DIGITAL PROCESSING
OF SIGNALS
Third Edition

DIGITAL PROCESSING OF SIGNALS

Theory and Practice
Third Edition

Maurice Bellanger

Conservatoire National des Arts et Métiers, Paris, France

Translated by
John C. C. Nelson, School of Electronic and Electrical Engineering,
University of Leeds, UK

JOHN WILEY & SONS, LTD

Chichester · New York · Weinheim · Brisbane · Singapore · Toronto

Library of Congress Cataloging-in-Publication Data

Bellanger, Maurice.
    [Traitement numérique du signal. English]
    Digital processing of signals / Maurice Bellanger ; translated by
John C.C. Nelson. - 3rd ed.
      p. cm.
    Includes bibliographical references.
    ISBN 0-471-97673-3 (alk. paper)
    1. Signal processing—Digital techniques. I. Title.
TK5102.9.B4513 2000   99-42711
621.382′ 2—dc21   CIP

British Library Cataloguing in Publication Data

A catalogue record for this book is available from the British Library

ISBN 0 471 97673 3

Produced from camera-ready copy supplied by the Authors
Printed and bound in Great Britain by Antony Rowe Ltd, Chippenham, Wiltshire
This book is printed on acid-free paper responsibly manufactured from sustainable forestry, in which at least two trees are planted for each one used for paper production.

CONTENTS

FOREWORD

In technology the most important advances and those with the most significant consequences are not always obvious to the eventual users of the product. Modern methods for the digital processing of signals fall into this category as their consequences are still not generally recognized, and they do not hit the headlines in the popular press.

It is interesting to reflect briefly on how these techniques have evolved. Signal processing by digital calculation in the widest sense of the term is certainly not a new idea in itself. When Kepler developed his laws of planetary motion from the series of observations by his father-in-law, Tycho-Brahé, he was actually studying digital signal processing, where the signal was the time series of Tycho-Brahé's observations of the planetary positions. However, it is only over the last few decades that digital signal processing has become a discipline in itself. The novel feature is that electrical signals can now be processed in real time using digital methods.

Before this could become possible a sequence of technical developments had to occur in many fields. The most important of these was undoubtedly the fundamental need for the data to be available as an electrical signal. This led to the development of a wide range of devices that are sometimes called data transducers. These range in complexity from radar to the simple strain gauge, which required a considerable amount of research into the physics of solids before it became possible.

It was then necessary to develop the technological tools which allow arithmetic operations to be carried out at the very high rates required by real time processing. This was achieved through extensive progress in microelectronics. Today, it is quite common to use a microprocessor weighing only a few grams and consuming milliwatts of electrical power, and with a mean time between failures of more than 10 years, to perform calculations which early computers took several hours to achieve, and often with frequent interruptions due to machine breakdowns. Indeed, the first electronic computer, ENIAC, was built only some forty years ago.

Finally, programming methods had to be refined and optimized because, whatever the impressive calculation capacities of modern microprocessors, their potential should not be wasted in performing redundant operations. The development of algorithms for fast Fourier transforms is one of the most striking examples of the importance of improved programming methods.

This convergence of technological progress in widely varying fields, whether dependent on physics, electronics or mathematics, has not occurred accidentally, To a certain degree, advances in one field created new demands which were met by advances in other disciplines. An examination of this relationship will undoubtedly have great implication for future studies in the history and epistemology of science and technology.

The immediate consequences of these developments are considerable. Analogue processing of electrical signals preceded digital processing, and will continue to be important in certain applications. However, the advantages of digital processing, paraphrased in the words 'accuracy' and 'reliability', have themselves encouraged applications far beyond the electronics and tele-communications fields from which they sprang. To quote only one example, X-ray tomography (or scanning) is based on a theorem derived by Radon in 1917. However, it was not until the developments outlined above were made that this new diagnostic medical tool could become a practical reality. It is certain that, in future, digital signal processing techniques will be involved in a wider range of products, including those used by the general public, who, while benefiting from the resulting advantages in price, performance and reliability, will not always be aware of the considerable infrastructure of research, technology and invention that was involved. This stage has already been reached in the domestic television market.

Almost inevitably, such a technological revolution leads to a further difficulty—user training. For the user, not only is a new tool involved but also often a new way of thinking. If care is not taken, this training stage could become a major obstacle in the introduction of new techniques. This is why this book by M. Bellanger is important. It is based on a course given for several years at the École Nationale Supérieure des Télécommunications and at the Institut Supérieur d'Électronique in Paris. With its textbook style and its extensive exercises, this book will be frequently used, and will undoubtedly play a role in speeding the evolution of this important technological advance.

P. AIGRAIN

PREFACE

Innovation requires an engineer to give material form to his knowledge and to determine the potential offered by new techniques which have been discovered and developed in research laboratories. The use of digital techniques in signal processing opens up a wide range of potential advantages such as precise systems design, equipment standardization and stability of performance characteristics, as well as ease of monitoring and control. However, these techniques involve a certain degree of abstraction and their practical application demands a basis of theoretical knowledge which is often considered to be more familiar, or at least more readily accessible, to the research worker than to the engineer. This can represent an obstacle to their utilization. It is the aim of this book to overcome this obstacle and to facilitate access to digital techniques by relating theory and practice, and by placing the most valuable results in this field within the reach of the working engineer.

The book is based on a course given for several years at the École Nationale Supérieure des Télécommunications and the Institut Supérieur d'Électronic in Paris. It is intended primarily for engineers. The author has attempted to provide a clear and concise presentation of the technical principles of digital signal processing, to compare their merits and to present the most useful results in a form directly usable both for the design of systems and for rapid evaluation in the realization of a project on a limited timescale. Theoretical treatment has been reduced to that which is necessary for a clear understanding and correct application of the results. The reader will find additional material which may be required in the references. Useful programs for design and system simulation are provided; they enable practical applications to be tackled directly. There are several exercises at the end of each chapter which are often based on real cases; these facilitate assimilation of the material in the chapter and familiarization with its application. Outline solutions and hints for these exercises have been provided at the end of the work.

This book is also intended for research workers for whom it will provide, in addition to a range of useful results, suggestions for the direction of their work by showing clearly the restrictions of practical technology. It also contains a number of results arising from research work by the author and his colleagues. In order to establish dialogue with research workers and to apply their discoveries with the minimum delay, engineers should become part of the scientific community and make their own contribution to research; by continuous contact with practical aspects they can not only evaluate and assimilate the results obtained by the researchers but also open up new approaches.

In comparison with its predecessors, this third edition includes several additions associated with new results and new applications. Some techniques have been replaced by others which are simpler or more efficient, particularly in the design of filters. Two aspects have been extended in accordance with the development of audiovisual signal processing which is associated with the rapid expansion of digital television and multimedia networks. These are filter banks for analysis and synthesis, which are the subject of a new chapter, and multi-carrier transmission techniques.

It should be emphasized that the work upon which this book is based originates from collaboration with, and the support of, the Centre National d'Études des Télécommunications which the author wishes to acknowledge. He would also like to express his gratitude to Mr J. Daguet, Technical Director of the Société Télécommunications Radioélectriques et Téléphoniques for his clear guidance and encouragement for many years. The author also sincerely thanks his colleagues for their contributions and the constant assistance which they have provided.

INTRODUCTION

Pocket calculators can be found today in every child's schoolbag. Thanks to a tiny electronic circuit, they are capable of performing the four basic numerical operations of addition, subtraction, multiplication and division. They can also store data in a memory, find percentages or even perform running totals. When such simple and familiar functions can be performed by tiny and inexpensive machines it is not surprising that they should extend their influence from their original area of information processing to every other aspect of technology. Instrumentation systems and industrial production equipment are examples of how the search for precision and safety has led to digitally controlled machine tools. However, it is in signal processing that one of the best illustrations of the opportunities for digital techniques is to be found. Indeed, these techniques have led to simplifications in design and implementation that are essential for the widespread development of this supposedly difficult and abstract field. They reduce the most complicated functions to a sequence of elementary operations.

The signal is the carrier of intelligence in these systems. It transmits the commands in the control equipment and conveys words or images over information networks. It is particularly fragile and has to be handled with great care. The processing that it undergoes is intended to extract information, modify the message that it carries or adapt it to the method of transmission. This is where digital techniques are involved. If a signal can be replaced by a set of numbers which are stored in a memory and represent its amplitude at suitably chosen instants of time, then the processing, even in its most complicated form, reduces to a sequence of arithmetic and logic operations on this set of numbers.

The conversion of a continuous analogue signal to a digital form is either performed by processors operating on recordings of the signal or carried out directly in the equipment which sends or receives the signal. The operations which follow this conversion are accomplished on suitably programmed digital computers. These techniques are applied in very different fields, and they are encountered in automation, industrial processes, aeronautics, radar systems, telecommunications, telemetry, medical instrumentation and geophysics.

Before introducing the various chapters of this book it is appropriate to give a more precise discussion of the type of processing that is involved. The term 'digital signal processing' is used to describe the complete set of operations, arithmetic calculations and numerical manipulations which are performed on the group of numbers representing the signal to be processed in order to produce another set of numbers representing the processed signal. Many different functions can be performed in this way: spectral analysis, linear or non-linear filtering, transcoding, modulation, detection, and estimation and extraction of parameters. Digital computers are used in all these applications and the processing obeys the laws of discrete systems. The numbers on which they operate can, in certain cases, be derived from a discrete process. However, more often they represent the amplitudes of samples of a continuous signal, and in this case the computer follows an analogue–digital converter and, eventually, precedes a digital–analogue converter. Signal digitization is of fundamental importance in the design of such systems and in the study of their operation, and it is necessary to examine sampling and coding. Distribution theory forms a concise approach, which is simple and effective for this analysis. After giving some of the background to Fourier analysis, distributions and the representation of signals, the first chapter presents the results which are of most use in the sampling and coding of a signal.

Digital processing began with the development of algorithms for the fast calculation of the discrete Fourier transform. This transform is the basis for the study of discrete systems and is the digital equivalent of the Fourier transform in the analogue case. It is the mechanism for passing from a discrete time space to a discrete frequency one. It appears naturally in spectral analysis using a step in the frequency domain which is a factor of the sampling frequency of the signal to be analysed.

Fast-calculation algorithms improve performance to such an extent that they allow real time calculations to be performed in many applications, provided certain basic conditions are fulfilled. Thus, the discrete Fourier transform not only forms a basic tool for determining the characteristics of a processing operation and in assessing its effect on the signal but gives a basis for the design of suitable equipment whenever a spectral analysis is involved: for example, in systems involving banks of filters or when, through the power of its algorithms, it leads to an advantageous approach for a filtering circuit. Chapters 2 and 3 are devoted to this subject and present the elementary properties of the fast algorithms and the mechanics of their application. They also outline a unified treatment of the algorithms and discuss possible developments. As for the system, the discrete Fourier transform computer is a discrete linear one, and is invariant in time.

Most of the present text is based on the study of one-dimensional time-invariant discrete linear systems. From the point of view of digital signal processing these are the most important, the most readily accessible and the most useful. Multi-dimensional systems, particularly those with two dimensions, have undergone a certain degree of development. They are applied, for example, to image analysis; however, their properties are generally deduced from those of one-dimensional systems, of which they are often only simple extensions.

Non-linear systems or systems which are variable in time either contain a large subset which has the properties of linearity and time invariance or one which can be analysed using the same techniques.

Linearity and time invariance involve the existence of a convolution relation which governs the operation of the system or involves a filter with the same properties. This convolution equation is defined using the response of the system to an elementary impulse, the impulse response. Thus, if $x(t)$ denotes the signal to be filtered and $h(t)$ is the impulse response of the filter, the filtered signal $y(t)$ is given by the equation:

$$y(t) = \int_{-\infty}^{\infty} h(\tau)x(t-\tau)\,d\tau$$

This type of equation, even though it directly represents the real operation of the filter, is of limited interest in practice. This is because it is difficult to determine the impulse response from the criteria which define the intended filtering operation. Also, an equation involving an integral does not allow the behaviour of the filter to be readily discovered or verified. The design is much more easily obtained in the frequency space because the Laplace or Fourier transforms allow access to a transformed space in which the convolution equations of amplitude–time space become simple products of functions. Fourier transformation produces the frequency response which corresponds to the impulse response of the system, and filtering amounts to producing this frequency response through the Fourier transform, or the spectrum, of the signal to be filtered.

In digital systems, which are discrete in character, convolution is performed by summation. The filter is defined by a set of numbers which forms its impulse response. Thus, if the set to be filtered is $x(n)$, the filtered set $y(n)$ is expressed by the following summation, where n and m are integers:

$$y(n) = \sum_{m} h(m)x(n-m)$$

Two cases are then possible. First, the summation can be performed over a finite number of terms, that is, all the $h(m)$ are zero except for a finite number of values of the integer variable m. The filter is said to have a finite impulse response. By reference to its construction, it is also said to be non-recursive, because its implementation does not require a feedback loop from the output to the input. It has a finite memory, as it only stores an elementary signal, such as an impulse, for a short time. The numbers $h(m)$ are called the coefficients of the filter, and define it completely. They can be calculated in a very simple direct way (for example, by performing the Fourier series expansion of the frequency response to be achieved). This type of filter has some very important basic characteristics. For example, it is possible to create a strictly linear phase response, i.e. a constant group delay. Signals whose components are in the pass band of the filter are not deformed when passing through it. This potential is used, for example, in data transmission systems or in spectral analysis.

Second, the summation can be performed over an infinite number of terms, as there are an infinite number of non-zero $h(m)$. The filter is said to have an infinite impulse response or to be of the recursive type, because it requires a feedback loop from the output to the input. Its operation is governed by an equation according to which an element of the output set $y(n)$ is calculated by the weighted summation of a certain number of elements of the input set $x(n)$ and of a certain number of elements of the preceding output set. For example, if L and K are integers, the operation of the filter can be defined by

$$y(n) = \sum_{l=0}^{L} a_l x(n-l) - \sum_{k=1}^{K} b_k y(n-k)$$

The numbers a_l $(l = 0, 1, \ldots, L)$ and b_k $(k = 0, 1, \ldots, L)$ are the coefficients. As with analogue filters, this type of filter cannot generally be analysed directly. It is necessary to change to a transformed space using, for example, the Laplace or Fourier transforms. However, for discrete systems a more suitable equivalent transform, the Z-transform, is available. A filter is described by its Z-transfer function, generally denoted by $H(Z)$, which involves the two sets of coefficients:

$$H(Z) = \frac{\sum_{l=0}^{L} a_l Z^{-l}}{1 + \sum_{k=1}^{K} b_k Z^{-k}}$$

The frequency response of the filter is obtained by replacing the variable Z in $H(Z)$ by the following expression, where f denotes the frequency and T is the sampling period of the signals:

$$Z = e^{j2\pi fT}$$

In this operation the imaginary axis in the Laplace plane corresponds to a circle of unit radius, centred on the origin in the Z-space. It is readily apparent that the frequency response of the filter defined by $H(Z)$ is a periodic function with the sampling frequency as its period. Another representation of the function $H(Z)$ is useful for the design of filters and involves the roots of the numerator, called the zeros of the filter, Z_l $(l = 1, 2, \ldots, L)$, and the roots of the denominator, called the poles, P_k $(k = 1, 2, \ldots, K)$:

$$H(Z) = a_0 \frac{\prod_{l=1}^{L} (1 - Z_l Z^{-1})}{\prod_{k=1}^{K} (1 - P_k Z^{-1})}$$

The term a_0 is a scale factor which defines the gain of the filter. The condition for the stability is expressed by the constraint that all the poles should be inside the unit circle. The positions of the poles and the zeros relative to the unit circle allow a very simple and commonly used representation of the filter characteristics.

A group of four chapters is devoted to the study of the properties of these digital filters. Chapter 4 discusses the properties of discrete time-invariant linear systems, reviews the principal properties of the Z-transform and gives the necessary elements for studying the filters. Chapter 5 discusses finite impulse response filters. Their properties are considered, together with the techniques for calculating the coefficients and the structures for their realization. As infinite impulse response filters are generally formed as a cascade arrangement of fundamental elements of the first and second order, Chapter 6 describes these elements and their properties. These elements considerably simplify the approach to this type of system and provide a set of results which are of use in practical applications. Chapter 7 describes the methods for calculating the coefficients of infinite impulse response filters and discusses the problems of their realization, with the limitations that are implied, and their consequences, in particular for round-off noise.

As infinite impulse response filters have properties comparable to those of continuous analogue filters it is natural to consider their realization in structures of the same type as are presently used in analogue filtering. This is the subject of Chapter 8, which describes ladder structures. A digression is made into switched-capacitor devices, which are not digital in the strict sense of the term but which are nevertheless of the sampled type and are very useful complements to digital filters. As a guide for the user, a resumé of the relative merits of the structures described is given at the end of the chapter.

Some equipment, as used, for example, in spectral analysis or in the telecommunications field, involves signals represented by a set of complex numbers. Analytic signals are an important example of particular practical interest. Their properties are discussed in Chapter 9, along with the design of devices for the production or processing of such signals. Some complementary concepts in filtering, such as the minimum phase shift condition, are also discussed in this chapter.

If the machines for digital signal processing are to operate in real time then they must operate at a rate which is closely related to the sampling frequency of the signals. Thus, their complexity depends on the number of operations to be performed and on the time interval available for their completion. The signal sampling frequency at the input or the output of the system is generally imposed by other constraints, but it can be varied within the system itself to match the properties of the signal and the processing, thus reducing the number of operations and the calculation rate. One simplification to the machines that can be of great importance is obtained by changing the sampling frequency to match the bandwidth of the useful signal throughout the processing. This is multirate filtering, which is discussed in Chapter 10. The effects on the processing characteristics are described, together with methods for their realization. Rules for their use and evaluation are provided. This technique yields particularly interesting results for narrow pass band filters or for banks of filters. In the latter case the system combines the discrete Fourier transform computer with a set of phase shift circuits.

Filter banks for the decomposition and reconstruction of signals have become a basic tool for compression. Their operation is described in Chapter 11 together with the computation methods and implementation structures.

Filters can be determined from a time specification, as would be the case, for example, when modelling the behaviour of a system. As the properties vary with time, the filters should adapt to match the evolution of the system to be modelled. This adaptation depends on an approximation criterion and is carried out at a rate up to the sampling rate of the system. Such a filter is said to be adaptive. Chapter 12 is devoted to simple adaptive filters, of the most common type, in which the coefficients vary according to the gradient algorithm. The approximation criterion is the least mean square of the error signal. After a brief review of the properties of random signals, and particularly their autocorrelation matrix whose eigenvalues play an important role, the gradient algorithm is introduced and the convergence conditions are studied. The two major adaptation parameters, the time constant and the residual error, are then analysed, together with the arithmetic complexity. Several implementation structures are discussed, and finally the important case of linear prediction is discussed.

Circuits and technological considerations are dealt with in Chapter 13. The elements required are an arithmetic unit to perform the calculations, an active memory to store the data, the intermediate results and the variable parameters, a read-only memory for fixed parameters such as the coefficients, and a control unit which co-ordinates the operation of the other components. The architecture of the system depends heavily on the arithmetic operator. This can perform operations on numbers presented either in serial form, with the binary digits (bits) representing the numbers at each input and output appearing one after the other, or in parallel form, where the bits are presented together simultaneously. The circuits are described in this chapter with examples of their construction using large-scale integration techniques for circuits which are specifically designed for digital signal processing. As the technology advances, the trend is towards the concentration of the functions and the use of programmable processors. Complexity parameters are introduced to give guidance in systems design or in the development of specific projects.

Telecommunications equipment represents a preferred area for the application of digital processing as transmission networks are progressively digitized and as digital computers are introduced at the terminals. Chapter 14 presents a group of typical applications illustrating the theory and the techniques presented in the book as a whole.

CHAPTER 1

SIGNAL DIGITIZING—SAMPLING AND CODING

The conversion of an analogue signal to a numerical form involves a double approximation. First, in time space, the signal function $s(t)$ is replaced by its values at integral time increments T, and is thus converted to $s(nT)$. This process is called sampling. Second, in amplitude space, each value of $s(nT)$ is approximated by a whole multiple of an elementary quantity. This process is called quantization. The approximate value thus obtained is then associated with a number. This process is called coding, a term often used to describe the whole process by which the value of $s(nT)$ is transformed to the number that represents it.

The effect on the signal of these two approximations will be analysed in this chapter and, to achieve this, two basic tools will be used: Fourier analysis and distribution theory.

1.1 FOURIER ANALYSIS

Fourier analysis is a method of decomposing a signal into a sum of individual components which can easily be produced and observed. The importance of this decomposition is that the response of a system to the signal can be deduced from these individual components using the superposition principle. These elementary component signals are periodic and complex, so that both the amplitude and phase of the systems can be studied. They are represented by a function $s_e(t)$ such that:

$$s_e(t) = e^{j2\pi ft} = \cos(2\pi ft) + j\sin(2\pi ft) \tag{1.1}$$

where f is the inverse of the period, that is, the frequency of the elementary signal.

Since the elementary signals are periodic it is clear that the analysis is simplified when the signal is itself periodic. This will be examined first, although it does not represent the most interesting case since a periodic signal is completely determined and carries practically no information.

1.1.1 Fourier series expansion of a periodic function

Let $s(t)$ be a function of a periodic variable t with period T, i.e. satisfying the relation:

1

$$s(t + T) = s(t) \tag{1.2}$$

Under certain conditions this function can be expanded in a Fourier series as:

$$s(t) = \sum_{n = -\infty}^{\infty} C_n e^{j2\pi nt/T} \tag{1.3}$$

The index n is an integer and the C_n, called the Fourier coefficients, are defined by:

$$C_n = \frac{1}{T} \int_0^T s(t) e^{-j2\pi nt/T} \, dt \tag{1.4}$$

In fact, the Fourier coefficients minimize the square of the difference between the function $s(t)$ and the series (1.3). Expression (1.4) is obtained by taking the derivative with respect to the index n coefficient of the quantity:

$$\int_0^T \left(s(t) - \sum_{m = -\infty}^{\infty} C_m e^{j2\pi mt/T} \right)^2 dt$$

and setting that derivative to zero.

Example: Figure 1.1 shows an example of a Fourier expansion of a function $i_p(t)$ composed of a train of impulses, each of width τ and amplitude a occurring at time intervals T. The time origin is taken as being at the centre of a pulse.

The coefficients C_n are given by:

$$C_n = \frac{1}{T} \int_{-\tau/2}^{\tau/2} a e^{-j2\pi nt/T} \, dt = \frac{a\tau}{T} \frac{\sin(\pi n\tau/T)}{\pi n\tau/T} \tag{1.5}$$

and the Fourier expansion is:

$$i_p(t) = \frac{a\tau}{T} \sum_{n = -\infty}^{\infty} \frac{\sin(\pi n\tau/T)}{\pi n\tau/T} e^{j2\pi nt/T} \tag{1.6}$$

The importance of this example for the study of sampled systems is readily apparent.

The properties of Fourier series expansions are given in Ref. [1]. One important property, expressed by the Bessel–Parseval equation, is that power

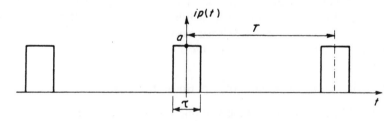

Fig. 1.1   Train of impulses

is conserved in the expansion of the signal:

$$\sum_{n=-\infty}^{\infty} |C_n|^2 = \frac{1}{T} \int_0^T |s(t)|^2 \, dt \tag{1.7}$$

The constituent elements resulting from the expansion of a periodic signal have frequencies which are integer multiples of $1/T$ (the inverse of the period). They form a discrete set in the space of all frequencies. In contrast, if the signal is not periodic the Fourier components form a continuous domain in the frequency space.

1.1.2 Fourier transform of a function

Let $s(t)$ be a function of t. Under certain conditions one can write:

$$s(t) = \int_{-\infty}^{\infty} S(f) e^{j2\pi f t} \, df \tag{1.8}$$

where

$$S(f) = \int_{-\infty}^{\infty} s(t) e^{-j2\pi f t} \, dt \tag{1.9}$$

The function $S(f)$ is the Fourier transform of $s(t)$. More commonly, $S(f)$ is called the spectrum of signal $s(t)$.

Example: To calculate the Fourier transform $I(f)$ of an isolated pulse $i(t)$ of width τ and amplitude a, centred on the time origin (Figure 1.2):

$$I(f) = \int_{-\infty}^{\infty} i(t) e^{-j2\pi f t} \, dt = a \int_{-\tau/2}^{\tau/2} e^{-j2\pi f t} \, dt$$

$$I(f) = a\tau \frac{\sin(\pi f \tau)}{\pi f \tau} \tag{1.10}$$

Figure 1.3 represents the function $I(f)$. This will be used frequently in what follows. It is important to note that it will be zero for non-zero frequencies which are whole multiples of the inverse of the pulse width. A table of this function is given in Appendix 1.

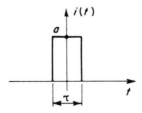

Fig. 1.2 Isolated impulse

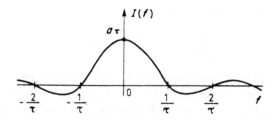

Fig. 1.3   Spectrum of an isolated impulse

This example shows clearly the correspondence between the Fourier coefficients and the spectrum. In effect, by comparing equations (1.6) and (1.10) it can be verified that, apart from the factor $1/T$, the coefficients of the Fourier series expansion of a train of pulses correspond to the values of the spectrum of the isolated pulse at frequencies which are whole multiples of the inverse of the period of the pulses.

In the case of a non-periodic function there is an expression similar to the Bessel–Parseval relation, but this time the energy in the signal is conserved, instead of the power:

$$\int_{-\infty}^{\infty} |S(f)|^2 \, df = \int_{-\infty}^{\infty} |s(t)|^2 \, dt \tag{1.11}$$

Let $s'(t)$ be the derivative of the function $s(t)$; its Fourier transform $S_d(f)$ is given by:

$$S_d(f) = \int_{-\infty}^{\infty} e^{-j2\pi ft} s'(t) \, dt = j2\pi f S(f) \tag{1.12}$$

Thus taking the derivative of a signal leads to multiplying its spectrum by $j2\pi f$.

One essential property of the Fourier transform (in fact this is the main reason for its use) is that it transforms a convolution into a simple product. Consider two time functions $x(t)$ and $h(t)$ with Fourier transforms $X(f)$ and $H(f)$, respectively. The convolution $y(t)$ is defined by:

$$y(t) = x(t) * h(t) = \int_{-\infty}^{\infty} x(t - \tau) h(\tau) \, d\tau \tag{1.13}$$

The Fourier transform of this product is:

$$Y(f) = \int_{-\infty}^{\infty} \left(\int_{-\infty}^{\infty} x(t - \tau) h(\tau) \, d\tau \right) e^{-j2\pi ft} \, dt$$

$$Y(f) = \int_{-\infty}^{\infty} h(\tau) e^{-j2\pi f\tau} \, d\tau \int_{-\infty}^{\infty} x(u) e^{-j2\pi fu} \, du = H(f) X(f)$$

Conversely, it can be shown that the Fourier transform of a simple product is a convolution product.

An interesting result can be derived from the above-mentioned properties. Let us consider the Fourier transform $II(f)$ of the function $i^2(t)$; because of equation (1.10) and (1.13), we obtain

$$II(f) = I(f) * I(f) = aI(f) \tag{1.14}$$

and therefore

$$\int_{-\infty}^{\infty} \frac{\sin(\pi\phi\tau)}{\pi\phi\tau} \cdot \frac{\sin(\pi(f-\phi)\tau)}{\pi(f-\phi)\tau} \, d\phi = \frac{1}{\tau} \frac{\sin(\pi f\tau)}{\pi f\tau}$$

Taking $f = n/\tau$, for any integer n,

$$\int_{-\infty}^{\infty} \frac{\sin(\pi\phi\tau)}{\pi\phi\tau} \cdot \frac{\sin(\pi(\phi\tau - n))}{\pi(\phi\tau - n)} \, d\phi = 0 \tag{1.15}$$

Thus the functions $\sin \pi(x - n)/[\pi(x - n)]$, with n integer, form a set of orthogonal functions.

The definition and properties of the Fourier transform can be extended to multivariate functions. Let $s(x_1, x_2, \ldots, x_n)$ be a function of n real variables: its Fourier transform is a function $S(\lambda_1, \lambda_2, \ldots, \lambda_n)$ defined by:

$$S(\lambda_1, \lambda_2, \ldots, \lambda_n) = \int\!\!\int_{\mathbb{R}^n} \cdots \int s(x_1, x_2, \ldots, x_n)$$
$$\times e^{-j2\pi(\lambda_1 x_1 + \lambda_2 x_2 + \cdots + \lambda_n x_n)} \, dx_1 \, dx_2 \cdots dx_n \tag{1.16}$$

If the function $s(x_1, x_2, \ldots, x_n)$ is separable, that is, if:

$$s(x_1, x_2, \ldots, x_n) = s(x_1)s(x_2) \cdots s(x_n)$$

then:

$$S(\lambda_1, \lambda_2, \ldots, \lambda_n) = S(\lambda_1)S(\lambda_2) \cdots S(\lambda_n)$$

The variables x_i $(1 \leqslant i \leqslant n)$ often represent distances (for example, for two dimensions), and in that case the λ_i are called spatial frequencies.

1.2 DISTRIBUTIONS

Mathematical distributions constitute a formal mathematical representation for the physical distributions found in experiment [1].

1.2.1 Definition

A distribution D is defined as a continuous linear function in the vector space $\mathscr{D}$ of functions defined in $\mathbb{R}^n$, indefinitely differentiable and having a bounded support.

With each function ϕ belonging to $\mathcal{D}$, the distribution D associates a complex number $D(\phi)$, which will also be denoted by $\langle D, \phi \rangle$, with the properties:

(1) $D(\phi_1 + \phi_2) = D(\phi_1) + D(\phi_2)$;
(2) $D(\lambda\phi) = \lambda D(\phi)$ where λ is a scalar;
(3) If ϕ_j converges to ϕ when j tends to infinity, the sequence $D(\phi_j)$ converges to $D(\phi)$.

Examples: (i) If $f(t)$ is a function which is summable over any bounded ensemble it defines a distribution D_f by:

$$\langle D_f, \phi \rangle = \int_{-\infty}^{\infty} f(t)\phi(t)\,\mathrm{d}t \tag{1.17}$$

(ii) If ϕ' denotes the derivative of ϕ, the function:

$$\langle D, \phi \rangle = \int_{-\infty}^{\infty} f(t)\phi'(t)\,\mathrm{d}t = \langle f, \phi \rangle \tag{1.18}$$

is also a distribution.

(iii) The Dirac distribution δ is defined by:

$$\langle \delta, \phi \rangle = \phi(0) \tag{1.19}$$

The Dirac distribution δ at a real point x is defined by:

$$\langle \delta(t - x), \phi \rangle = \phi(x) \tag{1.20}$$

This distribution is said to represent a mass of $+1$ at the point x.

(iv) Assume a pulse $i(t)$ of duration τ, with amplitude $a = 1/\tau$, centred on the origin. It defines a distribution D_i:

$$\langle D_i, \phi \rangle = \frac{1}{\tau} \int_{-\tau/2}^{\tau/2} \phi(t)\,\mathrm{d}t$$

For very small values of τ, this becomes:

$$\langle D_i, \phi \rangle \simeq \phi(0)$$

that is, the Dirac distribution can be regarded as the limit of the distribution D_i when τ tends towards 0.

1.2.2 Differentiation of distributions

The derivative $\partial D/\partial t$ of a distribution D is defined by the relation:

$$\left\langle \frac{\partial D}{\partial t}, \phi \right\rangle = -\left\langle D, \frac{\partial \phi}{\partial t} \right\rangle \tag{1.21}$$

To illustrate this, consider the Heaviside function Y, or single-step function, which is zero when $t < 0$ and $+1$ if $t \geqslant 0$:

$$\left\langle \frac{\partial Y}{\partial t}, \phi \right\rangle = -\left\langle Y, \frac{\partial \phi}{\partial t} \right\rangle = -\int_0^\infty \phi'(t)\,\mathrm{d}t = \phi(0) = \langle \delta, \phi \rangle \qquad (1.22)$$

As a result, the discontinuity in Y appears in the derivative as a point of unit mass.

This example illustrates the considerable practical interest of the notion of a distribution, which can extend to discontinuous functions a number of the concepts and properties of continuous functions.

1.2.3 The Fourier transform of a distribution

By definition, the Fourier transform of a distribution D is a distribution denoted by FD such that:

$$\langle FD, \phi \rangle = \langle D, F\phi \rangle \qquad (1.23)$$

By applying this definition to distributions with a point support we obtain:

$$\langle F\delta, \phi \rangle = \langle \delta, F\phi \rangle = \int_{-\infty}^\infty \phi(t)\,\mathrm{d}t = \langle 1, \phi \rangle \qquad (1.24)$$

Consequently, $F\delta = 1$. Similarly, $F\delta(t - a) = \mathrm{e}^{-\mathrm{j}2\pi fa}$.

A case which is fundamental to the study of sampling is that of the set (u) of Dirac distributions shifted by T and such that:

$$u(t) = \sum_{n=-\infty}^\infty \delta(t - nT) \qquad (1.25)$$

This set is a distribution of unit mass points separated on the abscissa by whole multiples of T. Its Fourier transform is:

$$Fu = \sum_{n=-\infty}^\infty \mathrm{e}^{-\mathrm{j}2\pi fnT} = U(f) \qquad (1.26)$$

and it can be shown that this sum is a point distribution.

An intuitive demonstration can be obtained from the Fourier series development of the function $i_p(t)$ formed by the set of separate pulses of duration T, with width τ, and amplitude $1/\tau$, centred on the time origin.

One can consider $u(t)$ as the limit of $i_p(t)$ as τ tends to zero:

$$u(t) = \lim_{\tau \to 0} i_p(t)$$

and by referring to relation (1.6) we find that:

$$\lim_{\tau \to 0} i_p(t) = \frac{1}{T} \sum_{n=-\infty}^\infty \mathrm{e}^{\mathrm{j}2\pi nt/T}$$

The following fundamental property is demonstrated in Ref. [2].

The Fourier transform of the time distribution represented by unit mass points separated by whole multiples of T is a frequency distribution of points of mass $1/T$ separated by whole multiples of $1/T$.

That is:

$$U(f) = \sum_{n=-\infty}^{\infty} e^{-j\pi f nT} = \frac{1}{T} \sum_{n=-\infty}^{\infty} \delta\left(f - \frac{n}{T}\right) \qquad (1.27)$$

This result will be used when studying signal sampling. The property of the Fourier transform whereby it exchanges convolution and multiplication applies equally to distributions.

Before considering the influences on the signal of the sampling and quantizing operations, it is useful to discuss the characteristics of those signals which are most often studied.

1.3 SOME COMMONLY STUDIED SIGNALS

A signal is defined as a function of time $s(t)$. This function can be either an analytic expression or the solution of a differential equation, in which case the signal is said to be determinist.

1.3.1 Determinist signals

Sine waves are the most frequently used signals of this type: for example,

$$s(t) = A \cos(\omega t + \alpha)$$

where A is the amplitude, $\omega = 2\pi f$ is the angular frequency and α is the phase of the signal.

Signals of this type are easy to reproduce and recognize at different points of a system. They allow the various characteristics to be visualized in a simple way. Moreover, as mentioned above, they serve as the basis for the decomposition of any determinist signal, through the Fourier transform.

If the system is linear and invariant in time it can be characterized by its frequency response $H(\omega)$. For each value of the frequency, $H(\omega)$ is a complex number whose modulus is the amplitude of the response. By convention, the function $\phi(\omega)$ such that:

$$H(\omega) = |H(\omega)| e^{-j\phi(\omega)} \qquad (1.28)$$

is defined as the phase. This convention allows the group delay $\tau(\omega)$, a positive function for real systems, to be expressed as

$$\tau(\omega) = \frac{d\phi}{d\omega} \qquad (1.29)$$

If the sinusoidal signal $s(t)$ is applied to the system then an output signal $s_r(t)$

is obtained such that:

$$s_r(t) = A|H(\omega)| \cos[\omega t + \alpha - \phi(\omega)] \qquad (1.30)$$

Once again this is a sinusoidal signal, and comparison with the applied signal allows the response of the system to be seen. The importance of this procedure (for example, for test operations) can readily be appreciated.

Determinist signals, meanwhile, do not give a good representation of real signals because they do not carry any information except by the fact of their presence. Real signals are generally represented by a random function $s(t)$. For testing and analysing systems, random signals are also used, but they must have particular characteristics which do not present undue complications for their generation and use. A study of such signals is given in Vol. 2 of Ref. [2].

1.3.2 Random signals

A random signal is defined at each instant of time, t, by a probability law for its amplitude. This law can be expressed by a probability density $p(x, t)$ defined by:

$$p(x, t) = \lim_{\Delta x \to 0} \frac{\text{Prob}[x \leqslant s(t) \leqslant x + \Delta x]}{\Delta x} \qquad (1.31)$$

It is stationary if its statistical properties are independent of time, that is, if the probability density is independent of time:

$$p(x, t) = p(x)$$

It is of second order if it possesses a first-order moment called the mean value, which is the mathematical expectation of $s(t)$, denoted by $E[s(t)]$ and defined by:

$$m_1(t) = E[s(t)] = \int_{-\infty}^{\infty} xp(x, t)\,\mathrm{d}x \qquad (1.32)$$

and a second-order moment, called the covariance:

$$E[s(t_1)s(t_2)] = m_2(t_1, t_2) = \int_{-\infty}^{\infty}\int_{-\infty}^{\infty} x_1 x_2 p(x_1, x_2; t_1, t_2)\,\mathrm{d}x_1\,\mathrm{d}x_2$$

where $p(x_1, x_2; t_1, t_2)$ is the probability density of a pair of random variables $[s(t_1), s(t_2)]$.

The stationarity can be limited to the moments of first and second order; then the signal is said to be stationary of order 2 or stationary in the wider sense. For such a signal:

$$E[s(t)] = \int_{-\infty}^{\infty} xp(x)\,\mathrm{d}x = m_1$$

The independence of time is translated for the probability density $p(x_1, x_2; t_1, t_2)$ as follows:

$$p(x_1, x_2; t_1, t_2) = p(x_1, x_2; 0, t_2 - t_1) = p(x_1, x_2; \tau)$$

where

$$\tau = t_2 - t_1$$

Only the difference between the two observation times is involved:

$$E[s(t_1)s(t_2)] = m_2(\tau) \tag{1.33}$$

The function $r_{xx}(\tau)$ such that:

$$r_{xx}(\tau) = E[s(t)s(t - \tau)] \tag{1.34}$$

is called the autocorrelation function of the signal.

A random signal $s(t)$ also has a time average m_T. This is a random variable defined by:

$$m_T = \lim_{T \to \infty} \frac{1}{T} \int_{-T/2}^{T/2} s(t)\, dt \tag{1.35}$$

The ergodicity of this average illustrates that it takes a particular value k with probability 1. For a stationary signal, the ergodicity of the time average implies equality with the average of the amplitudes at a given instant. In effect, we use the expectation of the variable m_T:

$$E[m_T] = k = E\left[\lim_{T \to \infty} \frac{1}{T} \int_{-T/2}^{T/2} s(t)\, dt \right] = \lim_{T \to \infty} \frac{1}{T} \int_{-T/2}^{T/2} E[s(t)]\, dt = m_1$$

This result has important consequences in practice as it provides a means of gaining access to the statistical properties of the signal at a given instant from observation over the time period. The ergodicity of the covariance in the stationary case is also very interesting because it leads to the relation:

$$r_{xx}(\tau) = \lim_{T \to \infty} \frac{1}{T} \int_{-T/2}^{T/2} s(t)s(t - \tau)\, dt \tag{1.36}$$

The autocorrelation function $r_{xx}(\tau)$ of the signal $s(t)$ is fundamental for the study of ergodic second-order stationary signals. Its principal properties are:

(1) It is an even function:

$$r_{xx}(\tau) = r_{xx}(-\tau)$$

(2) Its maximum is at the origin and corresponds to the power P of the signal:

$$r_{xx}(0) = E[s^2(t)] = P$$

(3) The power spectral density is the Fourier transform of the auto-correlation function:

$$\Phi_{xx}(f) = \int_{-\infty}^{\infty} r_{xx}(\tau) e^{j2\pi f \tau} d\tau = 2 \int_{0}^{\infty} r_{xx}(\tau) \cos(2\pi f \tau) d\tau$$

In effect, $r_{xx}(\tau) = s(\tau) * s(-\tau)$ and, if $S(f)$ denotes the Fourier transform of $s(t)$, we obtain:

$$\Phi_{xx}(f) = S(f)\overline{S(f)} = |S(f)|^2 \tag{1.37}$$

This last property is translated physically by the fact that the more rapidly varying the signal (i.e. the more its spectrum tends towards high frequencies), the narrower its autocorrelation function. In the limit, the signal is purely random and the function becomes zero for $\tau \neq 0$. This signal is called white noise and is described by:

$$r_{xx}(\tau) = P\delta$$

The spectral density is then constant:

$$\Phi_{xx}(f) = P$$

In fact, such a signal has no physical reality since its power is infinite, but it does offer a useful mathematical model for signals with a spectral density that is virtually constant over a wide frequency band.

1.3.3 Gaussian signals

Among the probability distributions that can be considered for a signal $s(t)$, one particular category is of special interest: normal or Gaussian distributions. In effect, normal random distributions preserve their normal character under any linear operation, such as convolution through a certain distribution, filtering, differentiation or integration. These random distributions are also frequently used for modelling real signals and for testing systems.

A random variable x is said to be Gaussian if its probability law has a density $p(x)$ which follows the normal or Gaussian law:

$$p(x) = \frac{1}{\sigma\sqrt{(2\pi)}} e^{-(x-m)^2/(2\sigma^2)} \tag{1.38}$$

The parameter m is the average of the variable x; the variance σ^2 is the second-order moment of the centred random variable $(x - m)$; σ is also called the standard deviation. The variable $(x - m)/\sigma$ has a zero average and a unit standard deviation. A tabulation is given in Appendix 2.

A random variable is characterized by its amplitude probability law and also by its moments m_n such that:

$$m_n = \int_{-\infty}^{\infty} x^n p(x) dx \tag{1.39}$$

These moments are the coefficients of the series expansion of the function $F(u)$, called the characteristic function of the random variable x and defined by:

$$F(u) = \int_{-\infty}^{\infty} e^{jux} p(x)\, dx \tag{1.40}$$

It is the inverse Fourier transform of the probability density $p(x)$ which is expressed by:

$$p(x) = \frac{1}{2\pi} \int_{-\infty}^{\infty} e^{-jux} F(u)\, du \tag{1.41}$$

From equation (1.40) the following series expansion is obtained:

$$F(u) = \sum_{n=0}^{\infty} \frac{(ju)^n}{n!} m_n \tag{1.42}$$

and for a centred Gaussian variable:

$$F(u) = e^{-\sigma^2 u^2/2} \tag{1.43}$$

The normal law can be generalized to multi-dimensional random variables [3]. The characteristic function of a k-dimensional Gaussian variable $x(x_1, x_2, \ldots, x_k)$ is expressed by:

$$F(u_1, u_2, \ldots, u_k) = \exp\left(-\frac{1}{2} \sum_{i=1}^{k} \sum_{j=1}^{k} r_{ij} u_i u_j \right) \tag{1.44}$$

with:

$$r_{ij} = E(x_i x_j)$$

The probability density is obtained through Fourier transformation. For two dimensions, we get:

$$p(x_1, x_2) = \frac{1}{2\pi\sigma_1\sigma_2\sqrt{(1-r^2)}} \exp\left\{ -\frac{1}{2(1-r^2)}\left[\frac{x_1^2}{\sigma_1^2} - \frac{2r x_1 x_2}{\sigma_1 \sigma_2} + \frac{x_2^2}{\sigma_2^2} \right] \right\} \tag{1.45}$$

where r is the correlation coefficient:

$$r = \frac{E(x_1 x_2)}{\sigma_1 \sigma_2}$$

A random signal $s(t)$ is said to be Gaussian if, for any set of k instants t_i $(1 \leqslant i \leqslant k)$, the k-dimensional random variable $s = [s(t_1), \ldots, s(t_k)]$ is Gaussian. According to equation (1.44), the probability law of that variable is completely defined by the autocorrelation function $r_{xx}(\tau)$ of the signal $s(t)$.

Example: The signal defined by the equations:

$$r_{xx}(\tau) = \sigma^2 e^{-|\tau|/(RC)} \tag{1.46}$$

$$p(x) = \frac{1}{\sigma\sqrt{(2\pi)}} e^{-x^2/(2\sigma^2)} \qquad (1.47)$$

is an approximation to white Gaussian noise which is used in the analysis of systems or for modelling signals. It is a stationary signal with a zero average and a spectral density which is not strictly constant but which corresponds to a uniform distribution filtered by an *RC* type low-pass filter. It is obtained by amplifying the thermal noise generated at the terminals of a resistance.

The Gaussian distribution can be obtained from a uniform distribution. Let $p(y)$ be the Rayleigh distribution:

$$p(y) = \frac{y}{\sigma^2} e^{-(y^2/2\sigma^2)}; \quad y \geqslant 0 \qquad (1.48)$$

and $p(x)$ the uniform distribution over the interval $[0, 1]$. Changing variables so that:

$$p(x)\,dx = p(y)\,dy$$

one gets:

$$p(y) = p(x)\frac{dx}{dy} = \frac{dx}{dy} = \frac{y}{\sigma^2} e^{-(y^2/2\sigma^2)}$$

and therefore

$$x = e^{-(y^2/2\sigma^2)}; \quad y = \sigma\sqrt{\left[2\ln\left(\frac{1}{x}\right)\right]} \qquad (1.49)$$

The Gaussian distribution is obtained by considering two independent variables x and y and the variable z given by:

$$z = y\cos 2\pi x \qquad (1.50)$$

With the help of the variable:

$$z' = y\sin 2\pi x$$

the following equations are obtained:

$$p(z, z')\,dz\,dz' = p(z)p(z')\,dz\,dz' = p(y)p(x)\,dx\,dy = p(z)p(z')y\,dy2\pi\,dx$$

Hence:

$$p(z)p(z') = \frac{1}{2\pi}\frac{1}{\sigma^2} e^{-(y^2/2\sigma^2)} = \frac{1}{2\pi\sigma^2} e^{-(z^2 + z'^2/2\sigma^2)}$$

and finally

$$p(z) = \frac{1}{\sigma\sqrt{2\pi}} e^{-(z^2/2\sigma^2)}$$

The method is often used to generate digital Gaussian signals.

1.3.4 Peak factor of a random signal

A random signal is defined by a probability law of its amplitude, such that this amplitude is often not bounded. This is the case for Gaussian signals, as can be seen from equation (1.38).

Signal processing can only be performed for a limited range of amplitudes, and restrictions are necessary. An important parameter is the peak factor defined to be the ratio of a certain amplitude A_m to the standard deviation σ. By convention, the amplitude A_m is that value which is exceeded for not more than 10^{-5} of the total time. This relationship is often expressed in decibels (dB) by f_c, where:

$$f_c = 20 \log_{10}(A_m/\sigma) \tag{1.51}$$

Following that convention, the peak factor for a Gaussian signal is 12.9 dB, and when this definition is applied to sinusoidal signal a peak factor of 3 dB results.

A stationary model used to represent telephone signals is formed by the random signal whose amplitude probability density obeys the following exponential, or Laplace, distribution:

$$p(x) = \frac{1}{\sigma\sqrt{2}} e^{-\sqrt{2}|x|/\sigma} \tag{1.52}$$

The peak factor rises in this case to 17.8 dB.

In conclusion, ergodic second-order stationary functions characterized by an amplitude probability distribution and an autocorrelation function allow us to model the majority of signals of interest and are widely used in system analysis.

In addition to the other possibilities for representing signals it is important to have some sort of global measure, so as to be able, for example, to follow a signal through the processing system. Such a measure can be obtained by defining norms on the function which represents the signal.

1.4 THE NORMS OF A FUNCTION

A norm is a positive real function which satisfies the relations:

$$\|x\| \geq 0; \qquad k\|x\| = \|kx\|$$

where k is real and positive.

One category of norms that is frequently employed is the set called L_p norms [4]. The L_p norm of a continuous function $s(t)$ defined over the interval $[0, 1]$ is denoted by $\|s\|_p$ and defined by:

$$\|s\|_p = \left[\int_0^1 |s(t)|^p \, dt \right]^{1/p} \tag{1.53}$$

Three values of p are of particular interest:

(1) $p = 1$:

$$\|s\|_1 = \int_0^1 |s(t)| \, dt \tag{1.53a}$$

(2) $p = 2$:

$$\|s\|_2^2 = \int_0^1 |s(t)|^2 \, dt \tag{1.53b}$$

which is the expression for the energy of the signal $s(t)$.

(3) $p = \infty$:

$$\|s\|_\infty = \max_{0 \leqslant t \leqslant 1} |s(t)| \tag{1.53c}$$

This is Chebyshev's norm.

Norms are also used in approximation techniques to measure the discrepancy between a function $f(x)$ and the function $F(x)$ being approximated. The approximation is made by least squares methods if the norm L_2 is used and in Chebyshev's sense if the norm L_∞ is used.

The L_p norms can be generalized by introducing a real positive weighting function of $p(x)$. The weighted L_p norm of the difference function $f(x) - F(x)$ is then written:

$$\|f(x) - F(x)\|_p = \left[\int_0^1 |f(x) - F(x)|^p p(x) \, dx \right]^{1/p} \tag{1.53d}$$

These expressions are applied when calculating filter coefficients and in realization problems, in particular for the scaling of internal data in memories and for noise estimation.

1.5 SAMPLING

Sampling consists of representing a signal function $s(t)$ by its value $s(nT)$ taken at whole multiples of the time interval T, called the sampling period. Such an operation can be analysed in a simple and concise way by using distribution theory. In fact, by definition, the distribution of unit masses at whole multiple points on the axis, with period T, associates with the function $s(t)$ the ensemble of its values $s(nT)$, where n is a whole number. Conforming to the notation given earlier, this distribution is denoted by $u(t)$ and is written:

$$u(t) = \sum_{n=-\infty}^{\infty} \delta(t - nT)$$

The sampling operation affects the spectrum $S(f)$ of the signal. By considering the fundamental relation (1.27) it appears that the spectrum $U(f)$ of the distribution $u(t)$ is formed of lines of amplitude $1/T$ at frequencies which are whole multiples of the sampling frequency $f_s = 1/T$. Thus $u(t)$ is expressed as a sum of elementary sinusoidal signals, collectively called carriers:

$$u(t) = \frac{1}{T} \sum_{n=-\infty}^{\infty} e^{j2\pi nt/T} \tag{1.54}$$

From this, the set of values for the signal $s(nT)$ corresponds to the product with the signal $s(t)$ of the ensemble of component signals which make up $u(t)$. That is, the operation of sampling is an amplitude modulation of the signal by an infinite number of carriers with frequencies which are whole multiples of the sampling frequency $f_s = 1/T$. Consequently, the sampled signal spectrum includes the function $S(f)$, called the base-band, as well as other image or sidebands, which correspond to the translation of the base-band by whole multiples of the sampling frequency. The operation of sampling and its influence on the signal spectrum are represented in Figure 1.4.

The sampled signal spectrum $S_s(f)$ is expressed as the convolution of $S(f)$ with $U(f)$, so that:

$$S_s(f) = 1/T \sum_{n=-\infty}^{\infty} S(f - n/T) \qquad (1.55)$$

It is important to note that the function $S_s(f)$ is periodic, that is, the sampling has introduced a periodicity into the frequency space, which constitutes a fundamental characteristic of the sampled signals.

The sampling operation as described above, which is called ideal sampling, may seem rather unrealistic in that it would be difficult in practice to manipulate or reconstitute an instantaneous signal value. Real samplers, or circuits which reconstitute samples, all possess a certain aperture time. In fact, it can be shown that sampling or the reconstitution of samples by pulses having a given width simply introduces a modification of the signal spectrum.

In effect, when sampling a signal $x(t)$ by a set of pulses separated by period T, with width τ and amplitude a, it is quite possible that a quantity σ_n is collected in the nth period, and this is written:

$$\sigma_n = a \int_{nT-\tau/2}^{nT+\tau/2} s(t)\,dt$$

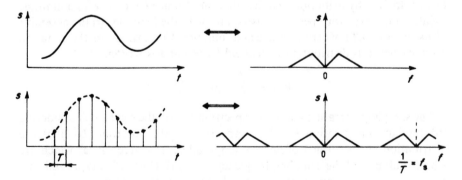

Fig. 1.4   *The spectral incidence of sampling*

This quantity expresses the result of the convolution of the signal $s(t)$ by the elementary pulse $i(t)$. The function which results in this case at the sampling time nT is the function $s*i$. That is, the sampled signal does not have $S(f)$ for its spectrum but the product:

$$S(f)a\tau\frac{\sin(\pi f\tau)}{\pi f\tau}$$

Similar reasoning applies in the case of reconstitution of samples with a duration τ. In fact, it is the convolution product of the samples $s(nT)$ with the elementary pulse $i(t)$ which is reconstituted. Thus we have the proposition:

Sampling or the reconstitution of samples by pulses with width τ can be treated as an ideal sampling or an ideal reconstitution, provided that the signal spectrum is multiplied by the spectrum of the elementary pulse train.

In practice, however, whenever τ is small in comparison with the period T the correction is negligible.

1.6 FREQUENCY SAMPLING

The type of sampling considered above is carried out on a time basis. Nevertheless, the general characteristics are also applicable to frequency sampling.

Let us calculate the spectrum of a periodic function $s_p(t)$ with period T. Such a function can be regarded as resulting from the convolution product of the function $s(t)$ which takes values of $s_p(t)$ over one period and is zero elsewhere,

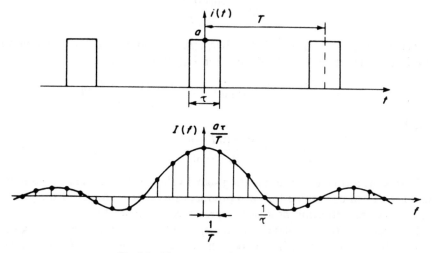

Fig. 1.5 The spectrum of a train of impulses

and the point distribution $u(t)$. This results in the following relation between the Fourier transforms:

$$S_p(f) = U(f)S(f) = \frac{1}{T} \sum_{n=-\infty}^{\infty} S\left(\frac{n}{T}\right)\delta\left(f - \frac{n}{T}\right) \qquad (1.56)$$

where we again meet the coefficients of the Fourier series expansion of the function $s_p(t)$. The case where $s(t)$ is a square pulse is represented in Figure 1.5.

It is apparent that the spectrum of the periodic function $s_p(t)$ is a set of lines forming a sampling of the spectrum of one period of the function. Sampling in frequency space corresponds to a periodicity in time space. This is a useful interpretation for digital analysis of spectra.

1.7 THE SAMPLING THEOREM

This theorem establishes the conditions under which the set of samples of a signal correctly represents that signal. A signal is said to be correctly represented by the set of its samples taken with the periodicity T, if it is possible, from this set of values, to completely reconstitute the original signal.

The sampling has introduced a periodicity of the spectrum in frequency space. To restore the original signal this periodicity must be suppressed, i.e. the image bands must be eliminated. This can be achieved using a low-pass filter with a transfer function $H(f)$ which is equal to $1/f_s$ up to the frequency $f_s/2$ and is zero at higher frequencies. At the output of such a filter a continuous signal appears, which can be expressed as a function of the values $s(nT)$. The square-wave impulse response $h(t)$ of the filter is written, using equation (1.10), as:

$$h(t) = \frac{\sin(\pi t/T)}{\pi t/T}$$

The signal at the output of the filter, $s(t)$, corresponds to the convolution product of the set $s(nT)$ with the function $h(t)$. It is written as:

$$s(t) = \int_{-\infty}^{\infty} \left[\sum_{n=-\infty}^{\infty} s(\theta)\delta(\theta - nT) \right] \frac{\sin \pi(t - \theta)/T}{\pi(t - \theta)/T} \, d\theta$$

and hence

$$s(t) = \sum_{n=-\infty}^{\infty} s(nT) \frac{\sin \pi(t/T - n)}{\pi(t/T - n)} \qquad (1.57)$$

This is the formula for interpolating signal values at points sited between the samples. It can be verified that it reproduces $s(nT)$ for multiples of the period T, and the reconstitution process is represented in Figure 1.6.

In order that the calculated signal $s(t)$ should be identical to the original signal, the spectrum $S(f)$ has to be identical to the spectrum of the original signal. As shown in Figure 1.6, this condition is satisfied if and only if the

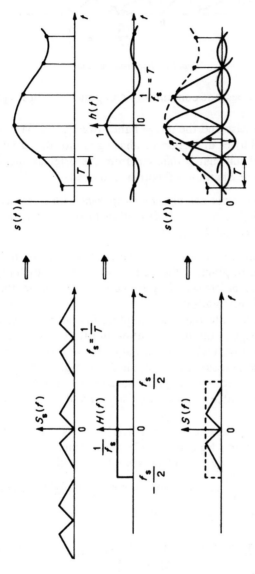

Fig. 1.6   Reconstruction of the signal after sampling

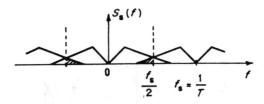

Fig. 1.7 Spectrum aliasing

original spectrum does not contain any components with frequencies greater than or equal to $f_s/2$.

If this is not the case and the image bands overlap the base-band, as shown in Figure 1.7, a foldover distortion or aliasing is introduced and the restoring filter produces a signal that is different from the original one. From this, the sampling theorem or Shannon's theorem is derived:

A signal which does not contain any component with a frequency equal to or greater than a value f_m is completely determined by the set of its values at regularly spaced intervals of period $T = 1/(2f_m)$.

Thus the sampling frequency of a signal is determined by the upper limit of its frequency band. In practice, the signal band is generally limited by filtering to a value below $f_s/2$ before sampling at frequency f_s, in order that the restoring filter can be practically realizable.

It is interesting to note that the sampling frequency is determined by the bandwidth of the signal; the reconstitution illustrated in Figure 1.6 was for a low-frequency signal with which a low-pass filter was associated. It is apparent that the same reasoning can also be applied to a signal occupying a restricted region in frequency space by using an associated band-pass filter. In particular, this property is applicable to modulated signals and is used in certain types of digital filter.

The result given at the end of Section 1.1.2 enables sampling to be presented from another viewpoint. Equation (1.57) shows that sampling corresponds to decomposition of the signal $s(t)$ in accordance with the set of orthogonal functions $\dfrac{\sin \pi(t/T - n)}{\pi(t/T - n)}$ and Shannon's theorem simply expresses the condition for this set to form the basis of decomposition of the signal.

1.8 SAMPLING OF SINUSOIDAL AND RANDOM SIGNALS

The properties given above can be clearly illustrated by sampling sinusoidal signals, whose features are of use in numerous applications.

1.8.1 Sinusoidal signals

Let the signal $s(t) = \cos(2\pi f t + \phi)$, with $0 \leqslant \phi \leqslant \pi/2$, be sampled with the period

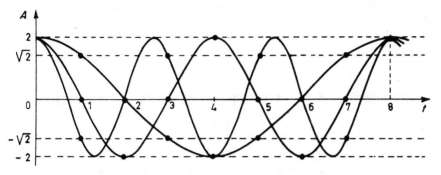

Fig. 1.8   Sampling the signals $\cos(2\pi N/8\,t)$

$T = 1/f_s = 1$. The samples are given by the set $s(n)$ such that:

$$s(n) = \cos(2\pi f n + \phi)$$

If the ratio $f/f_s = f$ is a rational number, this becomes:

$$f = N_1/N_2 \quad \text{with } N_1 \text{ and } N_2 \text{ as integers}$$

Then

$$s(n + N_2) = \cos[2\pi f(n + N_2) + \phi] = s(n)$$

The set $s(n)$ exhibits the periodicity N_2 and comprises at most N_2 different numbers. On the other hand, since the sampling frequency is more than twice the signal frequency, $N_1/N_2 < \frac{1}{2}$. The ensemble of N_2 different samples permits the representation of a number of sinusoidal signals equal to the largest whole number less than $N_2/2$. For example, if $N_2 = 8$, with the ensemble of numbers: $2\cos(2\pi n/8 + \phi)$, $(n = 0, 1, \ldots, 7)$, it is possible to represent samples of three sinusoidal signals:

$$2\cos\left(2\pi \frac{N_1}{8} t + \phi\right) \quad \text{with} \quad N_1 = 1, 2, 3$$

Figure 1.8 represents this sampling for $\phi = 0$, and in this particular case four numbers are sufficient: ± 2 and $\pm\sqrt{2}$.

If we then add to the three sinusoidal signals in Figure 1.8, the continuous signal with the value 1 and the oscillating signal $\cos(\pi t)$ with frequency $\frac{1}{2}$ and amplitude 1, the sampling of this sum gives zero values, except for points which are multiples of 8, where the value 8 is obtained as shown in Figure 1.9(a). The spectrum of this sum is obtained directly by applying the relation:

$$\cos x = \tfrac{1}{2}(e^{jx} + e^{-jx})$$

It is formed of lines with an amplitude of 1 at frequencies which are multiplies of $\frac{1}{8}$ (Figure 1.9(b)). This spectrum has already been studied in Section 1.2, and equation (1.27) again applies.

Spectrum analysers and digital frequency synthesizers use the fact that it is possible to produce a range of sinusoidal signals from a limited ensemble of numbers which are stored, for example, in a computer memory.

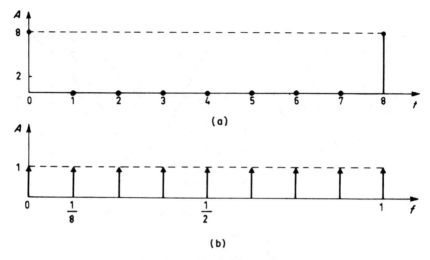

Fig. 1.9   (a) Sampling the signal:

$$s(t) = 1 + 2 \sum_{n=1}^{3} \cos(2\pi n/8t) + \cos(\pi t)$$

(b) corresponding spectrum

1.8.2 Discrete random signals

If the random signal $s(t)$ is sampled with the unit period $T = 1$, a discrete random signal results, which, by definition, has the same probability distribution of amplitude. The results obtained for the continuous case can be applied to the discrete one, particularly for the random signals which are second-order stationary and ergodic [5].

Thus the autocorrelation function of the discrete signal $s(n)$ is the sequence $r(n)$, such that:

$$r(n) = E[s(i)s(i - n)] \tag{1.58}$$

This is the sampling of the autocorrelation function $r_{xx}(\tau)$ of the continuous random signal defined by expression (1.34). Its Fourier transform gives the power spectrum density $\Phi_d(f)$ of the discrete signal, which is related to the spectral density $\Phi_{xx}(f)$ of the continuous signal by a relation similar to equation (1.55), that is:

$$\Phi_d(f) = \frac{1}{T} \sum_{n=-\infty}^{\infty} \Phi_{xx}\left(f - \frac{n}{T}\right) \tag{1.59}$$

If the sampling frequency is not large enough, or if the spectrum $\Phi_{xx}(f)$ spans an infinite domain, aliasing takes place.

The hypothesis of ergodicity for the discrete signal $s(n)$ leads to the relation:

$$r(n) = \lim_{N \to \infty} \frac{1}{2N + 1} \sum_{i=-N}^{N} s(i)s(i - n) \tag{1.60}$$

This relation gives the opportunity to extend the concept of autocorrelation function to deterministic signals. Then for a periodic signal with period N_0, the autocorrelation function is the sequence $r(n)$ given by:

$$r(n) = \frac{1}{N_0} \sum_{i=0}^{N_0-1} s(i)s(i - n) \tag{1.61}$$

It is a periodic sequence with the same period.

Example:

$$s(n) = A \sin\left(2\pi \frac{n}{N_0} \right)$$

$$r(n) = \frac{1}{N_0} \sum_{i=0}^{N_0} A^2 \sin\left(2\pi \frac{i}{N_0} \right) \sin 2\pi \left(\frac{i - n}{N_0} \right)$$

$$r(n) = \frac{A^2}{2} \cos\left(2\pi \frac{n}{N_0} \right)$$

The period of $r(n)$ is N_0 and $r(0)$ is the signal power.

A discrete random signal can also be defined directly. For example, if $r(n) = 0$ for $n \neq 0$, the signal $s(n)$ is a discrete white noise, and the spectral density is constant over the frequency interval $[-\frac{1}{2}, \frac{1}{2}]$. This signal has a physical reality: it is just a sequence of non-correlated random variables and it can be obtained through an algorithm which produces statistically independent numbers.

1.8.3 Discrete noise generation

It has been shown in Section 1.8.1 that the generation of digital sinusoidal signals is particularly simple. Such signals can be used to simulate noise, for example by addition of a large number of sine waves of different frequencies, constant amplitudes, having random or pseudo-random phases. This approach can lead to particularly efficient realizations, like that which has been standardized for measuring equipment in digital telephone transmission, and is as follows.

A pseudo-random sequence is created, which is a periodic sequence of $2^N - 1$ bits comprising approximately as many 'zeros' as 'ones' and which simulates a random sequences in which the bits are independent and have the probability $\frac{1}{2}$ of having a value of 0 or 1 (or in order to centre the variables, a value of $\pm\frac{1}{2}$).

If such a set is filtered (filtering in fact consists of a weighted summation), the numbers obtained after filtering obey a probability distribution which approximates the normal distribution.

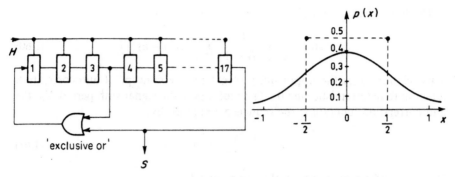

Fig. 1.10   *Pseudo-random generator and the probability distribution after filtering*

Pseudo-random sequences are studied in Ref. [6] and can be obtained easily using a suitably looped N-bit shift register. Figure 1.10 gives an example used in measuring equipment where $N = 17$. The generator polynomial is written:

$$g(x) = 1 + x^3 + x^{17} \tag{1.62}$$

The sequence contains $2^N - 1 = 131\,071$ bits, and it is periodic with a period $T = (2^N - 1)\tau$, where $\tau = 1/f_H$ denotes the clock period. The spectrum is formed of lines $1/T$ apart. For $f_H = 370\,\text{kHz}$, the distance between the lines is 2.8 Hz and there are 36 lines per 100 Hz.

By applying to this set a narrow band filter which passes only the band 450–550 Hz an approximately Gaussian signal is obtained. The signal has a peak factor of 10.5 dB and is an excellent test signal for digital transmission equipment. If the filtering is performed numerically, the set of numbers obtained can be used to test digital processing equipment.

1.9 QUANTIZATION

Quantization is the approximation of each signal value $s(t)$ by a whole multiple of an elementary quantity q which is called the quantizing step. If q is constant for all signal amplitudes, the quantization is said to be uniform. This operation is carried out by passing the signal through a device which has a staircase characteristic and produces the signal $s_q(t)$, as shown in Figure 1.11 for $q = 1$.

The way in which the approximation is made defines the centring of this characteristic. For example, the diagram represents the case (called rounding), where each value of the signal between $(n - \frac{1}{2})q$ and $(n + \frac{1}{2})q$ is rounded to nq. This approximation minimizes the power of the error signal. It is also possible to have approximation by default, which is obtained by truncation and consists of approximating by nq every value between nq and $(n + 1)q$; the characteristic is therefore displaced by $q/2$ towards the right on the abscissa.

The effect of this approximation is to superimpose on the original signal an

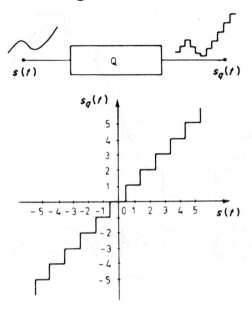

Fig. 1.11   The quantization operation

error signal $e(t)$ called the quantizing distortion or, more commonly, the quantizing noise. Thus:

$$s(t) = s_q(t) + e(t) \tag{1.63}$$

The rounding case is illustrated in Figure 1.12. The amplitudes at odd multiples of $q/2$ are called the decision amplitudes. The amplitude of the error signal lies between $-q/2$ and $q/2$ and power measures the degradation undergone by the signal.

When the variations in the signal are large relative to the quantizing step, that is, when quantization has been carried out sufficiently finely, the error signal is equivalent to an ensemble of elementary signals which are each formed from a straight-line segment (Figure 1.13). The power of such an elementary signal of width τ is written:

$$B = \frac{1}{\tau} \int_{-\tau/2}^{\tau/2} e^2(t)\,dt = \frac{1}{\tau} \left(\frac{q}{\tau}\right)^2 \int_{-\tau/2}^{\tau/2} t^2\,dt = \frac{q^2}{12} \tag{1.64}$$

The value obtained in this way, $B = q^2/12$, is a satisfactory estimate of the power of the quantizing noise in the majority of actual cases.

The spectral distribution of the error signal is more difficult to discern. The spectrum of the elementary error signal in Figure 1.13, $E_\tau(f)$, can be derived from its derivative. Using expressions (1.22) and then (1.12), the following is

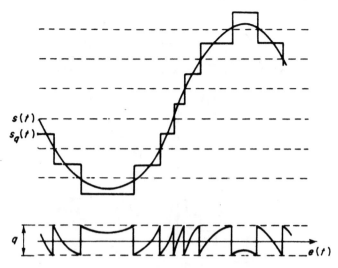

Fig. 1.12   *Quantization error*

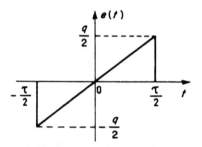

Fig. 1.13   *Elementary error signal*

obtained:

$$E_\tau(f) = \frac{1}{j2\pi f} q \left[\frac{\sin(\pi f \tau)}{\pi f \tau} - \cos(\pi f \tau) \right] \tag{1.65}$$

It appears that the largest part of the energy is found around the frequency $1/\tau$. Under these conditions the spectral distribution of the error signal depends both on the slope of the elementary signal, i.e. on the statistical distribution of the derivative $s'(t)$ of the signal, and on the size of the step q relative to the signal. Reference [7] gives the calculation of this spectrum for a noise signal and shows the spread as a function of frequency, when the quantizing step is sufficiently small, to be on a range several hundred times the width of the signal band. If the signal to be quantized is not random the spectrum of the error signal will be concentrated on certain frequencies, e.g. the harmonics of a sinusoidal signal.

When converting an analogue signal into digital form, quantization interferes, along with the sampling, as the two operations are carried out in succession. While sampling is normally carried out first, it is equally valid to carry out the quantization first and the sampling second, at a frequency f_s which is usually a little over twice the bandwidth of the signal. Under these conditions the error signal often has a spectrum which extends beyond the sampling frequency, and since it is actually the sum of the signal and the error signal which is sampled, aliasing occurs, and the whole energy of the error signal is recovered in the frequency band $(-f_s/2, f_s/2)$. In the majority of cases the spectral energy density of the quantizing noise is constant and the following statement can be made:

The noise produced during uniform quantization with a step q has a power which is generally expressed by $B = q^2/12$, and shows a constant spectral distribution in the frequency band $(-f_s/2, f_s/2)$.

It should be noted that the quantization of small signals (those with amplitude of the same order of magnitude as the step q) depends critically on the centring of the characteristic. For example, with the centring in Figure 1.11 a sinusoidal signal with an amplitude of less than $q/2$ is totally suppressed. It is possible, nevertheless, to suitably code these small signals by superimposing on them a large-amplitude auxiliary signal which is removed later in the process. Thus, the coding introduces limits on the small amplitudes of the signal. Equally, however, other limits appear for large amplitudes, as will be seen below.

1.10 THE CODING DYNAMIC RANGE

The signal which is sampled and quantized in amplitude is represented by a set of numbers which are almost always in binary form. If each number has N bits, the maximum number of quantized amplitudes that it is possible to represent is 2^N. Thus, the range of amplitudes that can be coded is subject to a double limitation: at low values it is limited by the quantum q and at larger values by $2^N q$. Any amplitude that exceeds this value cannot be represented and the signal is clipped. This results in degradation, so that, for example, if the signal is sinusoidal, harmonic distortion is introduced.

If the range of amplitudes to be coded covers the domain $[-A_m, A_m]$ we get:

$$A_m = 2^N q/2 \tag{1.66}$$

and, with rounding, the error signal $e(t)$ is:

$$|e(t)| \leq A_m 2^{-N}$$

By definition, the peak power of a coder is the power of that sinusoidal signal which has the maximum possible amplitude, A_m, which the coder will pass

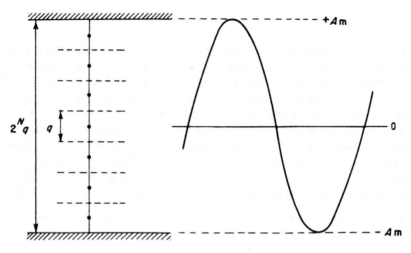

Fig. 1.14   Peak power of the coder

without clipping. It is expressed by:

$$P_c = \frac{1}{2}\left[\frac{2^N q}{2}\right]^2 = 2^{2N-3}q^2$$

Figure 1.14 illustrates this signal together with the quantizing step and the decision amplitudes.

The coding dynamic range is defined as the ratio between this peak power and the power of the quantizing noise; this is in fact the maximum value of the signal-to-noise ratio for a sinusoidal signal with uniform coding. The following formula expresses this dynamic range:

$$P_c/B = (S/B)_{max} = 2^{2N-3} \cdot 12 = \tfrac{3}{2} \cdot 2^{2N}$$

This can be expressed more conveniently in decibels as:

$$P_c/B = 6.02N + 1.76\,\text{dB} \tag{1.67}$$

This formula, which is of great practical use, relates the number of bits in the coding to the range of amplitudes which can be coded.

The signal to be coded is, in general, not sinusoidal. However, it is still possible to treat this case if an equivalent peak power can be defined for the signal, which can then be taken as the peak power of the coder. This occurs, for example, with multiplexed telephone signals. The case of random Gaussian signals is of particular importance because they conveniently represent a large number of signals which are encountered in practice. The maximum amplitude of the coder must then be correctly positioned relative to the signal amplitude, so that the distortion introduced be peak limitation remains within the imposed limits.

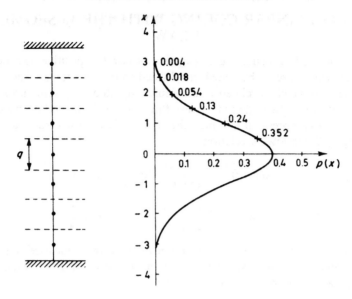

Fig. 1.15   Coding of a Gaussian signal

It is evident from the table given in Appendix 1 that the probability is less than 10^{-3} that a signal exceeds 3.4σ when it has a zero mean and a power of σ^2. Figure 1.15 gives an example of coding where $\sigma = q$. It would appear that the probability of clipping is less than 5×10^{-4} for the chosen parameters.

It is generally acknowledged that clipping distortion is negligible when the probability is less than 10^{-5}. Thus, it is sufficient to adjust the coder so that clipping occurs only for amplitudes greater than the root mean square signal by $20 \log(4.42) = 12.9\,\text{dB}$. This is the signal peak factor as defined in Section 1.3.

The peak power of the coder is then found at $12.9 - 3 = 9.9\,\text{dB}$ above the signal power. Thus the following proposition can be derived:

Distortion due to signal clipping is negligible for a random Gaussian signal if the peak power of the coder is 10 dB above the signal power.

This result is used not only in specifying analogue signal coders but also in digital processing for determining memory sizes and scaling internal data.

The coding dynamic range for a given number of bits can be considerably increased if coding is carried out with a quantizing step which varies with the amplitude of the signal. This is called non-linear coding. Numerous variations can be envisaged, but of particular importance is the 13-segment coding law which has been standardized by the International Telegraph and Telephone Consultative Committee (CCITT) for the coding of signals in telecommunications networks [8].

1.11 NON-LINEAR CODING WITH THE 13-SEGMENT A-LAW

In non-linear coding using the 13-segment A-law the positive and negative amplitudes which are to be coded are divided into seven ranges, each with a quantizing step which is related to an elementary step q by some power of 2. This operation can be regarded as resulting from a linear coding which was preceded by a compression in which the signal x is transformed to a signal y according to the following relations:

$$y = \text{sign}(x)\frac{1 + \log A|x|}{1 + \log A} \quad \text{for } \frac{1}{A} \leqslant |x| \leqslant 1$$

$$y = \text{sign}(x)\frac{A|x|}{1 + \log A} \quad \text{for } 0 \leqslant |x| \leqslant \frac{1}{A} \tag{1.68}$$

The parameter A controls the dynamic range increase; the value used is $A = 87.6$. The compression characteristic of the 13-segment A-law is shown in Figure 1.16.

The operation can be described as follows:

$$\text{if } 0 < x < \tfrac{1}{64}, \quad \text{then } y = 16x$$
$$\tfrac{1}{64} \leqslant |x| \leqslant \tfrac{1}{32} \qquad y = 8x + \tfrac{1}{8}$$
$$\tfrac{1}{32} \leqslant |x| \leqslant \tfrac{1}{16} \qquad y = 4x + \tfrac{1}{4}$$
$$\tfrac{1}{16} \leqslant |x| \leqslant \tfrac{1}{8} \qquad y = 2x + \tfrac{3}{8}$$
$$\tfrac{1}{8} \leqslant |x| \leqslant \tfrac{1}{4} \qquad y = x + \tfrac{1}{2}$$
$$\tfrac{1}{4} \leqslant |x| \leqslant \tfrac{1}{2} \qquad y = \tfrac{1}{2}x + \tfrac{5}{8}$$
$$\tfrac{1}{2} \leqslant |x| \leqslant 1 \qquad y = \tfrac{1}{4}x + \tfrac{3}{4}$$

This characteristic causes seven straight-line segments to appear in both the positive and negative quadrants; as the two segments which encompass the origin are colinear, the characteristic has a total of 13 segments.

Since the quantization of the y amplitudes is carried out with the quantum q, quantization of the x amplitudes near the origin is based on a quantum $q/16$, that is, the dynamic of the coder is increased by 24 dB. Amplitudes close to unity are less well quantized as the step is multiplied by 4. The power of the quantization noise is a function of the signal amplitude: it is necessary to calculate an average for each value and this interferes with the statistics of the signal.

Figure 1.17 gives the signal-to-noise ratio for a Gaussian signal as a function of the signal level after coding into 8 bits for linear and non-linear coding. The reference level for the signal (0 dB) is the peak power of the coder, and it is clearly apparent that the dynamic is extended by non-linear coding. For low amplitudes, quantization in fact corresponds to 12 bits. In practice, the signal can be coded by linear quantization into 12 bits, followed by a process which is very close to the conversion of an integer into a floating-point number:

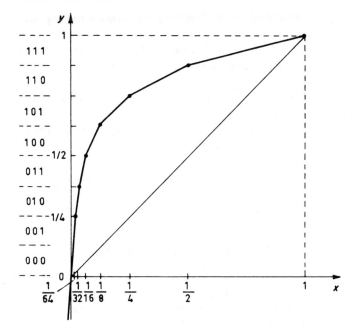

Fig. 1.16   The 13-segment compression A-law

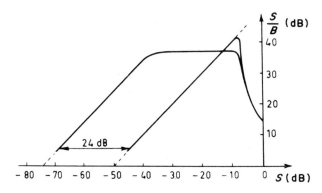

Fig. 1.17   Eight-bit linear and non-linear coding of a Gaussian signal

For example, the 12-bit number

$$+0\ 0\ 0\ 1\ 0\ 1\ 1\ 0\ 1\ 1\ 0$$

corresponds to the 8-bit one

$$+1\ 0\ 0\ 0\ 1\ 1\ 0$$

after application of the compression law.

The three bits which follow the sign give the code for the exponent; the four following bits give the position within the segment, or the mantissa. The difference between the conversion from a whole number to a floating-point number appears in the neighbourhood of the origin.

The implementation can use an array of gates arranged in parallel or a shift register combined with a 3-bit counter for serial realization. Equally, one could use memories holding the conversion table.

Another non-linear coding law is also widely used in telecommunications, the 15-segment μ-law, which follows the relation:

$$y = \text{sign}(x)\frac{\log(1 + \mu|x|)}{\log(1 + \mu)} \quad \text{for } -1 \leqslant x \leqslant 1 \tag{1.69}$$

The compression parameter is $\mu = 255$.

1.12 OPTIMAL CODING

The coding can be improved when the probability distribution $p(x)$ of the signal amplitude is known; for a given number of bits N an optional quantizing characteristic can be found, which minimizes the total quantizing distortion.

The signal amplitude range is divided into $M = 2^N$ subsets (x_{i-1}, x_i) with $-(M/2) + 1 \leqslant i \leqslant M/2$ and every subset is represented by a value y_i as shown in Figure 1.18. The optimization consists of determining the set of values x_i and y_i which minimize the error signal power E^2 expressed by:

$$E^2 = \sum_{i=-(M/2)+1}^{M/2} \int_{x_{i-1}}^{x_i} (x - y_i)^2 p(x)\,dx$$

Taking the derivative with respect to x_i and y_i, it can be shown that the

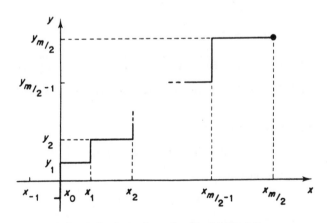

Fig. 1.18 Optimal quantization characteristic

following relations must hold:

$$x_i = \tfrac{1}{2}(y_i + y_{i+1}) \quad \text{for} \quad -\frac{M}{2} + 1 \leqslant i \leqslant \frac{M}{2} - 1$$

$$\int_{x_{i-1}}^{x_i} (x - y_i)p(x)\,dx = 0 \quad \text{for} \quad -\frac{M}{2} + 1 \leqslant i \leqslant \frac{M}{2} \tag{1.70}$$

$$p(x_{M/2}) = p(x_{-M/2}) = 0$$

These relations lead to the determination of the quantizing characteristic. If $p(x)$ is an even function $x_0 = 0$ and an iterative procedure is used starting with an *a priori* choice for y_1. If relation (1.70) is not satisfied for $M/2$, another initial choice is made for y_1 and so on [9].

Table 1.1 gives the error signal power obtained with a Gaussian signal of unit power, for several numbers of bits N, for the optimal coding and for uniform coding with the best scaling of the quantizing characteristic [9]. Table 1.2, taken from Ref. [10], gives for the same conditions the values which correspond to a signal probability density following expression (1.48).

The coding optimization can also be carried out with respect to the information content, by introducing the concept of entropy H defined as follows [2, Vol. 3]:

$$H = -\sum_i p_i \log_2(p_i) \tag{1.71}$$

with

$$-\frac{M}{2} + 1 \leqslant i \leqslant \frac{M}{2}$$

and where p_i designates the probability that the signal is in the amplitude subrange represented by y_i.

Table 1.1. CODING OF UNITARY GAUSSIAN SIGNAL

E^2	N	1	2	3	4	5
Optimal coding		0.3634	0.1175	0.03454	0.0095	0.0025
Uniform coding		0.3634	0.1188	0.03744	0.01154	0.00349
Entropy H		1	1.911	2.825	3.765	4.730

Table 1.2. CODING OF UNITARY LAPLACIAN SIGNAL

E^2	N	1	2	3	4	5
Optimal coding		0.5	0.1765	0.0548	0.0954	0.00414
Uniform coding		0.5	0.1963	0.0717	0.0254	0.0087

Considering that:

$$\sum_i p_i = 1$$

the entropy is zero when the amplitude is concentrated in a single subrange. It is maximal when the signal amplitude is uniformly distributed, when it takes the value H_{max} equal to the number of bits N of the coder:

$$H_{max} = \log_2 M = N \tag{1.72}$$

In fact the entropy measures the difference between a given distribution and the uniform distribution. The quantizing characteristic which maximizes the entropy is that which leads to amplitude subranges corresponding to a uniform probability distribution.

The last row in Table 1.1 shows that for a Gaussian signal the quantizing law which minimizes the error signal power leads to entropy values close to the maximum N.

1.13 QUANTITY OF INFORMATION AND CHANNEL CAPACITY

The results obtained for sampling and quantization can be used, inversely, to evaluate the quantity of information carried by a signal or to determine the capacity of a transmission channel.

A real channel of bandwidth f_m can carry $2f_m$ independent samples per second, as shown in Figure 1.4, by replacing τ with $2f_m$. The quantity of information per sample depends on the relative powers of the useful signal and the noise and their amplitude distributions.

Disregarding the amplitude distributions, which can be taken into account by coding operations, the number N of bits per sample can be expressed as a function of the ratio of the useful signal S to the noise power B.

Again using equation (1.66) gives:

$$\frac{A_m^2}{(q/2)^2} = 2^{2N} \tag{1.73}$$

Regarding the term A_m^2 as representing the maximum power of the whole signal, that is useful signal and noise, and $(q/2)^2$ the noise power:

$$\frac{S + B}{B} = 2^{2N} \tag{1.74}$$

With $C = 2f_m$, this gives:

$$C = f_m \log_2 \left(1 + \frac{S}{B}\right) \tag{1.75}$$

This is the classic equation giving the capacity C of a real channel of bandwidth f_m, expressed in bits per second.

1.14 BINARY REPRESENTATIONS

There are several ways of establishing the correspondence between the set of quantized amplitudes and the set of binary numbers which must represent them. As the signals to be coded have amplitudes which are generally both positive and negative, the preferred representations are those which preserve the sign information. The following are the most usual representations:

(1) Sign and absolute value;
(2) Offset binary;
(3) 1s-complement;
(4) 2s-complement.

The definitions and features of these representations are given in Ref. [11]; Table 1.3 defines them for three bits.

Table 1.3. BINARY REPRESENTATION FOR LINEAR CODE

Number	Sign and value	Offset binary	1s complement	2s complement
+3	0 1 1	1 1 1	0 1 1	0 1 1
+2	0 1 0	1 1 0	0 1 0	0 1 0
+1	0 0 1	1 0 1	0 0 1	0 0 1
+0	0 0 0	1 0 0	0 0 0	0 0 0
−0	1 0 0	—	1 1 1	—
−1	1 0 1	0 1 1	1 1 0	1 1 1
−2	1 1 0	0 1 0	1 0 1	1 1 0
−3	1 1 1	0 0 1	1 0 0	1 0 1
		(0 0 0)		(1 0 0)

The most useful representations for analogue-to-digital conversion are those of sign and absolute value and offset binary. The remaining two representations are mostly used in arithmetic circuits, and are given in detail below.

As mentioned in Section 1.11, non-linear coding brings a considerable increase of the dynamic range. Digital processing machines, and particularly general-purpose ones, often use floating-point representations in which each number has three parts: the sign bit, the mantissa and the exponent. The mantissa represents the fractional part and the exponent is a power of the base number; for example, with base 10: $+0.719 \times 10^5$.

The dynamic range extension comes from the multiplicative effect introduced by the exponent. For example, in base 2 for a 6-bit exponent and 16-bit mantissa the dynamic range is: $2^{26} \times 2^{16} = 2^{80} \simeq 10^{24}$, that is, 24 decimal numbers. Additional gain is achieved by choosing a base itself a power of two, such as 8 or 16, leading to octal or hexadecimal operations.

However, floating-point representations involve more complex arithmetic operations and circuits. The numbers produced by the coding are presented depending upon the technique employed, either in parallel form, i.e. N bits appear at N connection points at the same time, or in series, i.e. N bits appear successively at the same connection point, with the sign first and then with the data bits in decreasing significance. Reference [12] describes the technical principles of analogue-to-digital conversion.

APPENDIX 1
The function $I(x)$

The set of values:

$$I(n) = \frac{\sin(\pi(n/20))}{\pi(n/20)}$$

for $0 \leqslant n \leqslant 159$ when $n = k + 20N$.

THE SET OF NUMBERS

k	$N = 0$	$N = 1$	$N = 2$	$N = 3$	$N = 4$	$N = 5$	$N = 6$	$N = 7$
0	1	0	0	0	0	0	0	0
1	0.99589	−0.04742	0.02429	−0.01633	0.01229	−0.00986	0.00823	−0.00706
2	0.98363	−0.08942	0.04684	−0.03173	0.02399	−0.01929	0.01613	−0.01385
3	0.96340	−0.12566	0.06721	−0.04588	0.03482	−0.02806	0.02350	−0.02021
4	0.93549	−0.15591	0.08504	−0.05847	0.04455	−0.03598	0.03018	−0.02599
5	0.90032	−0.18006	0.10004	−0.06926	0.05296	−0.04287	0.03601	−0.03105
6	0.85839	−0.19809	0.11196	−0.07804	0.05989	−0.04850	0.04088	−0.03528
7	0.81033	−0.21009	0.12069	−0.08466	0.06520	−0.05301	0.04466	−0.03859
8	0.75683	−0.21624	0.12614	−0.08904	0.06880	−0.05606	0.04730	−0.04091
9	0.69865	−0.21682	0.12832	−0.09113	0.07065	−0.05769	0.04874	−0.04220
10	0.63662	−0.21221	0.12732	−0.09095	0.07074	−0.05787	0.04897	−0.04244
11	0.57162	−0.20283	0.12329	−0.08856	0.06910	−0.05665	0.04800	−0.04164
12	0.50455	−0.18921	0.11643	−0.08409	0.06581	−0.05406	0.04587	−0.03983
13	0.43633	−0.17189	0.10702	−0.07770	0.06099	−0.05020	0.04265	−0.03707
14	0.36788	−0.15148	0.09538	−0.06960	0.05479	−0.04518	0.03844	−0.03344
15	0.30011	−0.12862	0.08185	−0.06002	0.04739	−0.03914	0.03335	−0.02904
16	0.23387	−0.10394	0.06682	−0.04924	0.03989	−0.03298	0.02751	−0.02399
17	0.17001	−0.07811	0.05071	−0.03753	0.02980	−0.02470	0.02110	−0.01841
18	0.10929	−0.05177	0.03392	−0.02522	0.02007	−0.01667	0.01426	−0.01245
19	0.05242	−0.02544	0.01688	−0.01261	0.01006	−0.00837	0.00716	−0.00626

APPENDIX 2
The reduced normal distribution

$$f(x) = \frac{1}{\sqrt{2\pi}} e^{-x^2/2} \qquad P = \frac{2}{\sqrt{2\pi}} \int_\lambda^\infty e^{-x^2/2}\, dx \qquad P = \frac{2}{\sqrt{2\pi}} \int_\lambda^\infty e^{-x^2/2}\, dx$$

x	$10^5 \cdot f(x)$	$100\,P$	λ	λ	$100\,P$
0	39 894	100	0	0	100
0.2	39 104	95	0.0627	0.2	84.148
0.4	36 827	90	0.1257	0.4	68.916
0.6	33 322	85	0.1891	0.6	54.851
0.8	28 969	80	0.2533	0.8	42.371
1	24 197	75	0.3186	1	31,731
1.2	19 419	70	0.3853	1.2	23.014
1.4	14 973	65	0.4538	1.4	16.151
1.6	11 092	60	0.5244	1.6	10.960
1.8	7 895.	55	0.5978	1.8	7.186
2	5 399	50	0.6745	2	4.550
2.2	3 547	45	0.7554	2.2	2,781
2.4	2 239	40	0.8416	2.4	1.640
2.6	1 358	35	0.9346	2.6	0.932
2.8	792	30	1.0364	2.8	0.511
3	443	25	1.1503	3	0.270
3.2	238	20	1.2816	3.2	0.137
3.4	123	15	1.4395	3.4	0.067
3.6	61	10	1.6449	3.6	0.032
3.8	29	5	1.9600	3.8	0.014
4	13	1	2.5758	4	0.006
4.2	5.9	0.1	3.2905	4.5	0.00068
4.4	2.5	0.01	3.8906	5	0.000057
4.6	1	0.001	4.4172	5.5	0.000004
		0.0001	4.8916		
		0.00001	5.3267		

REFERENCES

[1] A. PAPOULIS, *The Fourier Integral and its Applications*, McGraw-Hill, New York, 1962.
[2] E. ROUBINE, *Introduction à la théorie de la communication*, 3 vols, Masson, Paris, 1970.
[3] W. B. DAVENPORT, *Probability and Random Processes*, McGraw-Hill, New York, 1970.
[4] J. R. RICE, *The Approximation of Functions*, vol. 1, Addison-Wesley, Reading, Mass., 1964.
[5] B. PICINBONO, *Principles of Signal's and Systems*, Artech House Inc., London, 1988.
[6] W. PETERSON, *Error Correcting Codes*, MIT Press, 1972.

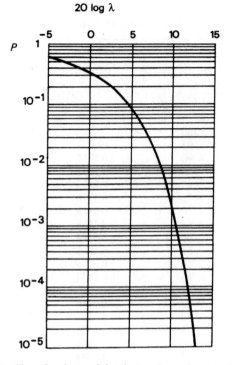

Fig. 1.19 The reduced normal distribution: P as a function of 20 log λ.

[7] W. B. BENNET, Spectra of quantized signals. *The Bell System Technical Journal*, July 1948.

[8] CCITT, Digital Networks—transmission systems and multiplexing equipment. *Yellow Book*, Vol. III, 3, Geneva, Switzerland, 1981.

[9] J. MAX, Quantizing for minimum distortion. *IRE Transactions on Information Theory*, **IT 6**, 7–12, 1960.

[10] M. D. PAEZ and T. H. GLISSON, Minimum mean-squared error quantization in Speech PCM and DPCM systems. *IEEE Transactions on Communications*, **COM 20**, 225–30, 1972.

[11] E. R. HNATEK, *A User's Handbook of Integrated Circuits*, John Wiley, New York, 1973

[12] A. VAN DER PLASSCHE, *Integrated analog-to-digital and digital-to-analog converters*, Kluwer, 1994.

EXERCISES

1 Consider the Fourier series expansion of the periodic function $i(t)$ of period T, which is zero throughout the period except for the range $-\tau/2 \leqslant t \leqslant \tau/2$, where it has a value of 1.

Give the value of the coefficients for $\tau = T/2$ and $\tau = T/3$.

Verify that the expansion leads to $i(0) = 1$, and draw the function when the expansion is limited to 5 terms.

2 Analyse the sampling at the frequency f_s of the signal $s(t) = \sin(\pi f_s t + \phi)$ when ϕ varies from 0 to $\pi/2$.

Examine the reconstitution of this signal from the samples.

3 Calculate the amplitude distortion introduced into a signal reconstituted by pulses with a width of half the sampling period.

4 A signal occupies the frequency band $[f_1, f_2]$. What conditions should be imposed on the frequency f_1 so that this signal can be sampled directly at a frequency between f_2 and $2f_2$?

5 Analyse the sampling of a signal given by:

$$s_i(t) = \sum_{n=1}^{3} \sin\left(2\pi \frac{n}{8}\frac{t}{T}\right)$$

and compare it with that of the signal:

$$s_r(t) = \sum_{n=1}^{3} \cos\left(2\pi \frac{n}{8}\frac{t}{T}\right)$$

Show by studying the spectra that the combination of the two sets of samplings forms the sampling of a complex signal.

6 Let $s(t)$ be the signal defined by:

$$s(t) = 1 + 2\sum_{k=1}^{3} \cos\left(2\pi\frac{kt}{8} + \phi_k\right) + \cos(\pi t + \phi_4)$$

This signal is sampled with the period $T = 1$. What is the maximum value of $s(n)$, where n is a whole number? Show that there exists an ensemble of values ϕ_k ($k = 1, 2, 3, 4$) which minimizes the maximum value of $s(n)$. Can this property be generalized?

7 A digital frequency synthesizer is constructed from read-only memory of 16 kbits with an access time of 500 ns. Knowing that the numbers which represent the samplings of sinusoidal signals total 8 bits, what are the characteristics of the synthesizer, and the frequency range and increment step that can be obtained?

8 What is the probability distribution of the amplitudes of the sinusoidal signal:

$$s(t) = A\cos\left(2\pi\frac{t}{T}\right)$$

Give its autocorrelation function. Give the autocorrelation function of a stationary random Gaussian function whose spectrum has a uniform distribution in the frequency band (f_1, f_2).

9 Calculate the spectrum of a set of impulses with width $T/2$, separated by T, the occurrence of each pulse having probability p. In particular, examine the case where $p = \frac{1}{2}$.

What happens to the spectrum if these pulses form a pseudo-random sequence with a length $2^4 - 1 = 15$ produced by a 4-bit shift register, following the polynomial $g(x) = x^4 + x + 1$?

10 A sinusoidal signal with frequency 1050 Hz is sampled at 8 kHz and coded in 10 bits. What is the maximum value of the signal-to-noise ratio? What is the value of the signal-to-quantization-noise ratio measured in the frequency band 300–500 Hz? What are the values if the sampling frequency is increased to 16 kHz?

11 The sinusoidal signal $\sin(2\pi t/8 + \phi)$ with $0 \leqslant \phi \leqslant \pi/2$ is sampled with period $T = 1$ and coded into 5 bits.

In the case where $\phi = 0$, calculate the power and the spectrum of the quantization noise. How does this spectrum appear as a function of the phase ϕ?

12 Consider a coding scale in which the quantizing step has a value q. What is the quantization of the signal $s_1(t) = \alpha q \sin(\omega_1 t)$ for $-1 \leqslant \alpha \leqslant 1$, as a function of the centring of the quantization characteristic? Show the envelope of the restored signal after decoding and narrow filtering around the frequency ω_1.

The signal $s_2(t) = 10q \sin \omega_2 t$ is superimposed on $s_1(t)$. Show the envelope of the restored signal under these conditions.

13 Assume a Gaussian signal is to be coded. How many bits would be required to have the signal-to-quantization-noise ratio greater than $50\,\mathrm{dB}$? Can this number be reduced if signal clipping is allowed for 1% of this time?

14 The signal $s(t) = A \sin(2\pi.810t)$ is coded into 8 bits. If the quantization step is q, trace the curve which shows the signal-to-quantization-noise ratio as a function of the amplitude A when this amplitude varies from q to $2^7 q$. Sketch the corresponding curve for non-linear coding following the 13-segment A-law.

15 Calculate the limits of the amplitude subranges for the optimal 2-bit coding of a unit Gaussian signal.

CHAPTER 2

THE DISCRETE FOURIER TRANSFORM

The discrete Fourier transform is introduced when the Fourier transform of a function is to be calculated using a digital computer. This type of processor can handle only numbers and in a quantity limited by the size of its memory. It follows that the Fourier transform

$$S(f) = \int_{-\infty}^{\infty} s(t) e^{-j2\pi f t} dt$$

must be adapted, by replacing the signal $s(t)$ with the numbers $s(nT)$ which represent a sampling of the signal, and by limiting to a finite value N the set of numbers on which the calculations are carried out. The calculation then provides numbers $S^*(f)$ defined by

$$S^*(f) = \sum_{n=0}^{N-1} s(nT) e^{-j2\pi f n T}$$

As the computer is limited in its processing power it can only provide results for a limited number of values of the frequency f, and it is natural to choose multiples of a certain frequency step Δf. Thus,

$$S^*(k\,\Delta f) = \sum_{n=0}^{N-1} s(nT) e^{-j2\pi n k\,\Delta f T}$$

The conditions under which the calculated values form a good approximation to the required values are examined below. An interesting simplifying choice is to take $\Delta f = 1/NT$. Then there are only N different values of $S^*(k/NT)$, which is a periodic set of period N since

$$S^*[(k+N)/NT] = S^*(k/NT)$$

On the other hand, the transform thus calculated appears as discrete values and, as shown in Section 1.6, this property is characteristic of the spectrum of periodic functions. Thus, the set $S^*(k/NT)$ is obtained by Fourier transformation of the set $s(nT)$, which is periodic with period NT. The discrete Fourier transform (DFT) and the inverse transform establish the relations between these two periodic sets.

The definition, properties, methods of calculation and applications of the discrete Fourier transform have been treated in numerous publications. Overviews are given in Refs [1–4].

2.1 DEFINITION AND PROPERTIES OF THE DISCRETE FOURIER TRANSFORM (DFT)

If two sets of complex numbers $x(n)$ and $X(k)$ which are periodic with period N are chosen, then the discrete Fourier transform and the inverse transform establish the following relationships between them:

$$X(k) = \frac{1}{N} \sum_{n=0}^{N-1} x(n) e^{-j2\pi nk/N} \tag{2.1}$$

$$x(n) = \sum_{k=0}^{N-1} X(k) e^{j2\pi kn/N} \tag{2.2}$$

The position of the scale factor $1/N$ is chosen so that the $X(k)$ are the coefficients of the Fourier series expansion of the set $x(n)$. This transformation has the following properties.

Linearity: If $x(n)$ and $y(n)$ are two sets with the same period and with transforms $X(k)$ and $Y(k)$, respectively, the set $v(n) = x(n) + \lambda y(n)$, where λ is a scalar, has the transform

$$V(k) = X(k) + \lambda Y(k)$$

A translation of the $x(n)$ implies a rotation of the phase of the $X(k)$. If the transform $X_{n_0}(k)$ of the set $x(n - n_0)$ is calculated, then

$$X_{n_0}(k) = \sum_{n=0}^{N-1} x(n - n_0) e^{-j2\pi nk/N} = X(k) e^{-j2\pi n_0 k/N}$$

A translation of the $x(n)$ from n_0 induces on $X(k)$ a rotation of the phase through an angle equal to $2\pi n_0 k/N$.

Symmetry: If the set $x(n)$ is real, the numbers $x(k)$ and $X(N - k)$ are complex conjugates:

$$\bar{X}(N - k) = \sum_{n=0}^{N} x(n) e^{j2\pi n(N-k)/N} = X(k)$$

If the set $x(n)$ is real and even, then so is the set $X(k)$. Indeed, if $x(N - n) = x(n)$ then, for example, for $N = 2P + 1$:

$$X(N - k) = x(0) + 2 \sum_{n=1}^{P} x(n) \cos\left(2\pi \frac{nk}{N}\right) = X(k)$$

If the set $x(n)$ is real and odd, the set $X(k)$ is purely imaginary. In this case: $x(N - n) = -x(n)$ and $x(0) = x(N) = 0$. For example, for $N = 2P + 1$ this becomes:

$$X(k) = -2j \sum_{n=1}^{P} x(n) \sin\left(2\pi \frac{nk}{N}\right) = -X(N - k)$$

It should be noted that $X(0) = X(N) = 0$.

As any real signal can always be decomposed into odd and even parts, these last two symmetry properties are important.

Circular convolution: The transform of a convolution product is equal to the product of the transforms.

If $x(n)$ and $h(n)$ are two sets with period N, the circular convolution $y(n)$ can be defined by the equation:

$$y(n) = \sum_{l=0}^{N-1} x(l)h(n-l) \tag{2.3}$$

This is a set which has the same period N. Its transform is written:

$$Y(k) = \sum_{n=0}^{N-1} \left[\sum_{l=0}^{N-1} x(l)h(n-l) \right] e^{-j2\pi nk/N}$$

$$= \sum_{l=0}^{N-1} x(l) \left[\sum_{n=0}^{N-1} h(n-l) e^{-j2\pi(n-l)k/N} \right] e^{-j2\pi lk/N}$$

$$Y(k) = \left(\sum_{n=0}^{N-1} h(n-l) e^{-j2\pi[(n-l)k/N]} \right) \left(\sum_{l=0}^{N-1} x(l) e^{-j2\pi lk/N} \right) = H(k)X(k) \tag{2.4}$$

This is an important property of the discrete Fourier transform. A direct application will be given later.

Parseval's relation: This relation states that the power of a signal is equal to the sum of the powers of its harmonics. Thus

$$\frac{1}{N} \sum_{n=0}^{N-1} x(n)\bar{x}(n) = \frac{1}{N} \sum_{n=0}^{N-1} x(n) \sum_{k=0}^{N-1} \bar{X}(k) e^{-j2\pi kn/N}$$

$$\frac{1}{N} \sum_{n=0}^{N-1} |x(n)|^2 = \sum_{k=0}^{N-1} \bar{X}(k) \frac{1}{N} \sum_{n=0}^{N-1} x(n) e^{-j2\pi kn/N} \tag{2.5}$$

$$\frac{1}{N} \sum_{n=0}^{N-1} |x(n)|^2 = \sum_{k=0}^{N-1} |X(k)|^2$$

However, the most important property of the discrete Fourier transform probably lies in the fact that it lends itself to efficient calculation techniques. This property has caused it to take a prominent position in digital signal processing.

2.2 FAST FOURIER TRANSFORM (FFT)

The equations defining the discrete Fourier transform provide a relationship between two sets of N complex numbers. This is conveniently written in matrix form by setting

$$W = e^{-j2\pi/N} \tag{2.6}$$

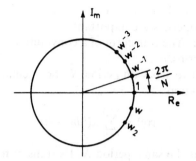

Fig. 2.1   *Co-ordinates of the coefficients of a discrete Fourier transform*

The co-ordinates of the numbers W^n, the coefficients of the discrete Fourier transform, appear on the unit circle in the complex plane as shown in Figure 2.1 and are the roots of the equation $Z^N - 1 = 0$.

The matrix equation for the direct transform is as follows:

$$
\begin{bmatrix} X_0 \\ X_1 \\ X_2 \\ \vdots \\ X_{N-1} \end{bmatrix} = \frac{1}{N} \begin{bmatrix} 1 & 1 & 1 & 1 & \cdots & 1 \\ 1 & W & W^2 & W^3 & \cdots & W^{N-1} \\ 1 & W^2 & W^4 & W^6 & \cdots & W^{2(N-1)} \\ \vdots & \vdots & \vdots & & & \vdots \\ 1 & W^{(N-1)} & W^{2(N-1)} & & \cdots & W^{(N-1)(N-1)} \end{bmatrix} \begin{bmatrix} x_0 \\ x_1 \\ x_2 \\ \vdots \\ x_{N-1} \end{bmatrix}
$$

For the inverse transform it is sufficient to remove the factor $1/N$ and to change W^n to W^{-n}.

The square matrix T_N of order N exhibits obvious features: rows and columns with the same index have the same elements and these elements are powers of a basic number such that $W^N = 1$. Significant simplifications can be envisaged under these conditions, leading to algorithms for fast calculation. A fast Fourier transform (FFT) is said to have been carried out when the discrete Fourier transform is calculated using such algorithms.

An important case occurs when N is a power of 2, because it leads to algorithms which are simple and particularly effective. These algorithms are based on a decomposition of the set to be transformed into a number of interleaved subsets. The case of interleaving in the time domain will be considered first, which leads to the so-called decimation-in-time algorithms.

2.2.1 Decimation-in-time fast Fourier transform

The set of elements $x(n)$ can be decomposed into two interleaved sets—those with even indices and those with odd ones. Using this decomposition, the first

$N/2$ elements of the set $X(k)$ can be written as

$$
\begin{bmatrix} X_0 \\ X_1 \\ X_2 \\ \vdots \\ X_{N/2-1} \end{bmatrix} = \begin{bmatrix} 1 & 1 & \cdots & 1 \\ 1 & W^2 & \cdots & W^{2(N/2-1)} \\ 1 & W^4 & \cdots & W^{4(N/2-1)} \\ \vdots & \vdots & & \vdots \\ 1 & W^{2(N/2-1)} & \cdots & W^{2(N/2-1)(N/2-1)} \end{bmatrix} \begin{bmatrix} x_0 \\ x_2 \\ x_4 \\ \vdots \\ x_{2(N/2-1)} \end{bmatrix}
$$

$$
+ \begin{bmatrix} 1 & 1 & \cdots & 1 \\ W & W^3 & \cdots & W^{N-1} \\ W^2 & W^6 & \cdots & W^{2(N-1)} \\ \vdots & \vdots & & \vdots \\ W^{N/2-1} & W^{3(N/2-1)} & \cdots & W^{(N/2-1)(N-1)} \end{bmatrix} \begin{bmatrix} x_1 \\ x_3 \\ x_5 \\ \vdots \\ x_{N-1} \end{bmatrix}
$$

If the matrix which multiplies the column vector of elements with even indices is denoted by $T_{N/2}$, then the matrix multiplying the vector of elements with odd indices can be factorized into the product $T_{N/2}$ and a diagonal matrix so that

$$
\begin{bmatrix} X_0 \\ X_1 \\ X_2 \\ \vdots \\ X_{N/2-1} \end{bmatrix} = T_{N/2} \begin{bmatrix} x_0 \\ x_2 \\ x_4 \\ \vdots \\ x_{2(N/2-1)} \end{bmatrix} + \begin{bmatrix} 1 & 0 & 0 & \cdots & 0 \\ 0 & W & 0 & \cdots & 0 \\ 0 & 0 & W^2 & \cdots & 0 \\ \vdots & \vdots & \vdots & & \vdots \\ 0 & 0 & \cdots & & W^{N/2-1} \end{bmatrix} T_{N/2} \begin{bmatrix} x_1 \\ x_3 \\ x_5 \\ \vdots \\ x_{N-1} \end{bmatrix}
$$

Similarly, for the last $N/2$ elements of the set $X(k)$, remembering that $W^N = 1$, one can write

$$
\begin{bmatrix} X_{N/2} \\ X_{N/2+1} \\ X_{N/2+2} \\ \vdots \\ X_{N-1} \end{bmatrix} = T_{N/2} \begin{bmatrix} x_0 \\ x_2 \\ x_4 \\ \vdots \\ x_{2(N/2-1)} \end{bmatrix} - \begin{bmatrix} 1 & 0 & 0 & \cdots & 0 \\ 1 & W & 0 & \cdots & 0 \\ 0 & 0 & W^2 & \cdots & 0 \\ \vdots & \vdots & \vdots & & \vdots \\ 0 & 0 & 0 & \cdots & W^{N/2-1} \end{bmatrix} T_{N/2} \begin{bmatrix} x_1 \\ x_3 \\ x_5 \\ \vdots \\ x_{N-1} \end{bmatrix}
$$

It is apparent that the calculation of $X(k)$ and $X(k + N/2)$ for $0 < k(N/2) - 1$ uses the same calculations with only a change in sign in the final sum. Hence the following diagram:

This shows that the calculation of a Fourier transform of order N reduces to the calculation of two transforms of order $N/2$, to which $M/2$ complex multiplications are added. By iteration through a number of steps given by $\log_2 N - 1 = \log_2(N/2)$, transforms of order of 2 are arrived at. These have the matrix

$$T_2 = \begin{bmatrix} 1 & 1 \\ 1 & -1 \end{bmatrix}$$

and no multiplications are required.

As each stage involves $N/2$ complex multiplications, the complete transformation requires M_c complex multiplications, where

$$M_c = (N/2)\log_2(N/2) \tag{2.7}$$

and A_c complex additions, where

$$A_c = N \log_2 N \tag{2.8}$$

In practice, the number of complex multiplications can again be reduced because some powers of W have certain properties. For example, $W^0 = 1$ and $W^{N/4} = -j$ do not require complex multiplications, and

$$W^{N/8} = \frac{\sqrt{2}}{2}(1 - j)$$

only require one complex half-multiplication each. Thus, three multiplications can be saved in the first stage, $3N/8$ can be eliminated in the penultimate one, and $2N/4$ in the last. The gain over all the stages is $5N/4 - 3$ and the minimum number of complex multiplications is given by

$$m_c = N/2[\log_2(N/2) - \tfrac{5}{2}] + 3 \tag{2.9}$$

It should be noted that all of these calculation reductions cannot always be easily implemented either in software or in hardware.

The matrix for the fourth-order transform is

$$T_4 = \begin{bmatrix} 1 & 1 & 1 & 1 \\ 1 & -j & -1 & +j \\ 1 & -1 & 1 & -1 \\ 1 & +j & -1 & -j \end{bmatrix} \tag{2.10}$$

The diagram for its reduction is shown in Figure 2.2. By convention, the arrows represent multiplications and the solid circles to the left of the elementary flow graphs, often called 'butterflies', represent an addition (upper one) and a subtraction (lower one). The eighth-order transform is represented in Figure 2.3.

It will be seen that in this treatment the indices of the $X(k)$ appear in natural order, while those of the $x(n)$ are in a permuted one. This permutation is caused by the successive interleavings and results in a reversal or inversion of the binary

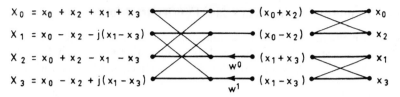

$$X_0 = x_0 + x_2 + x_1 + x_3$$
$$X_1 = x_0 - x_2 - j(x_1 - x_3)$$
$$X_2 = x_0 + x_2 - x_1 - x_3$$
$$X_3 = x_0 - x_2 + j(x_1 - x_3)$$

Fig. 2.2   *Transform of order 4 with decimation in time*

representation of the indices, which is often called 'bit-reversal'. For example, for $N = 8$,

$$x_0 \; (0 \; 0 \; 0) \text{ corresponds to } x_0 \; (0 \; 0 \; 0)$$
$$x_4 \; (1 \; 0 \; 0) \text{ corresponds to } x_1 \; (0 \; 0 \; 1)$$
$$x_2 \; (0 \; 1 \; 0) \text{ corresponds to } x_2 \; (0 \; 1 \; 0)$$
$$x_6 \; (1 \; 1 \; 0) \text{ corresponds to } x_3 \; (0 \; 1 \; 1)$$
$$x_1 \; (0 \; 0 \; 1) \text{ corresponds to } x_4 \; (1 \; 0 \; 0)$$
$$x_5 \; (1 \; 0 \; 1) \text{ corresponds to } x_5 \; (1 \; 0 \; 1)$$
$$x_3 \; (0 \; 1 \; 1) \text{ corresponds to } x_6 \; (1 \; 1 \; 0)$$
$$x_7 \; (1 \; 1 \; 1) \text{ corresponds to } x_7 \; (1 \; 1 \; 1)$$

The amount of data memory required to calculate a transform of order N is that needed to hold N complex positions. Indeed, the calculations are made on pairs of variables which undergo the operation represented by a butterfly and preserve their position in the ensemble of variables at the end of the operation, as is clearly shown in the diagrams. This is called 'in-place computation'. The inverse transform is simply obtained by changing the sign of the exponent of W. The factor $1/N$ can be introduced, for example, by halving the results of the additions and subtractions made in the butterflies. This allows scaling of the numbers in the memories.

This type of interleaving can also be applied to $X(k)$ when a similar algorithm is obtained. It leads to the so-called decimation-in-frequency algorithms.

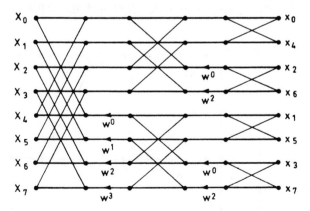

Fig. 2.3   *Transform of order 8 with decimation in time*

2.2.2 Decimation-in-frequency fast Fourier transform

The set of elements $X(k)$ can be decomposed into a pair of interleaved sets—one with even indices and the other with odd ones. For the even-index elements, since $W^N = 1$, the following situation occurs after elementary factorization:

$$
\begin{bmatrix} X_0 \\ X_2 \\ X_4 \\ \vdots \\ X_{2(N/2-1)} \end{bmatrix} = \begin{bmatrix} 1 & 1 & 1 & \cdots & 1 \\ 1 & W^2 & W^4 & \cdots & W^{2(N/2-1)} \\ 1 & W^4 & W^8 & \cdots & W^{4(N/2-1)} \\ \vdots & \vdots & \vdots & & \vdots \\ 1 & W^{2(N/2-1)} & & \cdots & W^{2(N/2-1)(N/2-1)} \end{bmatrix} \begin{bmatrix} x_0 + x_{N/2} \\ x_1 + x_{N/2+1} \\ x_2 + x_{N/2+2} \\ \vdots \\ x_{N/2-1} + x_{N-1} \end{bmatrix}
$$

For the elements with odd indices after a similar process, the corresponding equation is

$$
\begin{bmatrix} X_1 \\ X_3 \\ X_5 \\ \vdots \\ X_{N-1} \end{bmatrix} = \begin{bmatrix} 1 & W & W^2 & \cdots & W^{N/2-1} \\ 1 & W^3 & W^6 & \cdots & W^{3(N/2-1)} \\ 1 & W^5 & W^{10} & \cdots & W^{5(N/2-1)} \\ \vdots & \vdots & \vdots & & \vdots \\ 1 & W^{N-1} & W^{2(N-1)} & \cdots & W^{(N-1)(N/2-1)} \end{bmatrix} \begin{bmatrix} x_0 - x_{N/2} \\ x_1 - x_{N/2+1} \\ x_2 - x_{N/2+2} \\ \vdots \\ x_{N/2-1} - x_{N-1} \end{bmatrix}
$$

In this case the square matrix obtained is equal to the product of the matrix $T_{N/2}$ obtained for the elements with even indices, and the diagonal matrix whose elements are the powers W^k with $0 \leqslant k \leqslant N/2 - 1$. Thus,

$$
\begin{bmatrix} X_1 \\ X_3 \\ X_5 \\ \vdots \\ X_{N-1} \end{bmatrix} = \begin{bmatrix} 1 & 0 & 0 & \cdots & 0 \\ 0 & W & 0 & \cdots & 0 \\ 0 & 0 & W^2 & \cdots & 0 \\ \vdots & \vdots & \vdots & & \vdots \\ 0 & 0 & 0 & \cdots & W^{N/2-1} \end{bmatrix} \begin{bmatrix} x_0 - x_{N/2} \\ x_1 - x_{N/2+1} \\ x_2 - x_{N/2+2} \\ \vdots \\ x_{N/2-1} - x_{N-1} \end{bmatrix}
$$

The elements $x(k)$ with even and odd indices are calculated using the square matrix $T_{N/2}$ for the transform of order $N/2$ and the following diagram is obtained:

By adopting the same notation for the butterflies as was used in the preceding section, similar diagrams are obtained. Figure 2.4 shows the diagram for $N = 8$.

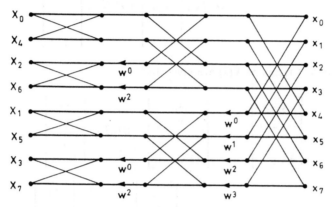

Fig. 2.4 *Transform of order 8 with decimation-in-frequency*

In decimation-in-frequency algorithms the number of calculations is the same as with time interleaving. The numbers $x(n)$ to be transformed appear in their natural order while the transformed numbers $X(k)$ are permuted.

The algorithms which have been obtained so far are based on a decomposition of the transform of order N into elementary second-order transforms which do not require multiplications. These algorithms are said to be radix 2 transforms. Other elementary transforms can also be used, and the most important is the radix 4 one, which uses the elementary matrix T_4.

2.2.3 Radix 4 FFT algorithm

This algorithm can be used when N is a power of 4. The set of numbers $x(k)$ is decomposed into four interleaved sets. Let us calculate the first $N/4$ values of $X(k)$ as an illustration of the decomposition. If $T_{N/4}$ denotes the square matrix of the transform of order $N/4$ and if D_i $(i = 1, 2, 3)$ is the diagonal matrix whose elements are the powers W^{ik} with $0 \leqslant k \leqslant N/4 - 1$, then

$$
\begin{bmatrix} X_0 \\ X_1 \\ X_2 \\ \vdots \\ X_{N/4-1} \end{bmatrix} = T_{N/4} \begin{bmatrix} x_0 \\ x_4 \\ x_8 \\ \vdots \\ x_{4(N/4-1)} \end{bmatrix} + D_1 T_{N/4} \begin{bmatrix} x_1 \\ x_5 \\ x_9 \\ \vdots \\ x_{N-3} \end{bmatrix}
$$

$$+ D_2 T_{N/4} \begin{bmatrix} x_2 \\ x_6 \\ x_{10} \\ \vdots \\ x_{N-2} \end{bmatrix} + D_3 T_{N/4} \begin{bmatrix} x_3 \\ x_7 \\ x_{11} \\ \vdots \\ x_{N-1} \end{bmatrix}$$

The next $N/4$ terms of $X(k)$ are given by:

$$\begin{bmatrix} X_{N/4} \\ X_{N/4+1} \\ \vdots \\ X_{N/2-1} \end{bmatrix} = T_{N/4} \begin{bmatrix} x_0 \\ x_4 \\ \vdots \\ x_{4(N/4-1)} \end{bmatrix} - jD_1 T_{N/4} \begin{bmatrix} x_1 \\ x_5 \\ \vdots \\ x_{N-3} \end{bmatrix}$$

$$- D_2 T_{N/4} \begin{bmatrix} x_2 \\ x_6 \\ \vdots \\ x_{N-2} \end{bmatrix} + jD_3 T_{N/4} \begin{bmatrix} x_3 \\ x_7 \\ \vdots \\ x_{N-1} \end{bmatrix}$$

This equation involves the same matrix calculations as the previous one with the addition of the multiplications by the elements of the second line of the matrix T_4. It can hence be shown that the calculation of the transform results in the diagram in Figure 2.5.

This type of transform is carried out in $\log_4 N - 1 = \log_4 (N/4)$ stages. Each stage requires $3(N/4)$ complex multiplications, which results in a total of M_{c4} multiplications, where

$$M_{c4} = \tfrac{3}{4} N \log_4 \left(\frac{N}{4} \right) \tag{2.11}$$

The number of complex additions A_{c4} is

$$A_{c4} = 2N \log_4 N \tag{2.12}$$

It is apparent that the number of additions is the same in radix 2 and in radix 4 algorithms, in contrast to the complex multiplications, where calculation in radix 4 algorithms results in a savings of over 25%. Figure 2.6 shows the complete diagram for $N = 16$.

Other radices can also be envisaged; for example, radix 8; in this case there are multiplications in the elementary matrix and the savings over radix 4 are negligible. Different radices can also be combined [5].

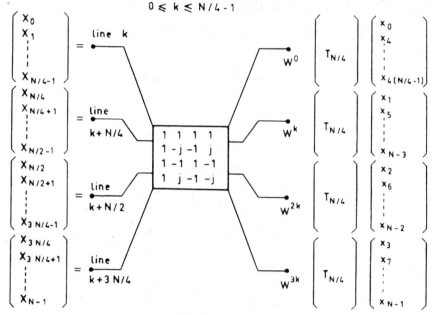

Fig. 2.5   Radix-4 transform

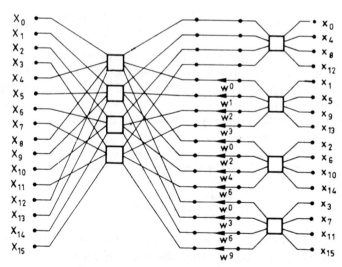

Fig. 2.6   Radix-4 transform of order 16

2.2.4 Split radix FFT algorithm

In a transform of order N the set of odd-index transformed values can be decomposed into two subsets:

$$X(4k+1) = \frac{1}{N} \sum_{n=0}^{N-1} x(n) W^n W^{4kn}$$

and:

$$X(4k+3) = \frac{1}{N} \sum_{n=0}^{N-1} x(n) W^{3n} W^{4kn}$$

Using definition (2.6) for W, the summations can also be written:

$$X(4k+1) = \frac{1}{N} \sum_{n=0}^{N/4-1} \left[\left[x(n) - x\left(n+\frac{N}{2}\right) \right] \right.$$
$$\left. - j\left[x\left(n+\frac{N}{4}\right) - x\left(n+\frac{3N}{4}\right) \right] \right] W^n W^{4nk} \qquad (2.13)$$

and:

$$X(4k+3) = \frac{1}{N} \sum_{n=0}^{N/4-1} \left[\left[x(n) - x\left(n+\frac{N}{2}\right) \right] \right.$$
$$\left. + j\left[x\left(n+\frac{N}{4}\right) - x\left(n+\frac{3N}{4}\right) \right] \right] W^{3n} W^{4kn} \qquad (2.14)$$

The set of even-index transformed values is:

$$X(2k) = \frac{1}{N} \sum_{n=0}^{N/2-1} \left[x(n) + x\left(n+\frac{N}{2}\right) \right] W^{2nk} \qquad (2.15)$$

The above equations show that the first step in the decimation-in-time approach can be replaced by the calculation of a transform of order $N/2$ and two transforms of order $N/4$. The split-radix algorithm is obtained through repetitive applications of the above procedure:

In the decimation-in-frequency approach the split-radix algorithm is based on the following decomposition:

$$X(k) = \sum_{n=0}^{N/2-1} x(2n) W^{2nk} + W^k \sum_{n=0}^{N/4-1} x(4n+1) W^{4nk}$$
$$+ W^{3k} \sum_{n=0}^{N/4-1} x(4n+3) W^{4nk} \qquad (2.16)$$

For a transform of order N the number of complex multiplications $M_{c2/4}(N)$ is given by the recurrence derived from the above equations:

$$M_{c2/4}(N) = M_{c2/4}\left(\frac{N}{2}\right) + 2M_{c2/4}\left(\frac{N}{4}\right) + \frac{N}{2} \qquad (2.17)$$

with the initial values $M(2) = M(4) = 0$. The value obtained in this way is slightly smaller than the same for a radix 4 algorithm.

As a reference, it is worth mentioning that the minimum number of non-trivial complex multiplications for a transform of order N, with $N = 2^m$, is given by $2m + 1 - 2m^2 + 4m - 8$ [5].

An additional point also is that the same algorithm can be used for the direct and the inverse transforms. In fact, it is sufficient to exchange the real and imaginary parts of the data at the input and output of a direct transform to achieve an inverse transform, because of the equality:

$$j\bar{x}(n) = \sum_{k=0}^{N-1} j\bar{X}(k) e^{-j2\pi kn/N}$$

and because the operation $j\bar{x}$ only consists of exchanging the real and imaginary parts of the complex number x.

The algorithms which have been presented for decimation-in-time and frequency and for radix 2 and 4 are elements of a large set of algorithms. A unified presentation of FFT algorithms is given in the next chapter, which allows the determination of the most appropriate for each application. In actual calculations, however, operations are carried out with limited precision. This results in some degradation of the signal.

2.3 DEGRADATION ARISING FROM WORDLENGTH-LIMITATION EFFECTS

The equipment used introduces limitations caused by the finite precision of arithmetic units and the limited capacity of memories. First, the coefficients are held in a memory, with a limited number of bits. Thus, the memory contents represent approximations to the actual coefficients, which are in general obtained by rounding. Second, as the calculation proceeds, roundings are performed so as to keep the data wordlengths within the capacities of the memory locations or of the arithmetic units. It is important to analyse the degradations introduced by these two types of wordlength limitation in order to be able to determine precisely the hardware which is necessary for producing a transform with a specific performance.

2.3.1 Effects of coefficient quantization

The coefficients actually used by the machine represent an approximation to the theoretical coefficients, which have values of the real and imaginary parts within the range $[-1, +1]$.

For the coefficient $e^{-j2\pi n/N}$, digitization into b_c bits involves a quantization error $\delta(n) = \delta_R(n) + j\delta_I(n)$ such that, if rounding is employed,

$$|\delta_R(n)| \leqslant 2^{-b_c} \quad \text{and} \quad |\delta_I(n)| \leqslant 2^{-b_c}$$

The calculation of each transformed number $X(k)$ from the data $x(n)$ is made with an error $\Delta(k)$ such that:

$$X(k) + \Delta(k) = \frac{1}{N} \sum_{n=0}^{N-1} x(n)[e^{-j2\pi nk/N} + \delta(nk)]$$

or

$$\Delta(k) = \frac{1}{N} \sum_{n=0}^{N-1} x(n)\delta(nk)$$

As the $x(n)$ and $X(k)$ are related by equation (2.2):

$$x(n) = \sum_{k=0}^{N-1} X(k)e^{j2\pi nk/N}$$

this becomes

$$\Delta(k) = \sum_{i=0}^{N-1} X(i)\varepsilon(i, k) \tag{2.18}$$

with

$$\varepsilon(i, k) = \frac{1}{N} \sum_{n=0}^{N-1} \delta(nk)e^{j2\pi ni/N}$$

Consequently, for the transformed number $X(k)$, rounding the coefficients of the transform introduces a perturbation $\Delta(k)$ obtained by summing the elementary perturbations, each of which is equal to the product of a transformed number by a factor representing its contribution. The transformed numbers interact with each other and are no longer strictly independent.

It is possible to calculate the $\varepsilon(i, k)$ for each transformation. In general, it is important to know the maximum value ε_m that the $|\varepsilon(i, k)|$ can have for a given order of transformation and for a given number of bits b_c.

The inequality $|\delta(n)| < 2^{-b_c}\sqrt{2}$, provides a maximum for ε_m as

$$\varepsilon_m \leqslant 2^{-b_c}\sqrt{2}$$

In practice, the values found for ε_m are much smaller than this maximum. For example, for $N = 64$, it is found that $\varepsilon_m \simeq 0.6 \times 2^{-b_c}$, and this value is also found for higher values of N [6].

2.3.2 Round-off noise in the FFT

The data are presented to a DFT calculator with a limited number of bits. At each operation of addition or multiplication this number of bits increases. In general, however, the number of bits available remains fixed within the machine.

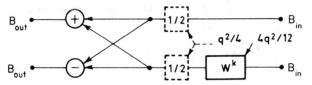

Fig. 2.7   *Butterfly of radix-2 fast Fourier transform*

This makes it necessary to limit the wordlength throughout the calculation.

This limitation is almost always made by eliminating the least significant bits by rounding. Moreover, as data overflow is not generally acceptable the scaling of the numbers in the memory must be carefully analysed.

Two simple but important cases will be examined for a radix 2 transform. The first is that of the direct transform, where, with a scale factor of $1/N$, it is sufficient to halve the results of the additions and subtractions at each butterfly operation to maintain a suitable scaling. The other case to be studied will be that in which the scaling selected at the beginning of the transform permits the whole of the calculation to be made without any risk of overflow. An example of this case could be the inverse transform.

In order to evaluate the power of the round-off noise it will be assumed that the machine stores the numbers in memories with a capacity of b_i bits for each real number and that the largest representable amplitude is unity. That is, the numbers take values within the range $[-1, +1]$ and the quantization step q has a value

$$q = 2 \times 2^{-b_i}$$

Figure 2.7 shows the schematic diagram of a butterfly with multiplication.

At the input of the butterfly, complex data are represented by a real and an imaginary part, each comprising b_i bits. After multiplication, the numbers are limited to b_i bits by rounding. On a real number this operation introduces noise with power estimated at $q^2/12$. In complex multiplication, which generally involves four real multiplications, the noise power introduced is $4(q^2/12)$. As the modulus of the coefficient W^k does not exceed unity, no scaling of the data in the memory is necessary.

Round-off noise with scaling

Scaling of the data is assumed to be done before the additions and subtractions of the butterfly. This is an unfavourable case but it is easy to realize. The real and imaginary parts are both divided by two by shifting, during which one bit is skipped. As this bit is either 0 or 1, with probability $\frac{1}{2}$, it can be assumed that noise of power $q^2/8$ results for each real number and $q^2/4$ for the complex one. Thus, the noise powers which were previously present have been divided by 4.

At the input to each butterfly the signal is affected by noise B_i and at the output by noise B_o. With scaling, the relation between B_i and B_o is given by

$$B_o = 2\left(\frac{q^2}{4} + \frac{B_i}{4}\right) + \frac{1}{4}\left(4\frac{q^2}{12}\right) = \tfrac{1}{2}q^2 + \tfrac{1}{2}B_i + \frac{q^2}{12}$$

The first stage of the transform does not involve multiplications, so if the noise at the input of the FFT calculator is neglected, the noise at the output of the first butterfly is

$$B_{o1} = \frac{q^2}{2}$$

The second stage in the decimation-in-time algorithm has multiplications by j, which do not involve rounding. The noise at the output under these conditions is

$$B_{o2} = 2\left(\frac{q^2}{4} + \frac{1}{4}B_{o1}\right) = \frac{q^2}{2} + \frac{q^2}{4}$$

and similarly:

$$B_{o3} = 2\left(\frac{q^2}{4} + \frac{1}{4}B_{o2}\right) + \frac{q^2}{12} = \frac{q^2}{2} + \frac{q^2}{4} + \frac{q^2}{8} + \frac{q^2}{12}$$

At the last stage of rank $\log_2(N)$:

$$B_{oT} = \frac{q^2}{2}\sum_{i=0}^{\log_2(N/2)}\frac{1}{2^i} + \frac{q^2}{12}\sum_{i=0}^{\log_2(N/8)}\frac{1}{2^i}$$

Finally, at the output of the transform, one can write

$$B_{oT} \simeq q^2$$

Thus, the result appears that in a transform involving scaling by dividing by 2 at each stage the power of the noise at each output can be estimated as

$$B_{oT} \simeq q^{2(1-b_i)} \tag{2.19}$$

Round-off noise without scaling

The powers of the noise at the input and output of a butterfly are related by

$$B_o = 2B_i + 4\frac{q^2}{12}$$

Considering that no noise is produced in the two first stages, the total noise at each output is given by:

$$B_{oT} = 4\frac{q^2}{12}\frac{N}{8}\sum_{i=0}^{\log_2(N/8)}\frac{1}{2^i}$$

whence:

$$B_{oT} \simeq N \frac{q^2}{12}$$

In a transform without scaling, the power of the noise at each output can be estimated as

$$B_{oT} = N \frac{2^{2(1-b_i)}}{12} \tag{2.20}$$

The same reasoning can be applied with comparable results to calculations involving other radices, in particular, radix 4.

In practice, the power of the noise should be related to the power of the signal and the most significant parameter is the signal-to-noise ratio. To determine how this parameter evolves under a transform the relation can be used which relates the power of the signal to the power of its spectrum:

$$\frac{1}{N} \sum_{n=0}^{N-1} |x(n)|^2 = \sum_{k=0}^{N-1} |X(k)|^2$$

The signal-to-noise ratio at the output of the transform depends on the distribution of the power between the $X(k)$. For example, in the calculation of the direct transform with scaling, under the hypothesis of uniform distribution of power between the $X(k)$, the power at each output is divided by N.

When S denotes the power of the signal at the input and when the noise at the input is neglected, the signal-to-noise ratio at the output of the transformation becomes:

$$(SNR)_{oT} = \frac{S}{N 2^{2(1-b_i)}} \tag{2.21}$$

Under the same conditions but without scaling, the power of the signal is multiplied by N and the ratio becomes:

$$(SNR)_{oT} = \frac{S}{2^{2(1-b_i)}/12} \tag{2.22}$$

The calculations made in this paragraph should be regarded only as approximations. They were made under the assumption of no correlation between the errors. Such an assumption is not always valid, particularly for transforms of low-order N. Such a simplified analysis, however, is sufficient in most applications; a more thorough analysis is given in Ref. [7]. The most direct application of the DFT is in spectrum analysis.

2.4 CALCULATION OF A SPECTRUM USING THE DFT

The calculation of a spectrum by the discrete Fourier transform requires that certain approximations and a suitable choice of parameters should be made to attain the desired performance. Before considering any applications it is therefore useful to look carefully at the function fulfilled by the discrete Fourier transform.

2.4.1 The filtering function of the DFT

Let us examine the relationship between the outputs $X(k)$ and the inputs $x(n)$, considered as the result of sampling a signal $x(t)$ with period T, which establishes the discrete Fourier transform. For $k = 0$ this is

$$X(0) = \frac{1}{N} \sum_{n=0}^{N-1} x(n)$$

The signal $X(0)$ thus defined results from the convolution of the signal $x(t)$ with the distribution $\phi_0(t)$ such that:

$$\phi_0(t) = \frac{1}{N} \sum_{n=0}^{N-1} \delta(t - nT)$$

The Fourier transform of this distribution is given by:

$$\Phi_0(f) = \frac{1}{N} \sum_{n=0}^{N-1} e^{-j2\pi nf/T} = \frac{1}{N} \frac{1 - e^{-j2\pi f/NT}}{1 - e^{-j2\pi fT}}$$

or

$$\Phi_0(f) = e^{-j\pi f(N-1)T} \Phi(f)$$

with

$$\Phi(f) = \frac{1}{N} \frac{\sin(\pi fNT)}{\sin(\pi fT)} \tag{2.23a}$$

Now, a convolution operation in time space corresponds to a product in frequency space, i.e. $X(0)$ is a signal obtained by filtering the input signal by the function $\Phi_0(f)$. Figure 2.8 shows the function $\Phi(f)$ and the function $\phi(t)$ of which it is the Fourier transform. The function $\Phi(f)$ is zero at points on the

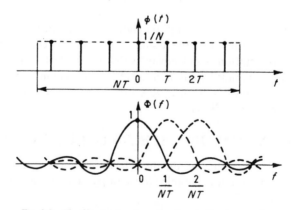

Fig. 2.8   The filtering function of a discrete Fourier transform

frequency axis which are whole multiples of $1/NT$, except for multiples of $1/T$. It is periodic and has a period $1/T$, conforming to the laws of sampling. It is simply the spectrum of a sampled impulse of width NT. Similarly, the output $X(k)$ has the corresponding function $\phi_k(t)$ such that:

$$\phi_k(t) = \frac{1}{N} \sum_{n=0}^{N-1} e^{-j2\pi nk/N} \delta(t - nT)$$

$$\Phi_k(f) = \frac{1}{N} \sum_{n=0}^{N-1} e^{-j2\pi nk/N} e^{-j2\pi fnT}$$

In compact form, after simplification, this becomes:

$$\Phi_k(f) = (-1)^k e^{-j\pi f(N-1)T} e^{j\pi k/N} \Phi\left(f + \frac{k}{NT}\right) \qquad (2.23b)$$

The output $X(k)$ provides the signal filtered according to the function $\Phi_0(f)$ but translated by k/NT along the frequency axis.

Thus, the discrete Fourier transform forms a set of N identical filters, or a bank of filters, distributed uniformly over the frequency domain at intervals of $1/NT$.

If the input signal is periodic, then, from the definition of the DFT, this bank of filters is frequency sampled at intervals of $1/NT$, and it should be noted that there is no interference between the outputs $X(k)$. Strictly speaking, this property is lost if the coefficients are rounded, as has been shown earlier.

The DFT function just pointed out also illustrates the problem of scaling numbers in the memory of an FFT calculator. Let us suppose that the numbers of transform $x(n)$ results from the sampling of a random signal for which the amplitude probability distribution has the variance σ^2. If the signal has a uniform distribution of its energy spectrum its power is uniformly distributed between the $X(k)$, and each has a variance equal to σ^2/N. By contrast, if the signal has a spectral distribution which can be concentrated on one $X(k)$, this $X(k)$ has the same probability distribution as the $x(n)$, in particular the variance σ^2. Scaling the numbers by dividing by 2 at each stage of an FFT calculation is suitable for the handling of such signals.

A different view of the filtering process is provided by observing that the outputs $X(k)$ of the DFT are the sums of the inputs $x(n)$ after phase shifting. In effect, the output $X(0)$ is the sum of the $x(n)$ with zero phase shift, the output $X(k)$ is the sum of $x(n)$ with phase shifts which are multiples of $2\pi(k/N)$, as shown in Figure 2.9. At each output, in-phase components of the resulting signals add, while the others cancel. For example, if the $x(n)$ are complex numbers with the same phase and modulus, all the $X(k)$ become zero except for $X(0)$. The input signal is thus found to be decomposed according to the base represented by the N vectors $e^{-j2\pi(k/N)}$ with $0 \leqslant k \leqslant N - 1$.

This result is useful in studying the banks of filters which include a DFT processor.

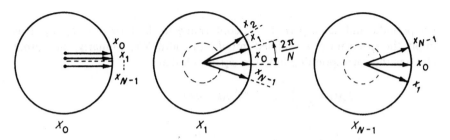

Fig. 2.9   Filtering by phase shift in a discrete Fourier transform

2.4.2 Spectral resolution

Spectrum analysis is used in many fields and is often performed on recorded data. By definition, the DFT establishes a relation between two periodic sets, the $x(n)$ and $X(k)$, each of which includes N different elements. In order to use it, it is necessary therefore to examine this double periodicity.

The periodicity in the frequency is introduced by the process of sampling the signal. The data to be processed are either in digital form, with sampling and coding being performed within the signal source, or in analogue form, which then has to be converted into numbers. The choice of sampling frequency $f_s = 1/T$ should be such that components of the signal with frequencies greater than or equal to $f_s/2$ are negligible, and, in any case, less than the tolerable error in the amplitude of the useful components. Fulfilment of this condition can be ensured by prefiltering the signal.

Periodicity in time is introduced artificially by assuming that the signal is repeated outside the time interval $\theta = NT$ which actually corresponds to the data being processed. Under these conditions, the DFT supplies a sampling of the spectrum with a frequency period Δf, equal to the inverse of the duration of the data, which constitutes the frequency resolution of the analysis. The relation $\Delta f = NT$ expresses the Heisenberg uncertainty principle for spectrum analysis. A more accurate analysis can be obtained by increasing the duration of the data collection, for example, by making it $N'T$ (with $N' > N$) with zero additional samples. The additional frequency samples are obtained simply from an interpolation of the others. This procedure is currently used to provide a number N' of data points which is a power of 2 and so to allow fast algorithms to be used. On the other hand, the fact that the signal is not formed solely of lines at frequencies which are multiples of $1/NT$ introduces interferences between the spectral components obtained. Indeed, the filter function $\Phi(f)$ of the DFT, which is given in Section 2.4.1, introduces ripples throughout the frequency band and, if the signal has a spectral component $S(f_0)$ at frequency f_0 such that $k/NT < f_0 < (k+1)/NT$, then:

$$X(k) = S(f_0)\Phi(k/NT - f_0), \quad 0 \leqslant k \leqslant N - 1 \tag{2.24}$$

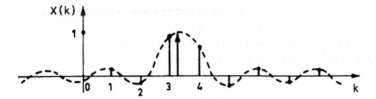

Fig. 2.10   Analysis of a signal with a frequency which is not a multiple of 1/NT

As a result, all the transform outputs can assume non-zero values as shown in Figure 2.10. Limitations thus appear for the resolution of the analyser. This effect can be reduced by modifying the filter function of the DFT by weighting the signal samples before transforming them.

This operation amounts to replacing the rectangular time window $\phi(t)$ by a function whose Fourier transform results in smaller ripples. Numerous functions are used, of which the simplest are the raised cosine window:

$$\phi(t) = \frac{1}{2}\left(1 + \cos 2\pi \frac{t}{NT}\right) \qquad (2.25)$$

and the Hamming window:

$$\phi(t) = 0.54 + 0.46 \cos 2\left(\pi \frac{t}{NT}\right) \qquad (2.26)$$

The latter function has 99.96% of its energy in the main lobe. The peak side lobe ripple is about 40 dB below the main lobe peak. Other time windows can be used and several efficient functions are introduced in Ref. [8].

Let $\Phi(f)$ be the spectrum of the time window $\phi(t)$ after sampling; the expression (2.24) can be extended to any signal with spectrum $S(f)$, using the definition of the convolution and taking into account the periodicity of $\Phi(f)$:

$$X(k) = \int_0^{1/T}\left[\frac{1}{T}\sum_{n=-\infty}^{\infty} S\left(u - \frac{n}{T}\right)\right]\Phi\left(\frac{k}{NT} - u\right) du \qquad (2.27)$$

The signal spectrum aliasing due to sampling with period T is apparent.

To cope better with the interferences between the calculated spectral components it is necessary to employ a bank of more selective filters, like that presented in Chapter 11.

The DFT can also be used indirectly in the calculation of convolutions.

2.5 FAST CONVOLUTION

The efficiency of fast Fourier transform algorithms leads to the use of the DFT in cases other than spectrum analysis and, in particular, in convolutions.

Although, in general, this approach is not the most efficient it can be useful in applications where an FFT processor is available.

One of the properties of the DFT is that the transform of a convolution product is equal to the product of its transforms. Given two sets $x(n)$ and $h(n)$ of period N, with transforms $X(k)$ and $H(k)$, the circular convolution

$$y(n) = \sum_{m=0}^{N-1} h(m)x(n-m)$$

is a set of the same period whose transform is written:

$$Y(k) = H(k)X(k)$$

Fast convolution consists of calculating the set $y(n)$ by applying a discrete Fourier transform to the set $Y(k)$. As one of the convolution terms is normally constant, the operation requires one DFT, one product and one inverse DFT. This technique is applied to sets of finite length. If $x(n)$ and $h(n)$ are two sets of N_1 and N_2 non-zero terms, the set $y(n)$ defined by

$$y(n) = \sum_{m=0}^{n} h(m)x(n-m)$$

is a set of finite length, having $N_1 + N_2 - 1$ terms. The fast convolution is applied by considering that the three sets $y(n)$, $x(n)$ and $h(n)$ have the period N such that $N \geqslant N_1 + N_2 - 1$. It is then sufficient to complete each set with a suitable number of zero terms. It is of particular interest if a power of 2 is chosen for N.

Nevertheless, in practice, convolution is a filtering operation, where the $x(n)$ represent the signal and the $h(n)$ the coefficients. The set of the $x(n)$ is much longer than that of the $h(n)$ and it is necessary to subdivide the calculation. To do this, the set of the $x(n)$ is regarded as a superposition of elementary sets $x_k(n)$ each of N_3 terms. That is,

$$x(n) = \sum_{k} x_k(n)$$

with $x_k(n) = x(n)$ for $kN_3 \leqslant n \leqslant (k+1)N_3 - 1$ and $x_k(n) = 0$ elsewhere.
We can then write:

$$y(n) = \sum_{m=0}^{n} h(m) \sum_{k} x_k(n-m)$$

$$y(n) = \sum_{k} \sum_{m=0}^{n} h(m)x_k(n-m) = \sum_{k} y_k(n)$$

Each set $y_k(n)$ contains $N_3 + N_2 - 1$ non-zero terms. Thus, the convolutions involve $N_3 + N_2 - 1$ terms. Figure 2.11 shows the sequence of operations. The sets $y_k(n)$ and $y_{k+1}(n)$ have $N_2 - 1$ terms which are superposed. The same operations can be performed by decomposing the set $x(n)$ into sets $x_k(n)$ such that $N_2 - 1$ terms are superposed.

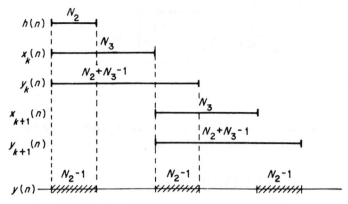

Fig. 2.11   Sequence of the operations in fast convolution

In this process the number of calculations to be made on each element of the set $y(n)$ increases as $\log_2(N_2 + N_3 - 1)$ and N_3 must not be chosen too large. Also, if $N_3 < N_2$, no terms in the set $y(n)$ can be obtained directly. Consequently, there is an optimal value for N_3. The number of memory locations needed increases as $N_3 + N_2 + 1$, and a good compromise is reached by taking for N_3 the first value above N_2 such that $N_3 + N_2 - 1$ is a power of 2.

2.6   CALCULATION OF A DFT USING CONVOLUTION

In certain applications only those operators which can form convolutions are available to calculate a DFT. This is the case for circuits using charge transfer devices which allow calculations to be made on the sampled signals in analogue form at speeds compatible with the frequencies met, for example, in radar applications.

The definition of a DFT can be written

$$X(k) = \frac{1}{N} \sum_{n=0}^{N-1} x(n) e^{-j2\pi(nk/N)}$$

By writing

$$nk = \tfrac{1}{2}[n^2 + k^2 - (n-k)^2] \quad \text{and} \quad W = e^{-j(2\pi/N)}$$

this becomes

$$X(k) = W^{k^2/2} \sum_{n=0}^{N-1} x(n) W^{n^2/2} W^{-(n-k)^2/2} \qquad (2.28)$$

This equation expresses the circular convolution product of the sets $x(n)W^{n^2/2}$ and $W^{-n^2/2}$. It follows that the calculation of $X(k)$ can be performed in three stages comprising the following operations:

(1) Multiply the data $x(n)$ by the coefficients $W^{n^2/2}$;
(2) Form the convolution product with the set of coefficients $W^{-n^2/2}$;
(3) Multiply the results by the coefficients $W^{k^2/2}$.

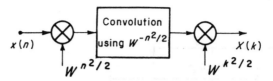

Fig. 2.12   Calculation of a discrete Fourier transform by convolution

This process is represented in Figure 2.12, and the method can be extended to the case where W is a complex number with a non-unit modulus [9].

2.7 IMPLEMENTATION

In order to implement DFT algorithms the following elements must be employed:

(1) A memory unit to store the input–output data and the intermediate results;
(2) A memory unit to store the coefficients of the transform;
(3) An arithmetic unit which can add and multiply complex numbers;
(4) A control unit to link the various operations.

These fundamental elements are found in every machine designed for digital signal processing, whether it uses hard-wired or programmed logic. Implementation of the FFT has two main characteristics:

(1) A large number of arithmetic calculations have to be performed;
(2) Permutations are required on the data, resulting in complicated calculations for the indices.

The constraints increase with the order of the transform.

The problem of implementation is to find efficient procedures for the algorithms described in the above sections. Various circuits and methods of organizing the calculations can be devised [10, 11]. For example, a simple circuit to implement the algorithm in Figure 2.4 is shown in Figure 2.13. The data $x(n)$ are presented in serial form and the transform is of order $N = 8$.

The circuit comprises a 4-bit register M_1, two 2-bit registers M_2 and two 1-bit registers M_3. The switches controlled, respectively, by the signals C_1, C_2 and C_3 permit suitable linkage of the calculations. The coefficients W^P must be applied to the multipliers at the appropriate moment. Figure 2.13 shows the control signals and the input and output signals.

The complex multiplier is implemented as a set of real multipliers and adders which carry out the operation:

$$c_R + jc_I = (a_R + ja_I)(b_R + jb_I)$$

as follows:

$$c_R = a_R b_R - a_I b_I; \quad c_I = a_R b_I + a_I b_R$$

There is another approach which leads to three multiplications instead of four

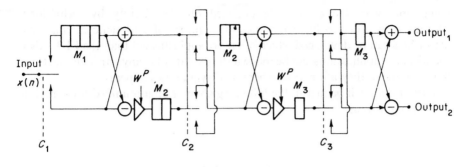

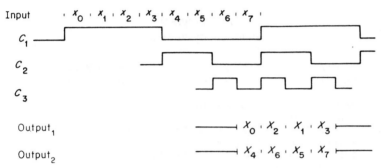

Fig. 2.13 *Serial implementation of radix-2 fast Fourier transform*

and five additions instead of two:

$$(a_R + ja_I)(b_R + jb_I) = [(a_R + a_I)b_R - a_I(b_R + b_I)] + j[(a_R + a_I)b_R + a_R(b_I - b_R)]$$
(2.29)

Moreover, if the number $b_R + jb_I$ is fixed, the number of additions reduces to three for every complex multiplication.

If the flow of data is continuous and the FFT calculations are to be performed on adjacent data blocks the fundamental rhythm of the calculations is that of the data, and it will be noted that the multipliers are inactive for half of the time. By processing consecutive blocks of data with a fundamental rhythm equal to half the rate of the data it is possible to optimize the circuit so as to use the multipliers to their maximum capacity.

Two points arise from this example. First, the circuit involves three stages, each with a different structure, which have different control signals. This is a drawback because we are looking for modular circuits, especially for large-scale integration. Second, the results are received in a permuted order, and an extra process is required to create the natural order of the indices.

When a degree of optimization is being sought, particularly to adapt the circuits to the algorithms, it is preferable to look for algorithms which are compatible with the constraints imposed on the circuity by technology. Moreover, in order to further reduce the number of arithmetic calculations, algorithms can be developed which are more sophisticated than the FFT. Also, significant simplifications can be made if the data display any particular features (for example, if they are real numbers or if symmetries appear).

These topics will be met in the following chapter after a unified presentation of FFT algorithms.

APPENDIX 1

FFT Program
(with binary inversion)

```
C         PROGRAM FFT1 (CALCULATION OF DISCRETE FOURIER TRANSFORM BY
C         FAST FOURIER TRANSFORM WITH BINARY INVERSION)
C
C         ORDER OF THE TRANSFORM :N
C         POWER OF TWO :N1
C         S :TABLE OF NUMBERS TO BE TRANSFORMED
C         W :TRANSFORM COEFFICIENTS
C         TYPE OF TRANSFORM:INVERSE IF INV=TRUE
          COMPLEX S, W
          LOGICAL INV
          DIMENSION S(2048)
          N=2048
          N1=11
          IMP=06
          INV=.FALSE.
C         TEST FUNCTION TO BE TRANSFORMED
          DO10I=1, N
    10    S(I)=(0.,0.)
          N32=N/32
          DO20I=1, N32
          S(I)=(1.,0.)
          S(N+1-I)=(1.,0.)
          S(N+1-N32)=(.5,0.)
          S(N32+1)=(.5,0.)
    20    CONTINUE
          CALL TFR (S, N, N1, INV)
          IF(INV)GOTO70
          FLN=FLOAT(N)
          DO80I=1,N
    80    S(I)=S(I)/FLN
    70    CONTINUE
C         PRINT RESULTS
          WRITE(IMP, 100)
   100    FORMAT (1H1,//,120(1H*),//)
          WRITE(IMP,201)
```

```
201    FORMAT(10X, 'FREQUENCY ',23X, 'FOURIER COEFF ', 13X,
       'AMPLITUDE',/)
       DO 50 I=1, 33
       I1=I-1
       R=CABS(S(I))
50     WRITE(IMP,200)I1,S(I),R
200    FORMAT(10X,I6,10X,3E20.4)
       WRITE(IMP, 102)
102    FORMAT(//,120(1H*),//)
       STOP
       END
       SUBROUTINE TFR(S, N, N1, INV)
C      S: TABLE OF NUMBERS TO BE TRANSFORMED
C      N: ORDER OF THE TRANSFORM
C      TYPE OF TRANSFORM: INVERSE IF INV=TRUE
       COMPLEX S(2048)
       LOGICAL INV
       COMPLEX W,W0,W00,AA,BB,S0
       DIMENSION W(1025)
     DIMENSION K(12), KK(14), JM (12)
C      BINARY INVERSION
       N2=2**(N1-1)
       IM=N1/2+0,1
       SING=-1
       IF(INV)SING=1.
       DO100I=1,IM
       IM1=N1-I
       D081I1=1.IM1
81     JM(I1)=2
       D082I1=Im1,12
82     JM(I1)=1
       J12M=JM(12)
       J11M=JM(11)
       J10M=JM(10)
       J9M=JM(9)
       J8M=JM(8)
       J7M=JM(7)
       J6M=JM(6)
       J5M=JM(5)
       J4M=JM(4)
       J3M=JM(3)
       J2M=JM(2)
       J1M=JM(1)
       D0100J12=1,J12M
       K(12)=J12-1
       D0100J11=1,J11M
       K(11)=J11-1
       D0100J10=1,J10M
       K(10)=J10-1
       D0100J9=1,J9M
       K(9)=J9-1
       D0100J8=1,J8M
       K(8)=J8-1
       D0100J7=1,J7M

       IF(I,EQ,1)GOTO94
       IM1=I-1
       D092I1=1,IM1
       KK(I1)=K(NBP+I1)
92     KK(N1+1-I1)=KK(I1)
94     KT1=1
       KT2=2**(N1-1)
       KT=KK(1)
       KI=KK(1)*KT2
       D0105J=2,N1
       KT1=KT1*2
       KT2=KT2/2+0.1
       KT=KT=KK(J)*KT1
105    KI=KI+KK(J)*KT2
       S0=S(KT+1)
       S(KT+1)=S(KI+1)
       S(KI+1)=S0
100    CONTINUE
C      CALCULATION OF EXPONENTIALS
       A=2.*3.14159265/N
       B=COS(A)
       C=SING*SIN(A)
       W0=CMPLX(B,C)
       W00=W0
       W(2)=W0
       D0190I=2,N2
       W00=W00*W0
190    W(I+1)=W00
       W(1)=1.
C      FAST ALGORITHM
       D080I=1,N1
       ND=2**(N1-I)
       NP1=2**(I-1)
```

```
      K(7)=J7-1                              NP=2*NP1
      DO100J6=1,J6M                          DO80KD=1,ND
      K(6)=J6-1                              KD0=(KD-1)*NP
      DO100J5=1,J5M                          DO80J=1,NP1
      K(5)=J5-1                              J0=J+KD0
      DO100J4=1,J4M                          NEXP=1+(J-1)*ND
      K(4)=J4-1                              AA=S(J0)
      DO100J3=1,J3M                          BB=S(J0+NP1)*W(NEXP)
      K(3)=J3-1                              S(J0)=AA+BB
      DO100J2=1,J2M                          S(J0,NP1)=AA-BB
      K(2)=J2-1                      80    CONTINUE
      DO100J1=1,J1M                          IF(INV)GOTO112
      K(1)=J1-1                              FLN=FLOAT(N)
      NBP=N1-2*I                             DO110I=1,N
      DO90I1=1,NBP                   110   S(I)=S(I)/FLN
90    KK(I1+I)=K(I1)                 112   CONTINUE
      KK(I)=1                                RETURN
      KK(N1+1-I)=0                            END
```

APPENDIX 2

FFT Program
(without binary inversion)

```
C        ORDER OF THE TRANSFORM :N                  SUBROUTINE TFF(T1,T2,W,N)
C        T1:TABLE OF NUMBERS TO BE TRANSFORMED      COMPLEX T1,T2,W
C        T2:WORK TABLE                              DIMENSION T1(2048),T2(2048)
C        W:TRANSFORM COEFFICIENTS                   W(1024)
C        TYPE OF TRANSFORM:INVERSE IF INV=TRUE      COMPLEX A,B
         COMPLEX T1, T2,W                           LOGICAL SW
         LOGICAL INV                                SW=.TRUE.
         DIMENSION T1(2048), T2(2048), W(1024)      N2=N/2
         N=2048                                     IH=N2
         IMP=06                               1   CONTINUE
         INV=.FALSE.                                IF(IH.EQ.0)GOTO2
C        TEST FUNCTION TO BE TRANSFORMED            J=0
         N16=N/16                             3   CONTINUE
         N161=N16+1                                L=J+N2
         DO20I=1,N16                                JK=J-MOD(J,IH)
20       T1(I)=(1.,0.)                              IP=J+JK
         DO30I=N161,N                               IQ=IP+IH
30       T1(I)=(0.,0.)                              IF(SW)GOTO4
1000     CONTINUE                                   A=T2(IP+1)
         CALL WEXP(INV,N,W)                         B=T2(IQ+1)*W(JK+1)
         CALL TFF(T1,T2,W,N)                        T1(J+1)=A+B
         IF(INV)GOTO70                              T1(L+1)=A-B
         FLN=FLOAT(N)                               GOTO5
         DO80I=1,N                            4   A=T1(IP+1)
80       T1(I)=T1(I)/FLN                            B=T1(IQ+1)*W(JK+1)
70       CONTINUE                                   T2(J+1)=A+B
```

```
C       PRINT RESULTS                         T2(L+1)=A-B
        WRITE(IMP,100)                    5   CONTINUE
100     FORMAT(1H1,//,120(1H*),//)            J=J+1
        WRITE(IMP,201)                        IF(J.LT.N2)GOTO3
201     FORMAT(10X,'FREQUENCY ',23X,          IH=IH/2
        'FOURIER COEFF ',15X,'AMPLITUDE',/)   SW=.NOT.SW
        DO50I=1,34                            GOTO1
        I1=I-1                            2   CONTINUE
        R=CABS(T1(I))                         IF(SW)RETURN
50      WRITE(IMP,200)I1,T1(I),R              DO6I=1,N
200     FORMAT(10X,I6,10X,3E20.4)         6   T1(I)=T2(I)
        WRITE(IMP,102)                        RETURN
102     FORMAT(//,120(1H*),//)                END
        STOP
        END
        SUBROUTINE WEXP(INV,N,W)
C       CALCULATION OF COMPLEX EXP.
        COMPLEX W
        DIMENSION W(1024)
        LOGICAL INV
        N2=N/2
        ARG=6.283185/N
        W(1)=(1.,0.)
        W(2)=CEXP(CMPLX(0.,-ARG))
        IF(INV)W(2)=CONJG(W(2))
        DO10I=3,N2
10      W(I)=W(2)*W(I-1)
        RETURN
        END
```

REFERENCES

[1] Special issue on FFT and applications. *IEEE Transactions*, **AU15**(2), 1967.

[2] A. OPPENHEIM and R. SCHAFER, *Digital signal processing*, Prentice Hall, Englewood Cliffs NJ, 1974, Chs 3 and 6.

[3] L. RABINER and B. GOLD, *Theory and application of digital signal processing*, Prentice Hall, Englewood Cliffs NJ, 1975, Chs 6 and 10.

[4] *Discreet Fourier Transform Programs*, Chapter 1, Programs for Digital Signal Processing, IEEE Press, John Wiley, New York, 1979.

[5] P. DUHAMEL and H. HOLLMANN, Split radix FFT algorithm. *Electronics Letters*, 5 January 1984, **20**, No. 1, 14–16.

[6] D. TUFTS and H. HERSEY, Effects of FFT Coefficient quantization on bin frequency response. *Proceedings of IEEE*, Jan 1972.

[7] TRAN-THONG and BEDE and LIU, Fixed point FFT error analysis. *IEEE Transactions*, **ASSP24**(6), 1976.

[8] A. EBERHARD, An optimal discrete window for the calculation of power spectra. *IEEE Trans.*, **AU-21**(1), 1973.

[9] L. R. RABINER, R. W. SCHAFER and C. M. RADER, The chirp Z-transform algorithm. *IEEE Transactions*, **AU17**, June 1969.

[10] H. L. GROGINSKY and G. A. WORKS, A pipeline fast Fourier transform. *IEEE Transactions on computers*, **19**, Nov 1970.

[11] B. GOLD and T. BIALLY, Parallelism in fast Fourier transform hardware. *IEEE Transactions*, **AU21**(1), 1973.

EXERCISES

1 Calculate the discrete Fourier transform of the set comprising $N = 16$ terms such that:

$$x(0) = x(1) = x(2) = x(14) = x(15) = 1$$
$$x(n) = 0 \quad \text{for} \quad 3 \leqslant n \leqslant 13$$

and of the set

$$x(0) = x(1) = x(2) = x(3) = x(4) = 1$$
$$x(n) = 0 \quad \text{for} \quad 5 \leqslant n \leqslant 15$$

Compare the results obtained. Carry out the inverse transform of these results.

2 Establish the diagram for the FFT algorithm of order 16 with time and frequency interleaving. What is the minimum number of multiplications and additions that are required?

3 Calculate the discrete Fourier transform of the set comprising $N = 128$ terms such that:

$$x(0) = x(1) = x(2) = x(126) = x(127) = 1$$
$$x(n) = 0 \quad \text{for} \quad 3 \leqslant n \leqslant 125$$

Compare the results with those in Exercise 1.

The set $X(k)$ obtained forms an approximation to the Fourier series expansion of a set of impulses. Compare the results obtained with the figures in the Table in Appendix 1, Chapter 1. Account for the differences.

4 We wish to develop a DFT of order 64 with a minimum of arithmetic operations. Determine the number of multiplications and additions required with radix 2, 4, and 8 algorithms.

5 Analyse the power of the rounding noise produced in a transform of order 32. Using the results in Section 2.3, show how the results vary at the different outputs. Calculate the distortion introduced by limiting the coefficients to 8 bits.

6 Show that each output of a DFT, $X(k)$, can be obtained from the inputs $x(n)$ by a recurrence relation. Calculate the number of multiplications that would be required.

7 Carry out a DFT of order 64 on data which are 16-bit numbers. Calculate the degradation of signal-to-noise ratio when a cascade of direct and inverse transforms is used on a 16-bit machine.

8 Assume that the bandwidth occupied by a signal for analysis is from 0 to 10 kHz. The spectral resolution required is 1 Hz. What length of recording is required in order to make such an analysis? What memory capacity is required to store the data assuming it is coded into 8 bits. Determine the characteristics of a computer capable of effecting such a spectral analysis: memory capacity, memory cycle, addition, and multiplication times.

9 Calculate the DFT of the set $x(n)$ which is defined by:

$$x(n) = \sin(2\pi n/3.5) + 0.2 \sin(2\pi n/6.5) \quad \text{with} \quad 0 \leqslant n \leqslant 15$$

The following windows are used to improve the analysis:

$$g(n) = \tfrac{1}{2}[1 - \cos(2\pi n/16)]$$

$$g(n) = 0.54 - 0.46 \cos\frac{2\pi n}{16} \quad \text{(Hamming)}$$

$$g(n) = 0.42 - 0.5\cos\frac{2\pi n}{16} + 0.08\cos\frac{4\pi n}{16} \quad \text{(Blakman)}$$

Compare the results.

10 Describe in detail the operation of the circuit in Figure 2.13. In particular, give the set of numbers at the output of each adder and subtractor. Show that a supplementary input memory is necessary to make the multipliers operate to their full capacity. Give the control signals in this case.

11 Generalize to radix 4 the implementation circuit given in Section 2.7. Give the detailed timing diagram.

Determine the necessary calculation power and the volume of memory for the data and coefficients if the transform is of order $N = 64$, if the data have 16 bits and appear at a rate of 8 kHz, and if the coefficients have 16 bits. Compare these results with the characteristics of a currently available microprocessor.

CHAPTER 3

OTHER FAST ALGORITHMS FOR THE FFT

Algorithms for the fast calculation of a discrete Fourier transform (DFT) are based on the factorization of the matrix of the transform. This factorization has already appeared in the decimation-in-time and decimation-in-frequency algorithms which were introduced in the preceding chapter and which are particular examples of a large group of algorithms.

In order to use these fast algorithms and thus to exploit to the full both the characteristics of the signals to be processed and the various technological possibilities, use has to be made of a suitable mathematical tool, the Kronecker product of matrices. By combining this product with the conventional product it is possible to factorize the matrix of the DFT in a simple way.

3.1 KRONECKER PRODUCT OF MATRICES

The Kronecker product is a tensor operation which is a generalization of the multiplication of a matrix by a scalar [1]. Knowing two matrices A and B with m and p rows and n and q columns respectively, the Kronecker product of A by B (written $A \times B$) is a new matrix with mp rows and nq columns which is obtained by replacing each element b_{ij} of the matrix B by the following array $b_{ij}A$:

$$
\begin{matrix}
b_{ij}a_{11} & b_{ij}a_{12} & \cdots & b_{ij}a_{1n} \\
\vdots & & & \vdots \\
b_{ij}a_{m1} & b_{ij}a_{m2} & \cdots & b_{ij}a_{mn}
\end{matrix}
$$

This product is generally not commutative:

$$A \times B \neq B \times A$$

As an example of the product, if the matrix B is

$$B = \begin{bmatrix} b_{11} & b_{12} \\ b_{21} & b_{22} \end{bmatrix}$$

the Kronecker product of the matrix A by the matrix B is

$$A \times B = \begin{bmatrix} b_{11}A & b_{12}A \\ b_{21}A & b_{22}A \end{bmatrix} \tag{3.1}$$

72

It will be noted in particular that the Kronecker product of the unit matrix I_N by the M-dimensional unit matrix I_M is equal to the unit matrix of dimension MN:

$$I_N \times I_M = I_{NM} \tag{3.2}$$

Similarly, the Kronecker product of a diagonal matrix by another diagonal matrix is once again a diagonal matrix.

The Kronecker product can be combined with conventional matrix products, and thus we have the following properties which will be used in the coming sections, provided that the dimensions are compatible:

(1) The Kronecker product of a product of matrices with the unit matrix is equal to the product of the Kronecker products of each matrix with the unit matrix:

$$(ABC) \times I = (A \times I)(B \times I)(C \times I) \tag{3.3}$$

(2) The product of Kronecker products is equal to the Kronecker product of the products:

$$(A \times B \times C)(D \times E \times F) = (AD) \times (BE) \times (CF) \tag{3.4}$$

(3) The inverse of a Kronecker product is equal to the Kronecker product of the inverses:

$$(A \times B \times C)^{-1} = A^{-1} \times B^{-1} \times C^{-1} \tag{3.5}$$

Transposition has the analogous property to inversion:

$$(A \times B \times C)^t = A^t \times B^t \times C^t \tag{3.6}$$

The transpose of the matrix of a Kronecker product is the Kronecker product of the transposes of the matrices.

These properties can be easily demonstrated using some simple examples; they are used to factorize matrices with redundant elements and, in particular, for DFT matrices [2]. Decimation-in-frequency will be considered first.

It should be noted that the scale factor $1/N$ is ignored throughout the rest of this chapter.

3.2 FACTORIZING THE MATRIX OF A DECIMATION-IN-FREQUENCY ALGORITHM

In the algorithms examined in the previous chapter one of the sets, either the input or the output, was permuted. The matrix which represents this algorithm is derived from the matrix T_N by permutation of the rows or the columns depending upon whether the decimation-in-frequency or decimation-in-time is considered [3].

Let T'_N denote the matrix corresponding to decimation-in-frequency. This is obtained by permutation of the rows of T_N as follows. The rows are numbered and each number is expressed in binary notation; then the binary numbers are reversed and the resulting number denotes the position of that row in the new matrix. For example, for $N = 8$ we obtain:

$$T_8 = \begin{bmatrix} 1 & 1 & 1 & 1 & 1 & 1 & 1 & 1 \\ 1 & W & W^2 & W^3 & -1 & -W & -W^2 & -W^3 \\ 1 & W^2 & -1 & -W^2 & 1 & W^2 & -1 & -W^2 \\ 1 & W^3 & -W^2 & W & -1 & -W^3 & W^2 & W \\ 1 & -1 & 1 & -1 & 1 & -1 & 1 & -1 \\ 1 & -W & W^2 & -W^3 & -1 & W & -W^2 & W^3 \\ 1 & -W^2 & -1 & W^2 & 1 & -W^2 & -1 & W^2 \\ 1 & -W^3 & -W^2 & -W & -1 & W^3 & W^2 & W \end{bmatrix} \begin{matrix} 0\,0\,0=0 \\ 0\,0\,1=1 \\ 0\,1\,0=2 \\ 0\,1\,1=3 \\ 1\,0\,0=4 \\ 1\,0\,1=5 \\ 1\,1\,0=6 \\ 1\,1\,1=7 \end{matrix}$$

$$T'_8 = \begin{bmatrix} 1 & 1 & 1 & 1 & 1 & 1 & 1 & 1 \\ 1 & -1 & 1 & -1 & 1 & -1 & 1 & -1 \\ 1 & W^2 & -1 & -W^2 & 1 & W^2 & -1 & -W^2 \\ 1 & -W^2 & -1 & W^2 & 1 & -W^2 & -1 & W^2 \\ 1 & W & W^2 & W^3 & -1 & -W & -W^2 & -W^3 \\ 1 & -W & W^2 & -W^3 & -1 & W & -W^2 & W^3 \\ 1 & W^3 & -W^2 & W & -1 & -W^3 & W^2 & -W \\ 1 & -W^3 & -W^2 & -W & -1 & W^3 & W^2 & W \end{bmatrix} \begin{matrix} 0\,0\,0=0 \\ 1\,0\,0=4 \\ 0\,1\,0=2 \\ 1\,1\,0=6 \\ 0\,0\,1=1 \\ 1\,0\,1=5 \\ 0\,1\,1=3 \\ 1\,1\,1=7 \end{matrix}$$

It will be noted that for $N = 2$ the matrix T'_2 is equal to T_2.

The matrix T'_N is factorized by finding the matrix $T'_{N/2}$ and the diagonal matrix $D_{N/2}$ whose elements are the numbers W^k with $0 \leqslant k \leqslant N/2 - 1$. Thus,

$$T'_N = \begin{bmatrix} T'_{N/2} & T'_{N/2} \\ T'_{N/2} D_{N/2} & -T'_{N/2} D_{N/2} \end{bmatrix}$$

This decomposition appears clearly for T'_8. If $I_{N/2}$ denotes the unit matrix of order $N/2$ we can write

$$T'_N = \begin{bmatrix} T'_{N/2} & 0 \\ 0 & T'_{N/2} \end{bmatrix} \begin{bmatrix} I_{N/2} & I_{N/2} \\ D_{N/2} & -D_{N/2} \end{bmatrix}$$

or

$$T'_N = \begin{bmatrix} T'_{N/2} & 0 \\ 0 & T'_{N/2} \end{bmatrix} \begin{bmatrix} I_{N/2} & 0 \\ 0 & D_{N/2} \end{bmatrix} \begin{bmatrix} I_{N/2} & I_{N/2} \\ I_{N/2} & -I_{N/2} \end{bmatrix}$$

By using the Kronecker products of the matrices, we obtain for T'_N:

$$T'_N = (T'_{N/2} \times I_2)\Delta_N(I_{N/2} \times T'_2) \tag{3.7}$$

where Δ_N is a diagonal square matrix of order N, in which the first $N/2$ elements have the value 1 and the subsequent elements are powers of W, W^k with $0 \leqslant k \leqslant N/2 - 1$.

The complete factorization is obtained by iteration:

$$T'_N = (T'_2 \times I_{N/2})(\Delta_4 \times I_{N/4})(I_2 \times T'_2 \times I_{N/4})$$

$$\dots\dots\dots\dots\dots\dots\dots\dots\dots\dots\dots\dots\dots\dots\dots$$

$$(\Delta_{N/2} \times I_2)(I_{N/4} \times T'_2 \times I_2)$$

$$\Delta_N(I_{N/2} \times T'_2)$$

or

$$T'_N = \prod_{i=1}^{\log_2 N} (\Delta_{2^i} \times I_{N/2^i})(I_{2^{i-1}} \times T'_2 \times I_{N/2^i}) \tag{3.8}$$

This expression shows that the transform is calculated in $\log_2(N)$ stages, each containing:

(1) One part involving the ordering of the data corresponding to the factor $(I_{2^{i-1}} \times I_{N/2^i})$, which contains only additions and subtractions;
(2) One part which involves the multiplications by the coefficients represented in the matrix $(\Delta_{2^i} \times T'_2 \times I_{N/2^i})$. The stage corresponding to $i = 1$ does not involve any multiplications. It can be verified that all the matrices indeed have the dimension N.

In order to see how factorization is generalized to radix 4 it is interesting to examine the matrix T'_{16}, which is obtained from T_{16} by the following permutation of the rows. The rows are numbered to base 4 and the order of the digits in the row numbers are reversed. The value obtained shows the number of the row in the new matrix. Following this permutation, we obtain $T_4 = T'_4$.

If D_4 denotes the diagonal matrix

$$D_4 = \begin{bmatrix} 1 & 0 & 0 & 0 \\ 0 & W & 0 & 0 \\ 0 & 0 & W^2 & 0 \\ 0 & 0 & 0 & W^3 \end{bmatrix}$$

The matrix of the transform of order 16 thus obtained is

$$T'_{16} = \begin{bmatrix} T_4 & T_4 & T_4 & T_4 \\ T_4 D_4 & T_4(-j)D_4 & T_4(-1)D_4 & T_4(+j)D_4 \\ T_4 D_4^2 & T_4(-1)D_4^2 & T_4 D_4^2 & T_4(-1)D_4^2 \\ T_4 D_4^3 & T_4(+j)D_4^3 & T_4(-1)D_4^3 & T_4(-j)D_4^3 \end{bmatrix}$$

$$T'_{16} = \begin{bmatrix} T_4 & 0 & 0 & 0 & I_4 & 0 & 0 & 0 \\ 0 & T_4 & 0 & 0 & 0 & D_4 & 0 & 0 \\ 0 & 0 & T_4 & 0 & 0 & 0 & D_4^2 & 0 \\ 0 & 0 & 0 & T_4 & 0 & 0 & 0 & D_4^3 \end{bmatrix}$$

$$\times \begin{bmatrix} I_4 & I_4 & I_4 & I_4 \\ I_4 & -jI_4 & -I_4 & +jI_4 \\ I_4 & -I_4 & I_4 & -I_4 \\ I_4 & +jI_4 & -I_4 & -jI_4 \end{bmatrix}$$

This expression is written in Kronecker product form as:

$$T'_{16} = (T_4 \times I_4)\Delta_{16}(I_4 \times T_4) \tag{3.9}$$

where Δ_{16} is a diagonal matrix in which the first four terms have the value 1, the next four terms W^k with $0 \leqslant k \leqslant 3$, and the subsequent terms $(W^2)^k$ and $(W^3)^k$ with $0 \leqslant k \leqslant 3$.

Factorization as Kronecker products forms the basis of algorithms which have various properties, notably the order of presentation and extraction of data and the linking of operations. It applies also to partial transforms which are of great practical importance.

3.3 PARTIAL TRANSFORMS

The transforms which have been studied in the above sections relate to sets of N numbers which may be complex. In a fine-spectrum analysis it can happen that the order of the transform N becomes very large while we are interested in knowing only a reduced number of points in the spectrum. The limitation of the calculation to useful single points can then permit a large saving.

Let us calculate the partial transform defined by the following equation, where r is a factor of N:

$$\begin{bmatrix} X_p \\ X_{p+1} \\ X_{p+2} \\ \vdots \\ X_{p+r-1} \end{bmatrix} = \begin{bmatrix} 1 & W^P & W^{2P} & \cdots & W^{(N-1)P} \\ 1 & W^{P+1} & W^{2(P+1)} & \cdots & W^{(N-1)(P+1)} \\ 1 & W^{P+2} & W^{2(P+2)} & \cdots & W^{(N-1)(P+2)} \\ \vdots & \vdots & & & \vdots \\ 1 & W^{(P+r-1)} & \cdots & \cdots & W^{(N-1)(P+r-1)} \end{bmatrix} \begin{bmatrix} x_0 \\ x_1 \\ x_2 \\ \vdots \\ x_{N-1} \end{bmatrix} \tag{3.10}$$

From the whole set of data one can form N/r subsets, each containing r terms:

$$(x_0, x_{N/r}, x_{2N/r}, \quad , \quad x_{(r-1)N/r})$$
$$(x_1, x_{(N/r)+1}, \quad , \quad x_{(r-1)N/r+1})$$
$$\cdots\cdots\cdots\cdots\cdots\cdots\cdots\cdots$$
$$(x_{(N/r)-1}, x_{(2N/r)-1}, \quad , \quad x_{N-1})$$

Assume D_r is the diagonal matrix of dimension r, whose elements are the powers of W, W^k with $0 \leqslant k \leqslant r - 1$.

The matrix of the partial transform can be separated into N/r submatrices which can each be applied to one of the sets which were defined earlier and the matrix equation of the transform is written:

$$[X]_{p,r} = \sum_{i=0}^{(N/r)-1} D_r^i T_r (W^P)^i D_r^{(N/r)} [x]_{i,r} \tag{3.11}$$

where $[X]_{p,r}$ denotes the set of r numbers X_k with $p \leqslant k \leqslant p + r - 1$ and $[x]_{i,r}$

the set of data x_k with $k = nN/r + i$ and $n = 0, 1, \ldots, r - 1$. The transform T_r is that of order r.

Consequently, if r is a factor of N, a partial transform relating to r points is calculated using N/r transforms of order r with which the appropriate diagonal matrices are associated.

If N and r are powers of 2, the number M_P of complex multiplications to be made is given by

$$M_P = \frac{N}{r}\left(\frac{r}{2}\log_2\frac{r}{2} + 2r\right) = N\left[\tfrac{1}{2}\log_2\left(\frac{r}{2}\right) + 2\right] \qquad (3.12)$$

This result is equally valid when it is the number of points to be transformed which is limited, as is often the case in spectrum analysis. A common example of a partial transform is that applied to real data.

3.3.1 Transform of real data and odd DFT

If the data to be transformed are real the properties listed in Chapter 2 show that the transformed numbers $X(k)$ and $X(N - k)$ are complex conjugates, that is, $X(k) = X(N - k)$. Thus, it is only necessary to calculate the set of the X_k with $0 \leqslant k \leqslant N/2 - 1$ and the above result can be applied:

$$[X]_{0,N/2} = T_{N/2}[x]_{0,N/2} + D_{N/2}\,T_{N/2}[x]_{1,N/2} \qquad (3.13)$$

In this particular case, the transform $T_{N/2}$ has only to be calculated once taking advantage of the following property of the discrete Fourier transform: if the set to be transformed x_k is purely imaginary, and the transformed set is such that:

$$X(k) = -\bar{X}(N - k)$$

Under these conditions the procedure for calculating the transform of a real set is as follows:

(1) Using the $x(k)$, form a complex set of $N/2$ terms $y(k) = x(2k) + jx(2k + 1)$ with $0 \leqslant k \leqslant N/2 - 1$.
(2) Calculate the transform $Y(k)$ of the set $y(k)$ with $0 \leqslant k \leqslant N/2 - 1$.
(3) Calculate the required numbers using the expression

$$X(k) = \frac{1}{2}\left[Y(k) + \bar{Y}\left(\frac{N}{2} - k\right)\right] + \tfrac{1}{2}je^{-j2\pi(k/N)}\left[\bar{Y}\left(\frac{N}{2} - k\right) - Y(k)\right] \qquad (3.14)$$

with $0 \leqslant k \leqslant N/2 - 1$.

If N is a power of 2, the number of complex multiplications Mc to be made is

$$Mc = \frac{N}{4}\log_2\left(\frac{N}{4}\right) + \frac{N}{2} = \frac{N}{4}\log_2 N \qquad (3.15)$$

Memory locations are required for N real numbers. An algorithm for real data

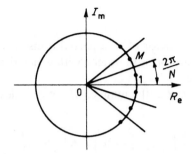

Fig. 3.1   Coefficients of the odd discrete Fourier transform

is described in detail in Ref. [4]. Another method of calculating the transforms of real numbers is to use odd transforms [5].

The odd discrete Fourier transform establishes by definition the following relations between two sets of N complex numbers $x(n)$ and $X(k)$:

$$X(k) = \frac{1}{N} \sum_{n=0}^{N-1} x(n) e^{-j2\pi(2k+1)n/(2N)} \tag{3.16}$$

$$x(n) = \sum_{k=0}^{N-1} X(k) e^{j2\pi(2k+1)n/(2N)} \tag{3.17}$$

The coefficients of this transform have as their co-ordinates the points M of a unit circle such that the vector OM makes an angle with the abscissa which is an odd multiple of $2\pi/2N$, as shown in Figure 3.1.

By setting $W = e^{-j(\pi/N)}$ the matrix of this transform is written:

$$T_N^I = \begin{bmatrix} 1 & W & W^2 & \cdots & W^{N-1} \\ 1 & W^3 & W^6 & \cdots & W^{3(N-1)} \\ 1 & W^5 & W^{10} & \cdots & W^{5(N-1)} \\ \vdots & \vdots & & & \vdots \\ 1 & W^{(2N-1)} & \cdots & \cdots & W^{(2N-1)(N-1)} \end{bmatrix}$$

If the $x(n)$ are real numbers, one can write

$$X(N-1-k) = \frac{1}{N} \sum_{n=0}^{N-1} x(n) e^{-j2\pi[2(N-1-k)+1]/(2N)}$$

$$= \frac{1}{N} \sum_{n=0}^{N-1} x(n) e^{j2\pi(2k+1)n/(2N)} \tag{3.18}$$

Thus,

$$X(N-1-k) = \overline{X(k)} \tag{3.19}$$

Consequently, since the $X(k)$ with even and odd indices are complex conjugates it is sufficient to calculate the $X(k)$ with even index in order to perform a transform on real numbers. Such a transform is the matrix T_R given by

$$T_R = \begin{bmatrix} 1 & W & W^2 & \cdots & W^{N/2} & \cdots & W^{N-1} \\ 1 & W^5 & W^{10} & \cdots & W^{5N/2} & \cdots & W^{5(N-1)} \\ \vdots & \vdots & & & \vdots & & \vdots \\ 1 & W^{2N-3} & \cdots & \cdots & W^{(2N-3)N/2} & \cdots & W^{(2N-3)(N-1)} \end{bmatrix}$$

Let $D_{N/2}$ be the diagonal matrix whose elements are W^k with $0 \leqslant k \leqslant N/2 - 1$, and let $T_{N/2}$ be the matrix of the transform of order $N/2$. Allowing for the fact that $W^{2N} = 1$ and $W^{N/2} = -j$, this becomes:

$$T_R = [T_{N/2}D, -jT_{N/2}D] = (T_{N/2}D) \times [1, -j] \tag{3.20}$$

The odd transform of the real data is then calculated by carrying out a transform of order $N/2$ on the set of complex numbers:

$$y(n) = \left[x(n) - jx\left(\frac{N}{2} + n\right) \right] W^n \quad \text{with} \quad 0 \leqslant n \leqslant \frac{N}{2} - 1 \tag{3.21}$$

The number of calculations is the same as in the method illustrated at the beginning of this section, but the structure is simpler. It should be noted that the transformed numbers give a frequency sampling of the signal spectrum represented by the $x(n)$, displaced by a half-step on the frequency axis.

An important case where significant simplifications are introduced is that of real symmetrical sets. Reductions in the calculations are illustrated by using the doubly odd transform [6].

3.3.2 The odd-time odd-frequency DFT

The odd-time odd-frequency discrete Fourier transform establishes by definition the following relations between two sets of N complex numbers $x(n)$ and $X(k)$:

$$X(k) = \frac{1}{N} \sum_{n=0}^{N-1} x(n) e^{-j2\pi(2k+1)(2n+1)/(4N)} \tag{3.22}$$

$$x(n) = \sum_{k=0}^{N-1} X(k) e^{j2\pi(2k+1)(2n+1)/(4N)} \tag{3.23}$$

The coefficients of this transform are based on the points M of a unit circle such that the vector $\overrightarrow{OM}$ forms an angle with the abscissa which is an odd multiple of $2\pi/4N$ as shown in Figure 3.2.

If the $x(n)$ are real numbers, expression (3.22) leads to:

$$X(N - 1 - k) = -\bar{X}(k)$$

Similarly, if the $X(k)$ are real numbers, then:

$$x(N - 1 - n) = -\bar{x}(n)$$

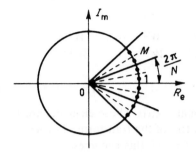

Fig. 3.2   Coefficients of the doubly odd discrete Fourier transform

By assuming as before that $W = e^{-j(\pi/N)}$, the matrix of the transform is written as:

$$
T_N^{II} = \begin{bmatrix}
W^{1/2} & W^{3/2} & W^{5/2} & \cdots & W^{N-1/2} \\
W^{3/2} & W^{9/2} & W^{15/2} & \cdots & W^{3(N-1/2)} \\
W^{5/2} & W^{15/2} & W^{25/2} & \cdots & W^{5(N-1/2)} \\
\vdots & \vdots & \vdots & & \vdots \\
W^{N-1/2} & W^{3(N-1/2)} & \cdots & \cdots & W^{(2N-1)(N-1/2)}
\end{bmatrix} \tag{3.24}
$$

This transform is factorized as follows:

$$
T_N^{II} = W^{1/2} \begin{bmatrix}
1 & 0 & 0 & \cdots & 0 \\
0 & W & 0 & \cdots & 0 \\
\vdots & \vdots & & & \vdots \\
\vdots & \vdots & & & \vdots \\
0 & 0 & & \cdots & W^{N-1}
\end{bmatrix}
\begin{bmatrix}
1 & 1 & 1 & \cdots & 1 \\
1 & W^2 & W^4 & \cdots & W^{2(N-1)} \\
1 & W^4 & W^8 & \cdots & W^{4(N-1)} \\
\vdots & & & & \vdots \\
1 & W^{2(N-1)} & & \cdots & W^{2(N-1)(N-1)}
\end{bmatrix}
$$

$$
\times \begin{bmatrix}
1 & 0 & 0 & \cdots & 0 \\
1 & W & 0 & \cdots & 0 \\
\vdots & \vdots & & & \vdots \\
0 & 0 & & \cdots & W^{N-1}
\end{bmatrix}
$$

That is,

$$
T_N^{II} = W^{1/2} D_N T_N D_N \tag{3.25}
$$

Let us consider the case where the set of data $x(n)$ is real and antisymmetric, i.e. $x(n) = -x(N-1-n)$. Then the same applies to the set $X(k)$. The set of the $x(n)$ for even n is equal to the set of the $x(n)$ for odd n, except for the sign. The situation is the same for the set of $X(k)$.

In order to calculate the transform it is sufficient in this case to carry out the calculations for the $x(2n)$ with $0 \leqslant n \leqslant N/2 - 1$ since the $X(k)$ are real numbers. Alternatively, it is sufficient to make the calculations on the $X(2k)$ with $0 \leqslant k \leqslant N/2 - 1$.

The corresponding matrix T_{RR} is written:

$$T_{RR} = W^{1/2} \begin{bmatrix} 1 & 0 & 0 & \cdots & & 0 \\ 0 & W^2 & 0 & \cdots & & 0 \\ \vdots & \vdots & & & & \vdots \\ 0 & 0 & \cdots & \cdots & & W^{2(N/2-1)} \end{bmatrix}$$

$$\begin{bmatrix} 1 & 1 & \cdots & 1 \\ 1 & W^2 & \cdots & W^{8[(N/2)-1]} \\ 1 & W^{16} & \cdots & \\ \vdots & \vdots & & \vdots \\ 1 & W^{8[(N/2)-1]} & W^{8[(N/2)-1](N/2)-1]} \end{bmatrix} \begin{bmatrix} 1 & 0 & \cdots & 0 \\ 0 & W^2 & \cdots & 0 \\ \vdots & \vdots & & \vdots \\ \vdots & \vdots & & \\ 0 & 0 & & W^{2[(N/2)-1]} \end{bmatrix}$$

Allowing for $W^{2N} = 1$, this becomes:

$$T_{RR} = W^{1/2} D_{N/2} \begin{bmatrix} T_{N/4} & T_{N/4} \\ T_{N/4} & T_{N/4} \end{bmatrix} D_{N/2} \tag{3.26}$$

and, as $W^{2N/4} = -j$, this calculation can be made with one performance of the operations represented by the matrix $T_{N/4}$ on the set of numbers $x(2n) - jx(2n + N/2)$ with $0 \leqslant n \leqslant N/4 - 1$. The $N/4$ numbers obtained are complex ones whose real parts form the set of the desired $X(2k)$ with $0 \leqslant k \leqslant N/4 - 1$. By carrying out the operation defined by T_{RR} for the transformed numbers of rank $2k + N/2$ with $0 \leqslant k \leqslant N/4 - 1$ it can be verified that we have obtained the earlier numbers multiplied by $-j$. That is, the imaginary part of the numbers obtained previously furnishes the set of the $X(2k + N/2)$. It follows that, if the doubly odd transform is applied to a real and antisymmetric set of N terms or to a symmetric set which becomes antisymmetric through a suitable change of sign, it is reduced to the equation

$$\left[X(2k) + jX\left(2k + \frac{N}{2}\right) \right] = W^{1/2} D_{N/4} T_{N/4} D_{N/4} \left[x(2n) - jx\left(2n + \frac{N}{2}\right) \right] \tag{3.27}$$

with $0 \leqslant k \leqslant N/4 - 1$, $0 \leqslant n \leqslant N/4 - 1$, and where $D_{N/4}$ is a diagonal matrix whose elements are W^{2i} with $0 \leqslant i \leqslant N/4 - 1$.

The number of complex multiplications Mc which are necessary is

$$Mc = \frac{N}{8} \log_2 \left(\frac{N}{8} \right) + 2 \frac{N}{4} = \frac{N}{8} \log_2 (2N) \tag{3.28}$$

Comparisons using the different transforms are given in Table 3.1 to illustrate the amount of calculation for each type.

The importance of the odd transforms can be easily seen. It should, however, be noted that other algorithms allow greater reductions to be made for real data and for symmetric real data [7], but these are not as simple to use, especially for practical implementations.

One feature of the doubly odd transform when applied to a real antisymmetric set is that it is identical to the inverse transform. Apart from the scale factor

Table 3.1. ARITHMETIC COMPLEXITIES OF THE VARIOUS FAST FOURIER TRANSFORMS

	Complex multiplications	Complex additions	Memory positions
Complex DFT	$\dfrac{N}{2}\log_2\left(\dfrac{N}{2}\right)$	$N\log_2 N$	$2N$
Odd DFT—real data	$\dfrac{N}{4}\log_2(N)$	$\dfrac{N}{2}\log_2\left(\dfrac{N}{2}\right)$	N
Doubly odd DFT— symmetrical real data	$\dfrac{N}{8}\log_2(2N)$	$\dfrac{N}{4}\log_2\left(\dfrac{N}{4}\right)$	$\dfrac{N}{2}$

$1/N$, there is no distinction, in this case, between the direct and the inverse transforms.

The Fourier transform of a real symmetric set is introduced, for example, when deriving the power spectrum density of a signal from its autocorrelation function.

3.3.3 Sine and cosine transforms

The transforms considered so far have complex coefficients. Discrete transforms of the same family can be obtained using the real and imaginary parts of the complex coefficients.

The following transforms can be defined:

(1) The cosine DFT (cos-DFT):

$$X_{CF}(k) = \frac{1}{N}\sum_{n=0}^{N-1} x(n)\cos\frac{2\pi nk}{N} \tag{3.29}$$

(2) The sine DFT (sin-DFT):

$$X_{SF}(k) = \frac{1}{N}\sum_{n=0}^{N-1} x(n)\sin\frac{2\pi nk}{N} \tag{3.30}$$

(3) The Discrete cosine Transform (DCT):

$$X_{DC}(0) = \frac{\sqrt{2}}{N}\sum_{n=0}^{N-1} x(n)$$

$$X_{DC}(k) = \frac{2}{N}\sum_{n=0}^{N-1} x(n)\cos\left(\frac{2\pi(2n+1)k}{4N}\right). \tag{3.31}$$

The inverse transform is given by:

$$x(n) = \frac{1}{\sqrt{2}}X_{DC}(0) + \sum_{n=1}^{N-1} X_{DC}(k)\cos\frac{2\pi(2n+1)k}{4N}$$

(4) The Discrete sine Transform (DST):

$$X_{DS}(k) = \sqrt{\left(\frac{2}{N+1}\right)} \sum_{n=0}^{N-1} x(n) \sin\left[\frac{2\pi(n+1)(k+1)}{2N+2}\right] \qquad (3.32)$$

Through algebraic manipulations, as in the previous sections, it is possible to establish relationships between the standard DFT and these transforms as well as among these transforms themselves.

For example, from the definitions we have:

$$DFT(N) = \cos\text{-}DFT(N) - j\sin\text{-}DFT(N)$$

Now, considering the cosine DFT [8]:

$$X_{CF}(k) = \sum_{n=0}^{N/2-1} x(2n) \cos\frac{2\pi nk}{N/2}$$

$$+ \sum_{n=0}^{N/4-1} [x(2n+1) + x(N-2n-1)] \cos\left[\frac{2\pi(2n+1)k}{4N/4}\right] \qquad (3.33)$$

it is clear that the cosine DFT of order N can be completed with the help of a cosine DFT of order $N/2$ and a DCT of order $N/4$; in concise form:

$$\cos DFT(N) = \cos DFT(N/2) + DCT(N/4)$$

Similarly, the DCT is expressed by:

$$X_{DC}(k) = \frac{2}{N} \sum_{n=0}^{N/2-1} \left[x(2n) \cos\frac{2\pi(4n+1)k}{4N} + x(2n+1) \cos\frac{2\pi[4(N-n-1)+1]k}{4N} \right]$$

and now, taking

$$y(n) = x(2n); \quad 0 \leqslant n \leqslant \frac{N}{2} - 1$$

$$y(N - n - 1) = x(2n + 1)$$

we have:

$$X_{DC}(k) = \frac{2}{N} \sum_{n=0}^{N-1} y(n) \cos\frac{2\pi(4n+1)k}{4N}$$

Expanding the cosine function yields, in concise form:

$$DCT(N) = \cos\frac{2\pi k}{4N} \cos DFT(N) - \sin\frac{2\pi k}{4N} \sin\text{-}DFT(N) \qquad (3.34)$$

Finally, the DCT of order N can be completed with the help of a DFT of the same order.

Among the transforms based on real coefficients, the Discrete Hartley transform (DHT) is worth mentioning. It is defined by [9]:

$$X_{DH}(k) = \frac{1}{N} \sum_{n=0}^{N-1} x(n) \left[\cos 2\pi \frac{nk}{N} + \sin 2\pi \frac{nk}{N} \right] \qquad (3.35)$$

and the inverse transform:

$$x(n) = \sum_{k=0}^{N-1} X_{DH}(k) \left[\cos 2\pi \frac{nk}{N} + \sin 2\pi \frac{nk}{N} \right]$$

The connection with the DFT is given by:

$$X(k) = \tfrac{1}{2}[X_{DH}(k) + X_{DH}(N-1-k) - j(X_{DH}(k) - X_{DH}(N-1-k))] \qquad (3.36)$$

The discrete sine and cosine transforms have been introduced for information compression, particularly in image processing. It is worth pointing out that, for images, they provide reasonably accurate approximations of the eigen-transform, which yields a signal representation with the minimum number of parameters.

3.3.4 The two-dimensional DCT

For a set of $(N \times N)$ real data the two-dimensional DCT (2D-DCT) is defined by:

$$X(k_1, k_2) = \frac{4e(k_1)e(k_2)}{N^2} \sum_{n_1=0}^{N-1} \sum_{n_2=0}^{N-1} x(n_1, n_2) \cos \frac{2\pi(2n_1+1)k_1}{4N} \cos \frac{2\pi(2n_2+1)k_2}{4N}$$

$$(3.37)$$

and:

$$x(n_1, n_2) = \sum_{k_1=0}^{N-1} \sum_{k_2=0}^{N-1} e(k_1)e(k_2) X(k_1, k_2) \cos \frac{2\pi(2n_1+1)k_1}{4N} \cos \frac{2\pi(2n_2+1)k_2}{4N}$$

with:

$$e(k) = \frac{1}{\sqrt{2}}; \quad k = 0$$

$$e(k) = 1; \quad k \neq 0$$

That transform is separable and it can be computed as follows:

$$X(k_1, k_2) = \frac{2}{N} e(k_2) \sum_{n_2=0}^{N-1} \cos \frac{2\pi(2n_2+1)k_2}{4N}$$

$$\times \left[\frac{2}{N} e(k_1) \sum_{n_1=0}^{N-1} x(n_1, n_2) \cos \frac{2\pi(2n_1+1)k_1}{4N} \right]$$

Thus, the 2D-transform can be computed using $2N$ times the 1D-DCT and the number of real multiplications is of the order of $N^2 \log_2(N)$. In fact, that amount can be reached, with the help of an algorithm based on the decomposition of a DCT of order N into two DCT of order $N/2$ [8]. It is even possible to reach the value $3/4\,N^2 \log_2 N$, through extension of the decimation technique and splitting the set of $(N \times N)$ data into subsets of $(N/2 \times N/2)$ data [10].

3.4 OTHER FAST ALGORITHMS

Fast Fourier transform algorithms form a technique for calculating a discrete Fourier transform of order N using a number of multiplications of the order of $N \log_2 N$. It has been shown in the previous sections that these algorithms have a relatively simple structure and offer sufficient flexibility for good adaptation to the operating constraints and to technological characteristics to be achieved. This explains why they are of great interest for practical applications.

Nevertheless, they are not the only method of fast calculation of a DFT, and algorithms can be elaborated which involve, at least in certain cases, a smaller calculation time or a lower number of multiplications, or which are applicable for values of the order N which are not necessarily powers of two.

A first approach consists of replacing the complex multiplications, which are costly in circuitry or time, by a set of operations which are simpler to put into operation. Reference [11] describes a technique which uses one property of the DFT mentioned in Section 2.4.1, namely the fact that multiplications by the coefficients W^k correspond to phase rotations.

The technique known as CORDIC (digital calculation with coordinate rotation) enables these rotations to be realized by linking simple operations: to rotate a vector (x, y) through an angle θ with an accuracy of $\theta/2^n$, a sequence of n elementary rotations of angle $d\theta_i$ is performed such that $\tan d\theta_i = 2^{-i}$ with $0 \leq i \leq n - 1$ and $-\pi/2 \leq \theta \leq \pi/2$. The coordinates x_i and y_i of the vector at iteration i yield the coordinates at iteration $i + 1$ by using the expressions:

$$x_{i+1} = x_i + \text{sign}[\theta_i] \cdot y_i 2^{-i}$$
$$y_{i+1} = y_i - \text{sign}[\theta_i] \cdot x_i 2^{-i}$$
$$\theta_{i+1} = \theta_i - \text{sign}[\theta_i] \cdot d\theta_i \tag{3.38}$$

The function sign $[\theta_i]$ is the sign of θ_i and $\theta_0 = -\theta$. These operations consist only of additions and shifts, they can be more advantageous than complex multiplication of the same precision.

One method which is particularly interesting is used to obtain a multiplication volume of order N, instead of $N \log_2 N$, for a DFT of order N. This method depends on factorizing the matrix T_N in a particular way. It is decomposed into a product of three factors:

$$T_N = B_N C_N A_N$$

where A_N is a J by N matrix, J is a whole number, C_N is a diagonal matrix of dimension J and B_N is an N by J matrix. The special feature of this factorization is that the elements of the matrices A_N and B_N are 0, 1, or -1. Under these conditions, the calculation requires only J multiplications. This decomposition is obvious for $J = N^2$; for example for $N = 3$, we obtain:

$$T_3 = \begin{bmatrix} 1 & 1 & 1 & 0 & 0 & 0 & 0 & 0 & 0 \\ 0 & 0 & 0 & 1 & 1 & 1 & 0 & 0 & 0 \\ 0 & 0 & 0 & 0 & 0 & 0 & 1 & 1 & 1 \end{bmatrix}$$

$$\times \begin{bmatrix} 1 & & & & & & & \\ & 1 & & & & & O & \\ & & 1 & & & & & \\ & & & 1 & & & & \\ & & & & W & & & \\ & & & & & W^2 & & \\ & & & & & & 1 & \\ & O & & & & & W^2 & \\ & & & & & & & W \end{bmatrix} \begin{bmatrix} 1 & 0 & 0 \\ 0 & 1 & 0 \\ 0 & 0 & 1 \\ 1 & 0 & 0 \\ 0 & 1 & 0 \\ 0 & 0 & 1 \\ 1 & 0 & 0 \\ 0 & 1 & 0 \\ 0 & 0 & 1 \end{bmatrix}$$

With some low values of N, certain factorizations are available in which J is of the order of N, in which case there are the same number of multiplications. In order to generalize this property and to illustrate a suitable factorization of T_N it is necessary to perform a permutation of the data before and after transformation. For example, for $N = 12$, by assuming

$$X' = \begin{bmatrix} X_0 \\ X_3 \\ X_6 \\ X_9 \\ X_4 \\ X_7 \\ X_{10} \\ X_1 \\ X_8 \\ X_{11} \\ X_2 \\ X_5 \end{bmatrix} \quad \text{and} \quad x' = \begin{bmatrix} x_0 \\ x_9 \\ x_6 \\ x_3 \\ x_4 \\ x_1 \\ x_{10} \\ x_7 \\ x_8 \\ x_5 \\ x_2 \\ x_{11} \end{bmatrix}$$

and by using the Kronecker products of the matrices, it can be shown that:

$$X' = (T_3 \times T_4)x'$$

Similarly, if N has L factors such that

$$N = N_L N_{L-1} \cdots N_1$$

It can be shown that:

$$X' = (T_{N_L} \times T_{N_{L-1}} \times \cdots T_{N_1})x' \tag{3.39}$$

By using the factorization defined earlier for the matrices T_{N_L} and the algebraic properties of the Kronecker products, this becomes

$$X' = (B_{N_L} \times B_{N_{L-1}} \times \cdots \times B_{N_1}(C_{N_L} \times C_{N_{L-1}} \times \cdots \times C_{N_1})$$
$$\times (A_{N_L} \times A_{N_{L-1}} \times \cdots \times A_{N_1})x'$$

This result defines a types of algorithm called the Winograd algorithm.

It can be clearly seen that the algorithm of order N is deduced from algorithms of order N_i with $1 \leqslant i \leqslant L$. Herein lies the importance of algorithms with small numbers of multiplications for small values of N. Reference [12] gives algorithms for $N = 2, 3, 4, 5, 7, 8, 9$, and 16, where the number of multiplications is of the order of N, as shown in Table 3.2. In the multiplication column, the figures in parentheses give the number of multiplications by coefficients different from 1. Further, these are complex multiplications which correspond to two real multiplications. The number of additions is comparable to that for FFT algorithms.

Table 3.2. ARITHMETIC COMPLEXITIES OF LOW ORDER
WINOGRAD ALGORITHMS

Order of the DFT	Multiplications		Additions
2	2	(0)	2
3	3	(2)	6
4	4	(0)	8
5	6	(5)	17
7	9	(8)	36
8	8	(2)	26
9	11	(10)	44
16	18	(10)	74

The algorithms for low values of N are obtained by calculating the Fourier transform as a set of correlations:

$$X_k = \sum_{n=1}^{N-1} (x_n - x_0)W^{nk}; \quad k = 1, \ldots, N-1$$

and by using the algebraic properties of the set of exponents of W which are defined modulo N.

For example, for $N = 4$, the sequence of operations is as follows:

$$t_1 = x_0 + x_2, \qquad t_2 = x_1 + x_3$$
$$m_0 = 1(t_1 + t_2), \qquad m_1 = 1(t_1 - t_2)$$
$$m_2 = 1(x_0 - x_2), \qquad m_3 = j(x_1 - x_3)$$

$$X_0 = m_0$$
$$X_1 = m_2 + m_3$$
$$X_2 = m_1$$
$$X_3 = m_2 - m_3$$

For $N = 8$:

$$t_1 = x_0 + x_4, \quad t_2 = x_2 + x_6, \quad t_3 = x_1 + x_5,$$
$$t_4 = x_1 - x_5, \quad t_5 = x_3 + x_7, \quad t_6 = x_3 - x_7,$$
$$t_7 = t_1 + t_2, \quad t_8 = t_3 + t_5,$$
$$m_0 = 1(t_7 + t_8), \qquad m_1 = 1(t_7 - t_8),$$
$$m_2 = 1(t_1 - t_2), \qquad m_3 = 1(x_0 - x_4),$$
$$m_4 = \cos\left(\frac{\pi}{4}\right)(t_4 - t_6), \quad m_5 = j(t_3 - t_5),$$

$$m_6 = j(x_2 - x_6), \qquad m_7 = j\sin\left(\frac{\pi}{4}\right)(t_4 + t_6),$$

$$s_1 = m_3 + m_4, \quad s_2 = m_3 - m_4, \quad s_3 = m_6 + m_7, \quad s_4 = m_6 - m_7$$
$$X_0 = m_0, \quad X_1 = s_1 + s_3, \quad X_2 = m_2 + m_5, \quad X_3 = s_2 - s_4,$$
$$X_4 = m_1, \quad X_5 = s_2 + s_4, \quad X_6 = m_2 - m_5, \quad X_7 = s_1 - s_3.$$

Finally, Winograd algorithms generally introduce a reduction in the amount of computation. This can be large when compared to FFT algorithms. The situation is similar for other algorithms, such as those using the polynomial transforms [13].

These techniques have a large application potential and are of considerable importance in certain cases. However, it should be noted that they may require a larger memory capacity and a more complicated sequence of operations, which results in an increase in the size of the system's control unit or in the volume of the program memory.

Another attractive path towards the optimization of processing and machines is to use transforms operating in finite fields, called number theoretic transforms.

3.5 NUMBER-THEORETIC TRANSFORMS

Fourier transformation involves making arithmetic operations on the field of complex numbers. The machines which carry out these operations generally use binary representations which are approximations to the data and the coefficients. The precision of the calculations is a function of the number of bits available in the machine.

In practice, a machine with B bits carries out its operations on the set with 2^B integers: $0, 1, \ldots, 2^B - 1$. In this set the usual laws of addition and multiplication cannot be applied, as shifting and truncation must be introduced, which lead to approximations in the calculations, as was shown in Section 2.3.

The first condition to be fulfilled to ensure exact calculations in a set E is that the product or the sum of two elements of the set E also belongs to this set. This condition is satisfied in the set of integers $0, 1, \ldots, M-1$ if the calculations are made modulo M. By appropriate selection of the modulus M it is possible to define transformations with properties which are comparable to those of the DFT and which allow error-free calculation of convolutions with fast calculation algorithms.

The definition of such transformations rests on the algebraic properties of integers modulo M, for certain values of M. They are called number-theoretic transforms.

The choice of the modulus M is governed by the following considerations:

(1) Simplicity of the calculations in the modular arithmetic. In principle, modular arithmetic implies a division by the modulus M. This division is trivial for $M = 2^m$. It is very simple for $M = 2^m \pm 1$, because the result is obtained by adding a carry-bit (1s complement arithmetic) or subtracting it.

(2) The modulus must be sufficiently large. The result of the convolution must be capable of representation without ambiguity in the modulo M arithmetic. For example, a convolution with 32 terms with 12-bit data and 8-bit coefficients requires $M > 2^{25}$.

(3) Suitable algebraic properties. The set of modulo M integers should have algebraic properties allowing the definition of transformations comparable to the DFT.

First, there should be periodic elements in order that the fast algorithms can be elaborated; it must have an element α such that:

$$\alpha^N = 1$$

A transformation can then be defined by the expression:

$$X(k) = \sum_{n=0}^{N-1} x(n)\alpha^{nk} \tag{3.40}$$

For the existence of the inverse transformation which is defined by the expression:

$$x(n) = N^{-1} \sum_{k=0}^{N-1} X(k)\alpha^{-nk} \tag{3.41}$$

it is first necessary for N and the powers of α to have inverses.

It can be shown that N has an inverse modulo M if N and M are prime relative to each other. The element α should be prime relative to the M and of order N, i.e. $\alpha^N = 1$.

The existence of the inverse transformation implies a further condition for the α^i, that

$$\sum_{k=0}^{N-1} \alpha^{ik} = N\delta(i) \quad \text{with} \quad \begin{array}{l} \delta(i) = 1 \text{ if } i = 0 \text{ modulo } N \\ \delta(i) = 0 \text{ if } i \neq 0 \text{ modulo } N \end{array}$$

This condition reflects the fact that each element $(1 - \alpha^i)$ must have an inverse. It can be shown that all of the conditions for the existence of a transformation and its inverse reduce to the following one. For each prime factor P of M, N must be a factor of $P - 1$. Thus, if M is prime, N must divide $M - 1$.

Fast algorithms can be elaborated if N is a composite number, in particular, if N is a power of 2. These algorithms are similar to those of the FFT:

$$N = 2^m$$

The calculations to be performed in the transformation are considerably simplified in the particular case when $\alpha = 2$.

Finally, an interesting choice for the modulus M is

$$M = 2^{2^m} + 1$$

where M is prime. These are the Fermat numbers.

A transform based on the Fermat numbers is defined as follows:

(1) Modulus $M = 2^{2^m} + 1$;
(2) Order of the transform: $N = 2^{m+1}$;
(3) Direct transform:

$$X(k) = \sum_{n=0}^{N-1} x(n) 2^{nk}$$

(4) Inverse transform:

$$x(n) = (2^t) \sum_{k=0}^{N-1} X(k) 2^{-nk} \quad \text{with} \quad t = 2^{m+1} - m - 1$$

Example: $m = 3$, $2^m = 8$, $M = 257$, $N = 16$, $t = 12$.

This transform allows the calculation of convolutions of real numbers, as with the discrete Fourier transform, but with the following advantages.

(a) The result is obtained without approximation;
(b) The operations relate to real numbers;
(c) The calculation of the transform and its inverse does not require any multiplication. The only multiplications which remain are in the transformed space.

This technique does, nevertheless, have some significant limitations. As the calculations are exact, the modulus M should be sufficiently large, resulting in numbers which are long.

The relations between the parameters M and N given above require that the calculations are carried out with a number of bits B of the order of $N/2$. That is, the number of terms in the convolution is approximately twice the number of bits of data in the calculation. The application is consequently restricted to convolutions which contain a small number of terms.

The field for the application of number theoretic transforms can be widened by employing numbers other than the Fermat numbers, or by treating the convolutions with many terms as two dimensional convolutions [14]. Reference [15] describes a practical example.

REFERENCES

[1] M. C. PEASE, *Methods of matrix algebra*, Academic Press, New York, 1965.

[2] C. S. BURRUS and T. W. PARKS, *DFT/FFT and convolution algorithms*, John Wiley, New York, 1985.

[3] H. SLOATE, Matrix representations for sorting and the fast Fourier transform. *IEEE Transactions*, **CAS21**(1), 1974.

[4] G. BERGLAND, A fast Fourier transform algorithm for real valued series. *Communications of the ACM*, **11**(10), 1968.

[5] J. L. VERNET, Real signals FFT by means of an odd discrete Fourier transform. *Proceedings of the IEEE*, October 1971.

[6] G. BONNEROT and M. BELLANGER, Odd-time odd-frequency DFT for symmetric real-valued series. *Proceedings of the IEEE*, March 1976.

[7] H. ZIEGLER, A fast transform algorithm for symmetric real valued series. *IEEE Transactions*, **AU20**(5), 1972.

[8] M. VETTERLI and H. J. NUSSBAUMER, Simple FFT and DCT algorithms with reduced number of operations. *Signal Processing*, **6**, No. 4, August, 267–78, 1984.

[9] R. N. BRACEWELL, The fast Hartley transform. *Proc. IEEE*, **22**, August 1984, 1010–18.

[10] M. A. HAQUE, A two-dimensional fast cosine transform, *IEEE Transactions*, **ASSP33**(6), 1532–39, 1985.

[11] A. DESPAIN, Very fast Fourier transform algorithms hardware for implementation. *IEEE Transactions on Computers* **28**(5), 1979.

[12] H. SILVERMAN, An introduction to programming the Winograd Fourier transform algorithm. *IEEE Transactions*, **ASSP25** (2), 1977.

[13] H. NUSSBAUMER, *Fast Fourier Transform and Convolution Algorithms*, Springer Verlag, Berlin, 1981.

[14] R. C. AGRAWAL and C. S. BURRUS, Number theoretic transforms to implement fast digital convolution. *Proceedings of the IEEE*, **63**, 1975.

[15] J. H. MACCLELLAN, Hardware realization of a Fermat number transform. *IEEE Transactions*, **ASSP24**, June 1976.

EXERCISES

1 Find the Kronecker product $A \times I_3$ for the matrix A such that:

$$A = \begin{bmatrix} a_{11} & a_{12} \\ a_{21} & a_{22} \end{bmatrix}$$

and the unit matrix I_3 of dimension 3.

Find the product $I_3 \times A$ and compare it with the above.

2 By taking square matrices of dimension 2, verify the properties of the Kronecker products given in Section 3.1.

3 Give the factorizations of the matrix of a DFT of order 64 on base 2, base 4, and base 8, following the procedure given in Section 3.2.

Calculate the required number of multiplications for each of these three cases and compare the results.

4 Using the decimation-in-time approach, factorize the matrix of the DFT of order 12. What is the minimum number of multiplication and additions required? Write a computer program for the calculation.

5 Factorize the matrix of a DFT of order 16 on base 2 for the following two cases:

(a) When the data appear at both input and output in natural order;
(b) When the stages in the calculation are identical.

For the latter case, devise an implementation scheme involving the use of shift registers as memories.

6 Calculate a discrete Fourier transform of order 16 relating to real data.
Give the algorithm which uses a DFT of order 8 for this calculation.
Give the algorithm based on the odd DFT.
Compare these algorithms and the numbers of operations.

7 Calculate a DFT of order 12 by using a factorization of the type given in Section 3.4 and with the given permuations for the data.
Evaluate the number of operations and compare it with the values found in Exercise 4.

8 To perform a circular convolution of the two sets $x = (2, -2, 1, 0)$ and $h = (1, 2, 0, 0)$ a number-theoretic transform of modulus $M = 17$ and coefficient $\alpha = 4$ is used.
As $N = 4$, verify that $\alpha^N = 1$. Give the matrices of the transformation and the inverse transformation. Prove that the desired result is the set $y = (2, 2, -3, 2)$.

CHAPTER 4

TIME-INVARIANT DISCRETE LINEAR SYSTEMS

Discrete linear systems which are invariant in time form a very important area for the digital processing of signals—digital filters with fixed characteristics. These systems are characterized by the fact that their behaviour is governed by a convolution equation. Their properties are analysed using the Z-transform, which plays the same role in discrete systems as the Laplace or Fourier transforms in continuous systems. In this chapter the elements which are most useful for studying such systems will be briefly introduced. To supplement this, reference should be made to Refs [1–5].

4.1 DEFINITION AND PROPERTIES

A discrete system is one which converts a set of input data $x(n)$ into a set of output data $y(n)$. It is linear if the set $x_1(n) + ax_2(n)$ is converted to the set $y_1(n) + ay_2(n)$, and it is invariant in time if the set $x(n - n_0)$ is converted to the set $y(n - n_0)$ for any integer n_0.

Assume $u_0(n)$ is a unit set as given in Figure 4.1 and defined by:

$$u_0(n) = 1 \quad \text{for} \quad n = 0$$
$$u_0(n) = 0 \quad \text{for} \quad n \neq 0 \tag{4.1}$$

The set $x(n)$ can be decomposed into a sum of suitably shifted unit sets:

$$x(n) = \sum_{m=-\infty}^{\infty} x(m)u_0(n - m) \tag{4.2}$$

Further, if $h(n)$ is the set forming the response of the system to the unit set $u_0(n)$, then $h(n - m)$ corresponds to $u_0(n - m)$ because of the time invariance. Linearity then implies the following relation:

$$y(n) = \sum_m x(m)h(n - m) = \sum_m h(m)x(n - m) = h(n) * x(n) \tag{4.3}$$

This is the convolution equation which represents the linear time invariant (LTI) system. Such a system is completely defined by the values of the set $h(n)$, which is called the impulse response of the system.

This system has the property of causality if the output with index $n = n_0$ depends only on inputs with indices $n < n_0$. This property implies that $h(n) = 0$

93

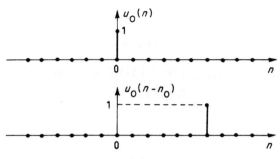

Fig. 4.1   *Unit series*

for $n < 0$, and the output is given by:

$$y(n) = \sum_{m=0}^{\infty} h(m)x(n-m) \tag{4.4}$$

An LTI system is stable if each input with a bounded amplitude has a corresponding bounded output. A necessary and sufficient condition for stability is given by the inequality

$$\sum_n |h(n)| < \infty \tag{4.5}$$

To show that the condition is necessary, it is sufficient to apply to the system the input set $x(n)$ such that:

$$\begin{aligned} x(n) = &+1 \quad \text{if} \quad h(n) \geqslant 0 \\ &-1 \quad \text{if} \quad h(n) < 0 \end{aligned}$$

Then, for $n = 0$, we obtain

$$y(0) = \sum_m |h(m)|$$

If the inequality (4.5) is not satisfied, $y(0)$ is not bounded and the system is not stable. If the input set is bounded, i.e.

$$|x(n)| \leqslant M \quad \text{for all } n$$

then we have

$$|y(n) \leqslant \sum_m |h(m)| \, |x(n-m)| \leqslant M \sum_m |h(m)|$$

If the inequality (4.5) is satisfied, then $y(n)$ is bounded and the condition is sufficient.

In particular, the LTI system defined by the response

$$h(m) = a^m \quad \text{with} \quad m \geqslant 0$$

is stable for $|a| < 1$. The properties of LTI systems will be studied using the Z-transform.

4.2 THE Z-TRANSFORM

The Z-transform, $X(Z)$, of the set $x(n)$ is defined by the following relation:

$$X(Z) = \sum_{n=-\infty}^{\infty} x(n)Z^{-n} \tag{4.6}$$

Z is a complex variable and the function $X(Z)$ has a convergence region which, in general, is an annular ring centred on the origin, with radii R_1 and R_2. That is, $X(Z)$ is defined for $R_1 < |Z| < R_2$. The values R_1 and R_2 depend on the set $x(n)$. If the set $x(n)$ represents the samples of a signal taken with period T, its Fourier transform is written:

$$S(f) = \sum_{n=-\infty}^{\infty} x(n)e^{-j2\pi f n T}$$

Consequently, for $Z = e^{j2\pi f T}$ the Z-transform of the set $x(n)$ coincides with its Fourier transform. That is, the analysis of a discrete system can be performed with the Z-transform, and, in order to find a frequency response, it is sufficient to replace Z by $e^{j2\pi f T}$.

This transform has an inverse. Assuming Γ is a closed contour containing the origin and all singular points, or poles, of $X(Z)$, we can write

$$Z^{m-1}X(Z) = \sum_{n=-\infty}^{\infty} x(n)Z^{m-1-n} = x(m)Z^{-1} + \sum_{n \neq m} Z^{m-1-n}x(n)$$

and from the theory of residues:

$$x(m) = \frac{1}{2\pi j} \int_{\Gamma} Z^{m-1}X(Z)\,dZ \tag{4.7}$$

For example, if $X(Z) = 1/(1 - pZ^{-1})$, direct application of the above equation results in

$$x(n) = p^n \quad \text{for} \quad n \geq 0$$
$$0 \quad \text{for} \quad n < 0$$

Similarly, for $X(Z)$ defined by:

$$X(Z) = \sum_{i=1}^{N} \frac{a_i}{1 - p_i Z^{-1}}$$

there is a corresponding set $x(n)$ such that:

$$x(n) = \sum_{i=1}^{N} a_i p_i^n \quad \text{for} \quad n \geq 0$$

$$x(n) = 0 \quad \text{for} \quad n < 0$$

A stability condition appears quite simply from observing that the set $x(n)$ is bounded if and only if $|p_i| < 1$ for $1 \leqslant i \leqslant N$, i.e. the poles of $X(Z)$ are inside the unit circle.

In these examples the terms of the set $x(n)$ can be obtained directly by series expansion. When $X(Z)$ is a rational fraction a very simple method of obtaining the first values of the set $x(n)$ is by direct division. For example, if

$$X(Z) = \frac{1 + 2Z^{-1} + Z^{-2} + Z^{-3}}{1 - Z^{-1} - 8Z^{-2} + 12Z^{-3}}$$

direct division gives:

$$X(Z) = 1 + 3Z^{-1} + 12Z^{-2} + 25Z^{-3} + \cdots$$

and thus:

$$x(0) = 1; \quad x(1) = 3; \quad x(2) = 12; \quad x(3) = 25$$

The Z-transformation has the property of linearity. Also, the Z-transform of the delayed set $x(n - n_0)$ is written:

$$X_{n_0}(Z) = Z^{-n_0} X(Z) \tag{4.8}$$

These two properties are used to calculate the Z-transform, $Y(Z)$, of the set $y(n)$ obtained at the output of a discrete linear system by convolution of the sets $x(n)$ and $h(n)$ which have transforms $X(Z)$ and $H(Z)$, respectively.

By calculating the Z-transform for the two components of the convolution equation (4.3):

$$y(n) = \sum_m h(m)x(n - m)$$

we have

$$Y(Z) = \sum_m h(m)Z^{-m}X(Z) = H(Z)X(Z) \tag{4.9}$$

Consequently, the Z-transform of a convolution product is the product of the transforms. The function $H(Z)$ is called the Z-transfer function of the LTI system being considered.

The Z-transform of the product of two sets $x_3(n) = x_1(n)x_2(n)$ is the function $X_3(Z)$ defined by:

$$X_3(Z) = \frac{1}{2\pi j} \int_\Gamma X_1(v)X_2\left(\frac{Z}{v}\right)v^{-1}\,dv \tag{4.10}$$

The integration contour is inside the region of convergence of the functions $X_1(v)$ and $X_2(Z/v)$.

When applied to causal sets the one-sided Z-transform is introduced. The one-sided Z-transform of the set $x(n)$ is written:

$$X(Z) = \sum_{n=0}^{\infty} x(n)Z^{-n} \tag{4.11}$$

The properties are the same as for the transform defined by equation (4.6), except for delayed sets where the transform of the set $x(n - n_0)$ is written:

$$X_{n_0}(z) = \sum_{n=0}^{\infty} x(n - n_0)z^{-n} = z^{-n_0}X(z) + \sum_{n=1}^{n_0} x(-n)z^{-(n_0-n)} \qquad (4.12)$$

The value of this transform in the study of system response is that account can be taken of the initial conditions and that the transient response can be exposed. It also allows determination of the extreme values of the set $x(n)$ from the $x(z)$. The initial value $x(0)$ is written:

$$x(0) = \lim_{Z \to \infty} X(Z) \qquad (4.13)$$

and the final value is:

$$x(\infty) = \lim_{Z \to 1} (Z - 1)X(Z) \qquad (4.14)$$

More on the Z-transform and its applications can be found in Ref. [6]. The above results can be applied to the calculation of the power of discrete signals.

4.3 ENERGY AND POWER OF DISCRETE SIGNALS

Let us calculate the energy E of a signal represented by the set $x(n)$, whose Z-transform is written $X(Z)$. By definition:

$$E = \sum_{n=-\infty}^{\infty} |x(n)|^2$$

The set $x_3(n)$ defined by:

$$x_3(n) = |x(n)|^2$$

can be regarded as the product of two sets $x_1(n)$ and $x_2(n)$ such that:

$$x_1(n) = x(n); \quad x_2(n) = \bar{x}(n)$$

The transform $X_3(Z)$ is calculated from the functions $X_1(Z)$ and $X_2(Z)$ using formula (4.10) given in the preceding section for the Z-transform of the product of two sets:

$$X_3(Z) = \frac{1}{2\pi j} \int_{\Gamma} X_1(v)X_2\left(\frac{Z}{v}\right)\frac{dv}{v}$$

The evaluation of $X_3(Z)$ at the point $Z = 1$ leads to the equation

$$X_3(1) = \sum_{n=-\infty}^{\infty} |x(n)|^2 = \frac{1}{2\pi j} \int_{\Gamma} X_1(v)X_2\left(\frac{1}{v}\right)\frac{dv}{v}$$

If Γ is the unit circle, $1/v = \bar{v}$ and consequently:

$$X_2\left(\frac{1}{v}\right) = X_2(\bar{v}) = \sum_{n=-\infty}^{\infty} \bar{x}(n)(\bar{v})^{-n} = \bar{X}(v)$$

Since $v = e^{j2\pi f}$, we have

$$E = \sum_{n=-\infty}^{\infty} |x(n)|^2 = \int_{-1/2}^{1/2} |X(e^{j2\pi f})|^2 \, df \tag{4.15}$$

This is the Bessel–Parseval relation given in Section 1.1.1 which expresses the conservation of energy for discrete signals. The energy of the signal is equal to the energy of its spectrum.

The calculations above provide an expression which is useful for the norm $\|X\|_2$ of the function $X(f)$. Indeed, by definition,

$$\|X\|_2^2 = \int_{-1/2}^{1/2} |X(f)|^2 \, df$$

This becomes:

$$\|X\|_2^2 = \frac{1}{2\pi j} \int_{|Z|=1} X(Z)X(Z^{-1}) \frac{dZ}{Z} \tag{4.16}$$

If $X(Z)$ is a holomorphic function of the complex variable in a domain which contains the unit circle, the integral is calculated by the method of residues and directly yields the value of $\|X\|_2$ which is also the L_2 norm of the discrete signal $x(n)$.

Let us now assume we want to calculate the energy E_y of the signal $y(n)$ at the output of the LTI system with impulse response $h(n)$, to which the signal $x(n)$ is applied.

The signal $x(n)$ is assumed to be determinist. Using equation (4.15), by setting $\omega = 2\pi f$ we can write:

$$E_y = \frac{1}{2\pi} \int_{-\pi}^{\pi} |Y(e^{j\omega})|^2 \, d\omega$$

Equation (4.9) directly provides the following result:

$$E_y = \frac{1}{2\pi} \int_{-\pi}^{\pi} |H(e^{j\omega})|^2 |X(e^{j\omega})|^2 \, d\omega \tag{4.17}$$

These results also apply to random signals.

4.4 FILTERING OF RANDOM SIGNALS

If the signal $x(n)$ is random and has a moment of order 1, $E[x(n)]$, the expectation value of the output $y(n)$ for the LTI system can be calculated as

$$E[y(n)] = \sum_m h(m)E[x(n-m)] \tag{4.18}$$

If the expectation value of $x(n)$ is stationary, then so is that of $y(n)$, provided that the system is stable, that is, it satisfies equation (4.5).

If the input signal $x(n)$ is stationary and of the second order, with an autocorrelation function $r_{xx}(n)$, then the autocorrelation function $r_{yy}(n)$ of the output of the LTI system can be calculated. Using the equation of definition (1.58) we have:

$$r_{yy}(n) = E[y(i)y(i-n)] = \sum_m h(m)E[x(i-m)y(i-n)]$$

If the correlation function $r_{xy}(n)$ between $x(n)$ and $y(n)$ is

$$r_{xy}(n) = E[x(i)y(i-n)] \tag{4.19}$$

we can write

$$r_{yy}(n) = h(n) * r_{xy}(n) \tag{4.20}$$

Then,

$$r_{xy}(n) = \sum_m h(m)E[x(i)x(i-n-m)] = h(-n) * r_{xx}(n)$$

Finally, one obtains

$$r_{yy}(n) = h(n) * h(-n) * r_{xx}(n) \tag{4.21}$$

Then, the Z-transforms $\Phi_{xx}(Z)$ and $\Phi_{yy}(Z)$ are related by

$$\Phi_{yy}(Z) = H(Z)H(Z^{-1})\Phi_{xx}(Z) \tag{4.22}$$

This expression can provide a more useful approach than equation (4.21) for calculating the autocorrelation function of the output signal by inverse transformation. With the Fourier transform, one has

$$r_{yy}(n) = \frac{1}{2\pi} \int_{-\pi}^{\pi} |H(e^{j\omega})|^2 \Phi_{xx}(e^{j\omega}) e^{jn\omega} \, d\omega \tag{4.23}$$

In particular, the output power of the signal can be written:

$$P_y = r_{yy}(0) = \frac{1}{2\pi} \int_{-\pi}^{\pi} |H(e^{j\omega})|^2 \Phi_{xx}(e^{j\omega}) \, d\omega \tag{4.24}$$

This is equivalent to equation (4.17) for random signals. If the signal $x(n)$ is assumed to be a white noise with variance σ_x^2, the variance σ_y^2 of the output signal $y(n)$ is given by:

$$\sigma_y^2 = \frac{\sigma_x^2}{2\pi} \int_{-\pi}^{\pi} |H(e^{j\omega})|^2 \, d\omega \tag{4.25}$$

or, using the equality (4.15):

$$\sigma_y^2 = \sigma_x^2 \left[\sum_n h^2(n) \right] \tag{4.26}$$

These results are of great practical importance and will be frequently used in later sections (for example, when evaluating the powers of the round-off noise in the filters).

4.5 SYSTEMS DEFINED BY DIFFERENCE EQUATIONS

The most interesting of the LTI systems are those where the input and output sets are related by a linear difference equation with constant coefficients. On the one hand, they represent simple examples, and on the other, they form an excellent representation of many natural systems.

A system of this type of order N is defined by the equation

$$y(n) = \sum_{i=0}^{N} a_i x(n-i) - \sum_{i=1}^{N} b_i y(n-i) \tag{4.27}$$

By applying the Z-transform to the two sides of this equation and by denoting the transforms of the sets $y(n)$ and $x(n)$ by $Y(Z)$ and $X(Z)$, we obtain:

$$Y(Z) = \sum_{i=0}^{N} a_i Z^{-i} X(Z) - \sum_{i=1}^{N} b_i Z^{-1} Y(Z) \tag{4.28}$$

Hence:

$$Y(Z) = H(Z)X(Z)$$

with:

$$H(Z) = \frac{a_0 + a_1 Z^{-1} + \cdots + a_N Z^{-N}}{1 + b_1 Z^{-1} + \cdots + b_N Z^{-N}} \tag{4.29}$$

The transfer function $H(Z)$ is a rational fraction. The a_i and b_i are the coefficients of the system. Some coefficients can be zero, as is the case, for example, when the two summations of expression (4.27) have different numbers of terms. To find the frequency response it is sufficient to replace the variable Z by $e^{j2\pi f}$ in $H(Z)$.

The function $H(Z)$ is written in the form of a quotient of two polynomials $N(Z)$ and $D(Z)$ of degree N which have N roots Z_i and P_i, respectively, with $1 \leqslant i \leqslant N$.

By using these roots, another expression for $H(Z)$ appears:

$$H(Z) = \frac{N(Z)}{D(Z)} = a_0 \frac{\prod_{i=1}^{N}(1 - Z_i Z^{-1})}{\prod_{i=1}^{N}(1 - P_i Z^{-1})} \tag{4.30}$$

where a_0 is a scale factor. Thus, we can write:

$$H(Z) = a_0 \frac{\prod_{i=1}^{N}(Z - Z_i)}{\prod_{i=1}^{N}(Z - P_i)} \tag{4.31}$$

In the complex plane, Z is the affix of a moving point M, P_i and Z_i $(1 \leqslant i \leqslant N)$

are affixes of the poles and zeros of the function $H(Z)$. Then

$$Z - Z_i = MZ_i \, e^{j\theta i} \quad \text{and} \quad Z - P_i = MP_i \, e^{j\theta i}$$

Consequently, the transfer function is expressed by

$$H(Z) = a_0 \prod_{i=1}^{N} \frac{MZ_i}{MP_i} e^{j\sum_{i=1}^{N}(\theta_i - \phi_i)} \tag{4.32}$$

From this, a graphic interpretation in the complex plane can be developed. The frequency response of the system is obtained when the moving point M lies on the unit circle. Figure 4.2 shows the example of a system of order $N = 2$.

The modulus of the transfer function is thus equal to the ratio of the product of the distances between the moving point M and the roots Z_i to the product of the distances between M and the poles P_i. The phase is equal to the difference between the sum of the angles between the vectors $\overrightarrow{P_iM}$ and the real axis, and the sum of the angles between the vectors $\overrightarrow{Z_iM}$ and the real axis, following the convention introduced in Chapter 1. This graphic interpretation is frequently used in practice because it offers a very simple visualization of the frequency response of a system.

Analysis of a system using its frequency response corresponds to steady-state operation. It is adequate only in so far as transient phenomena can be neglected. If this is not the case, initial conditions have to be introduced, to represent, for example, the status of the equipment and the contents of its memory at switch-on.

Consider, the behaviour for values with index $n \geqslant 0$ of the system defined by equation (4.27) to which the set $x(n)$ ($x(n) = 0$ for $n < 0$) is applied. The $y(n)$ are completely determined if the values $y(-i)$ with $1 \leqslant i \leqslant N$ are known. These values correspond to the initial conditions, and, in order to introduce them, the one-sided Z-transform has to be used.

The one-sided Z-transform is applied to both sides of equation (4.27) by

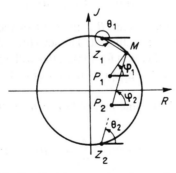

Fig. 4.2   Graphical interpretation of a transfer function

assuming that the input $x(n)$ is causal, that is, that $x(n) = 0$ for $n < 0$. Allowing for equation (4.12) which gives the transform $Y_i(Z)$ of the delayed set $Y(n - i)$:

$$Y_i(Z) = Z^{-i}Y(Z) + \sum_{n=1}^{i} y(-n)Z^{-(i-n)}$$

we obtain:

$$Y(Z) = \sum_{i=0}^{N} a_i Z^{-i} X(Z) - \sum_{i=1}^{N} b_i Z^{-i} Y(Z) - \sum_{i=1}^{N} b_i \sum_{n=1}^{i} y(-n)Z^{-(i-n)}$$

or,

$$Y(Z) = H(Z)X(Z) - \frac{\sum_{i=1}^{N} b_i \sum_{n=1}^{i} y(-n)Z^{-(i-n)}}{1 + \sum_{i=1}^{N} b_i Z^{-i}} \tag{4.33}$$

The system response with index $n, y(n)$, is obtained by series expansion or by inverse transformation.

It should be noted that the values $y(-i)$ represent the state of the system at switch-on, provided that only the set of output numbers is contained in the system memory. However, this often contains other internal variables which can be introduced into the analysis for generalization and to provide other features relating in particular to the implementation.

4.6 STATE VARIABLE ANALYSIS

The state of a system of order N at instant n is defined by a set of at least N internal variables represented by a vector $U(n)$ called the state vector. Its operation is governed by the relations between this state vector and the input and output signals. The behaviour of a linear system to which the input set $x(n)$ is applied and which provides the output set $y(n)$ is described by the following pair of equations, called state equations [7]:

$$\begin{aligned} U(n + 1) &= AU(n) + Bx(n) \\ y(n) &= C^t U(n) + dx(n) \end{aligned} \tag{4.34}$$

A is called the matrix of the system, B is the control vector, C the observation vector, and d is the transition coefficient. The set $x(n)$ is the innovation and $y(n)$ is the observation. The justification for these titles will appear later, particularly in Chapter 11. The matrix A is a square matrix of order N. The vectors B and C are N-dimensional.

The state of the system at time n is obtained from the initial state at time zero by the equation:

$$U(n) = A^n U(0) + \sum_{i=1}^{n} A^{n-i} Bx(i - 1) \tag{4.35}$$

Consequently, the behaviour of such a system depends on successive powers of the matrix A.

The Z-transfer function of the system is obtained by taking the Z-transform of the state equations (5.34). Thus:

$$(ZI - A)U(Z) = BX(Z)$$
$$Y(Z) = C^t U(Z) + dX(Z)$$

and consequently:

$$H(Z) = C^t(ZI - A)^{-1}B + d \tag{4.36}$$

The poles of the transfer function thus obtained are the values of Z for which the determinant of the matrix $(ZI - A)$ is zero, that is, they are the roots of the characteristic polynomial of A. Consequently, the poles of the transfer function are the eigenvalues of the matrix A and have absolute values less than unity to ensure stability. This result agrees with the equation for the operation of the system (4.35). Indeed, by diagonalizing the matrix A it can be seen that it is the condition under which the vector $U(n) = A^n U(0)$ tends to zero as n tends to infinity, a situation which corresponds to the free evolution of the system from the initial state $U(0)$.

Examination of the transfer function of the system (4.36) shows by another route that, when a system is specified by the input–output equation, there is some latitude in the choice of the state parameters. Indeed, only the eigenvalues of the matrix A are imposed, and the matrix of the system can be replaced by another matrix $A' = M^{-1}AM$, which has the same eigenvalues. Then, in order to preserve the same output set, using equation (4.35), the following criteria are necessary:

$$A' = M^{-1}AM; \quad C'^t = C^t M; \quad B' = M^{-1}B$$

The matrix A can also be replaced by its transpose A^t. The system is then described by a system of equations, parallel to that of equation (4.34), corresponding to the state vector $V(n)$, such that:

$$V(n + 1) = A^t V(n) + Cx(n) \tag{4.37}$$
$$y(n) = B^t V(n) + dx(n)$$

This state representation produces another method of realization.

The results obtained in this section are used subsequently for studying the properties of, and for finding the structures for, the realization of LTI systems.

REFERENCES

[1] L. R. RABINER and B. GOLD, *Theory and application of digital signal processing*, Prentice Hall, Englewood Cliffs NJ, 1975.

[2] A. V. OPPENHEIM and R. W. SCHAFER, Digital signal processing, Prentice Hall, Englewood Cliffs NJ, 1974, Ch. II.

[3] J. LIFERMAN, *Les systèmes discrets*, Masson, Paris, 1975.

[4] R. BOITE and H. LEICH, *Les filtres numériques: analyse et synthèse des filtres unidimensionnels*, Masson, Paris, 1980.

[5] J. MAX and OTHERS, *Méthodes et techniques de traitement du signal*, Masson, Paris, 1981.

[6] E. I. JURY, *Theory and application of the Z-transform method*, John Wiley, New York, 1964.

[7] J. E. CADZOW, *Discrete time systems*, Prentice Hall, Englewood Cliffs NJ, 1973.

EXERCISES

1 Assume an LTI system with an impulse response $h(n)$ such that:

$$h(n) = 1 \quad 0 \leqslant n \leqslant 3$$
$$0 \quad \text{elsewhere}$$

Calculate the response $y(n)$ to the set $x(n)$ given by

$$x(n) = a^n \quad \text{with } a = 0.7 \text{ for } 0 \leqslant n \leqslant 5$$
$$0 \quad \text{elsewhere}$$

Calculate the response to the set:

$$x(n) = \cos(2\pi n/8) \quad \text{for } 0 \leqslant n \leqslant 7$$
$$0 \quad \text{elsewhere}$$

2 Show that the Z-transform of the causal sequence $x(n)$ defined by:

$$x(n) = nT e^{-anT} \quad \text{for } n \geqslant 0$$
$$0 \quad \text{for } n < 0$$

is

$$X(Z) = \frac{T e^{-aT} Z^{-1}}{(1 - e^{-aT} Z^{-1})^2}$$

Calculate the inverse transform of $\log(Z - a)$, $Z/\{(Z - a)(Z - b)\}$ and establish the conditions on a and b such that the resulting sequence converges.

3 Calculate the Z-transform for the impulse response

$$h(n) = r^n \frac{\sin[(n + 1)\theta]}{\sin(\theta)} \quad n \geqslant 0$$
$$0 \quad n < 0$$

What is the domain of convergence of the function obtained? Plot its poles and zeros in the Z-plane.

4 Assume an LTI system whose transfer function $H(Z)$ is written:

$$H(Z) = \frac{1}{1 - 1.6Z^{-1} + 0.92Z^{-2}}$$

and to which is applied a signal with a uniform spectrum and unit power. Calculate the power of the signal at the output of the system and give the spectral distribution.

5 Use the one-sided Z-transform to calculate the response of the system defined by the difference equation

$$y(n) = x(n) + y(n-1) - 0.8y(n-2)$$

with the initial conditions $y(-1) = a$ and $y(-2) = b$, to the set $x(n)$ defined by

$$x(n) = e^{jn\omega} \quad \text{for } n \geqslant 0$$
$$0 \quad \text{for } n < 0$$

Show the response due to the initial conditions and the steady state response.

CHAPTER 5

FINITE IMPULSE RESPONSE (FIR) FILTERS

Digital finite impulse response (FIR) filters are discrete linear time-invariant systems in which an output number, representing a sample of the filtered signal, is obtained by weighted summation of a finite set of input numbers, representing samples of the signal to be filtered. The coefficients of the weighted summation constitute the impulse response of the filter and only a finite number of them take non-zero values. This filter is of the 'finite memory' type, that is, it determines its output as a function of input data of limited age. It is frequently called a non-recursive filter because, unlike the infinite impulse response filter, it does not require a feedback loop in its implementation.

The properties of finite impulse response filters will be illustrated by two simple examples.

5.1 FIR FILTERS

Let us consider a signal $x(t)$ represented by its samples $x(nT)$, taken at frequency $f_s = 1/T$, and examine the effect on its spectrum of replacing the set $x(nT)$ by the set $y(nT)$ defined by the equation:

$$y(nT) = \tfrac{1}{2}[x(nT) + x((n-1)T)] \tag{5.1}$$

This set is also obtained by sampling the signal $y(t)$ such that:

$$y(t) = \tfrac{1}{2}[x(t) + x(t-T)]$$

If $Y(f)$ and $X(f)$ denote the Fourier transforms of the signals $y(t)$ and $x(t)$, then

$$Y(f) = \tfrac{1}{2}X(f)(1 + e^{-j2\pi fT})$$

Such an operation corresponds to the transfer function

$$H(f) = Y(f)/X(f)$$

where

$$H(f) = e^{-j\pi fT}\cos(\pi fT) \tag{5.2}$$

This is called cosine filtering, and it conserves the zero frequency component and eliminates that at $f_s/2$, as can be readily verified.

In the expression for $H(f)$ the complex term $e^{-j\pi fT}$ represents a delay $\tau = T/2$ which is the propagation time of the signal through the filter.

106

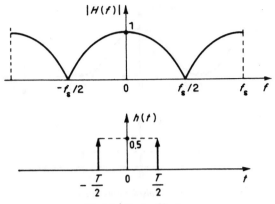

Fig. 5.1   Cosine filtering

The impulse response $h(t)$ which corresponds to the filter transfer function $|H(f)|$ is written:

$$h(t) = \frac{1}{2}\left[\delta\left(t + \frac{T}{2}\right) + \delta\left(t - \frac{T}{2}\right)\right]$$

Figure 5.1 shows the characteristics of the filter.

Another simple operation associates the set $x(nT)$ with the set $y(nT)$ by:

$$y(nT) = \tfrac{1}{4}[x(nT) + 2x[(n-1)T] + x[(n-2)T]] \tag{5.3}$$

As in the previous case, this equation conserves the component with zero frequency and eliminates the one with frequency $f_s/2$. This corresponds to the transfer function:

$$H(f) = \tfrac{1}{4}(1 + e^{-j2\pi f 2T} + e^{-j2\pi f 2T}) = e^{-j2\pi f T}\tfrac{1}{2}(1 + \cos 2\pi f T) \tag{5.4}$$

This is a raised cosine filter. Its propagation delay is $\tau = T$, and $|H(f)|$ corresponds to the impulse response $h(t)$ such that:

$$h(t) = \tfrac{1}{4}\delta(t + T) + \tfrac{1}{2}\delta(t) + \tfrac{1}{4}\delta(t - T)$$

This is a more selective low-pass filter than the preceding one and it is evident that an even more selective filtering function can be obtained merely by increasing the number of terms in the set $x(nT)$ over which the weighted summation is carried out.

These two examples have served to illustrate the following properties of FIR filters.

(1) The input set $x(n)$ and the output set $y(n)$ are related by an equation of the following type (the defining relation):

$$y(n) = \sum_{i=0}^{N-1} a_i x(n-i) \tag{5.5}$$

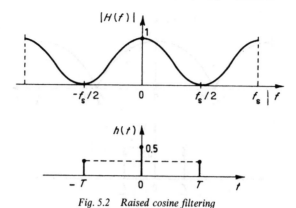

Fig. 5.2   Raised cosine filtering

The filter defined in this way comprises a finite number N of coefficients a_i. If it is regarded as a discrete system its response $h(i)$ to the unit set is:

$$h(i) = a_i \quad \text{if } 0 \leqslant i \leqslant N - 1$$
$$0 \quad \text{elsewhere}$$

That is, the impulse response is simply the set of the coefficients.

(2) The transfer function of the filter is

$$H(f) = \sum_{i=0}^{N-1} a_i e^{-j2\pi f i T} \tag{5.6}$$

or, expressed in terms of Z:

$$H(Z) = \sum_{i=0}^{N-1} a_i Z^{-i} \tag{5.7}$$

(3) The function $H(f)$, the frequency response of the filter, is periodic with period $f_s = 1/T$. The coefficients a_i $(0 \leqslant i \leqslant N - 1)$ form the Fourier series expansion of this function.

The Bessel–Parseval relation given in Section 1.1.1 allows the following to be written:

$$\sum_{i=0}^{N-1} |a_i|^2 = \frac{1}{f_s} \int_0^{f_s} |H(f)|^2 \, df \tag{5.8}$$

(4) If the coefficients are symmetric the transfer function can be written as the product of two terms, of which one is a real function and the other a complex number with modulus 1, representing a constant propagation delay τ which is a whole multiple of half the period of the sampling. Such a filter is said to have a linear phase.

5.2 PRACTICAL TRANSFER FUNCTIONS AND LINEAR PHASE FILTERS

A digital filter which processes numbers representing samples of a signal taken with period T has a periodic frequency response of period $f_s = 1/T$. As a consequence, this function $H(f)$ can be expanded by Fourier series:

$$H(f) = \sum_{n=-\infty}^{\infty} \alpha_n e^{+j2\pi fnT} \qquad (5.9)$$

with

$$\alpha_n = \frac{1}{f_s} \int_0^{f_s} H(f) e^{-j2\pi fnT} \, df \qquad (5.10)$$

The coefficients α_n of the expansion are, except for a constant factor, the samples taken with period T of the Fourier transform of the function $H(f)$ over a frequency interval of width f_s. As they form the impulse response the condition (4.5) for the stability of the filter implies that the α_n tend to zero when n tends to infinity. Consequently, the function $H(f)$ can be approximated by an expansion reduced to a limited number of terms:

$$H(f) \simeq \sum_{n=-P}^{Q} \alpha_n e^{j2\pi fnT} = H_L(f)$$

where P and Q are finite integers. The approximation is improved as the numbers increase.

The property of causality, which translates the fact that in a real filter the output cannot precede the input in time, implies that the impulse response $h(n)$ is zero for $n < 0$. Using relations (5.5) and (5.6), if the filter is causal, then $Q = 0$ and we obtain:

$$H_L(f) = \sum_{n=0}^{P} a_n e^{-j2\pi fnT}$$

As a result, any causal digital filtering function can be approximated by the transfer function of an FIR filter.

The fact that transfer functions with linear phase can be realized is an important property which is exploited in spectral analysis or in data transmission applications. For FIR filters this capability leads to a reduction in complexity, because it implies a symmetry in the coefficients which is used to halve the number of multiplications needed to produce each output number.

By definition, a linear phase filter has the following frequency response:

$$H(f) = R(f)e^{-j\phi(f)} \tag{5.11}$$

where $R(f)$ is a real function and the phase $\phi(f)$ is a linear one: $\phi(f) = \phi_0 + 2\pi f \tau$, when τ is a constant giving the propagation delay through the filter.

The impulse response of this filter is written:

$$h(t) = e^{-j\phi_0} \int_{-\infty}^{\infty} R(f)e^{j2\pi f(t-\tau)} df \tag{5.12}$$

By assuming ϕ_0 to be zero and by decomposing the real function $R(f)$ into an even part $P(f)$ and an odd part $I(f)$, this becomes:

$$h(t + \tau) = 2 \int_0^{\infty} P(f)\cos(2\pi ft) df + 2j \int_0^{\infty} I(f)\sin(2\pi ft) df$$

If the condition is imposed that the function $h(t)$ must be real, this becomes:

$$h(t + \tau) = 2 \int_0^{\infty} P(f)\cos(2\pi ft) df$$

This relation shows that the impulse is symmetric about the point $t = \tau$ on the time axis. Such a condition is satisfied in a filter with real symmetric coefficients. Two configurations are available, depending upon whether the number of coefficients N is even or odd:

(1) $N = 2P + 1$: the filter has a propagation time $\tau = PT$. The transfer function is

$$H(f) = e^{-j2\pi fPT}\left[h_0 + 2\sum_{i=1}^{P} h_i \cos(2\pi fiT)\right] \tag{5.13}$$

(2) $N = 2P$: the filter has a propagation delay $\tau = (P - \frac{1}{2})T$. The transfer function is

$$H(f) = e^{-j2\pi f(P-1/2)T}2\left(\sum_{i=1}^{P} h_i \cos[2\pi f(i - \frac{1}{2})T]\right) \tag{5.14}$$

The filter coefficients h_i form the response of the digital filter to the unit set. They can also be regarded as samples, taken with period T, of the continuous impulse response $h(t)$ of a filter which has the same frequency response as the digital filter in the range $(-T/2, T/2)$ but has no periodicity on the frequency axis. Figures 5.3 and 5.4 illustrate this for odd and even N.

These filters have the even function $P(f)$ in their frequency response. With real coefficients, a frequency response can also be obtained which corresponds to $I(f)$, the odd part of $R(f)$.

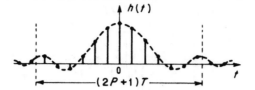

Fig. 5.3 *Symmetrical filter with odd order*

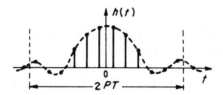

Fig. 5.4 *Symmetrical filter with even order*

As the function $h(t)$ must be real, this category of filter has the transfer function:

$$H(f) = -\,\mathrm{j}\,\mathrm{e}^{-\mathrm{j}2\pi f\tau}I(f) = \mathrm{e}^{-\mathrm{j}(\pi/2)}\,\mathrm{e}^{-\mathrm{j}2\pi f\tau}I(f)$$

A fixed phase shift $\phi_0 = \pi/2$, which corresponds to a quadrature, is added to the frequency proportional phase shift. This possibility is important in certain types of modulation and will be examined later. The impulse response is zero at the point $t = \tau$ on the time axis and is antisymmetric about it. Figures 5.5 and 5.6 show the configurations when the number of coefficients N is even or odd.

(1) $N = 2P + 1$: the filter has a propagation delay $\tau = PT$

$$H(f) = -\,\mathrm{j}\,\mathrm{e}^{-\mathrm{j}2\pi f\tau}2\sum_{i=1}^{P} h_i \sin(2\pi fiT) \tag{5.15}$$

(2) $N = 2P$: this becomes $\tau = (P - \tfrac{1}{2})T$

$$H(f) = -\,\mathrm{j}\,\mathrm{e}^{-\mathrm{j}2\pi f\tau}2\sum_{i=1}^{P} h_i \sin[2\pi f(i - \tfrac{1}{2})T] \tag{5.16}$$

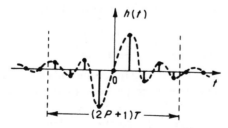

Fig. 5.5 *Antisymmetrical filter with odd order*

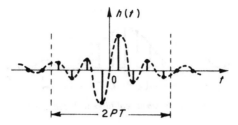

Fig. 5.6   Antisymmetrical filter with even order

As $h_0 = 0$, the transfer function has the same form in both cases. It is not difficult to envisage that fixed phase shifts other than $\phi_0 = 0$ and $\phi_0 = \pi/2$ can be obtained for filters with complex coefficients.

The calculation of the coefficients of FIR filters will now be considered, under the assumption of linear phase, which applies to a large number of applications and when specifications are given for the frequency response.

5.3 CALCULATION OF COEFFICIENTS BY FOURIER SERIES EXPANSION FOR FREQUENCY SPECIFICATIONS

The coefficients h_i must be determined so that the filter satisfies any imposed conditions. These conditions can be defined in time space, for example by the data of the impulse response itself, by the limits within which it should fall, or by some other type of constraint, as will be shown later. In fact, the calculation of the coefficients for specifications in time space is often related to adaptive filtering and is discussed in Chapter 12. The conditions are often defined in frequency space and the transfer function must approximate a given function and must remain within limits which constitute the specification.

For a low-pass filter, for example, the absolute value of the transfer function is required to approximate the value 1 with precision δ_1, in the frequency band $(0, f_1)$, called the pass band, and the value 0 with precision δ_2 in the band $(f_2, f_s/2)$, which is called the stop (or rejection) band. The corresponding limits are represented in Figure 5.7. The range $\Delta f = f_2 - f_1$ is called the transition band and the steepness of the cut-off is described by the parameter R_c such that:

$$R_c = \frac{f_1 + f_2}{2(f_2 - f_1)} \tag{5.17}$$

A very simple method of obtaining the coefficients h_i is to expand the periodic function $H(f)$ in a Fourier series to produce an approximation

$$h_i = \frac{1}{f_s} \int_0^{f_s} H(f) e^{-j2\pi i f/f_s} \, df$$

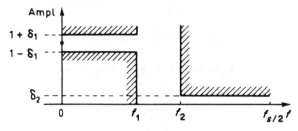

Fig. 5.7   Mask for a low-pass filter

In the case of a low-pass filter corresponding to the mask given in Figure 5.7, expression (1.5) leads to:

$$h_i = \frac{f_1 + f_2}{f_s} \cdot \frac{\sin \pi i[(f_1 + f_2)/f_s]}{\pi i[(f_1 + f_2)/f_s]} \tag{5.18}$$

The table given in Appendix 1 of Chapter 1 can then be used to provide a first estimation of the coefficients of an FIR filter. The optimized values are, in fact, not very different.

For the filter to be realizable it is necessary to limit the number of coefficients to N. This operation reduces to multiplying the impulse response $h(t)$ by a time window $g(t)$ so that:

$$g(t) = 1 \quad \text{for} \quad -\frac{NT}{2} \leqslant t \leqslant \frac{NT}{2}$$

$$0 \quad \text{elsewhere}$$

Using equation (1.10), the Fourier transform of this function is written:

$$G(f) = NT \frac{\sin(\pi f NT)}{\pi f NT} \tag{5.19}$$

Figure 5.8 shows these functions.

The real filter, with a limited number of coefficients N, has the following convolution product $H_R(f)$ as its transfer function:

$$H_R(f) = \int_{-\infty}^{\infty} H(f')G(f - f')\,df'$$

By limiting the number of coefficients, ripples are introduced and the steepness of the cut-off of the filter is limited as shown in Figure 5.9, which corresponds to an ideal low-pass filter with a cut-off frequency f_c.

The ripples depend on those of the function $G(f)$ and in order to reduce their amplitude it is sufficient to choose as the time window functions whose spectra introduce smaller ripples than that of the rectangular window given above. This

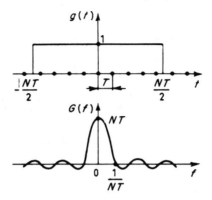

Fig. 5.8   Rectangular window

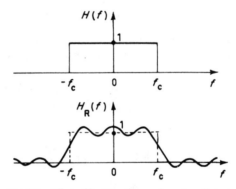

Fig. 5.9   Effect of limiting the number of coefficients

situation has been encountered in Section 2.4.2 for spectrum analysis and the same functions can be employed. One example is the Hamming window, which is defined as:

$$g(t) = 0.54 + 0.46\cos(2\pi i/NT) \quad \text{for } |t| \leqslant NT/2$$
$$0 \qquad\qquad\qquad\qquad\quad \text{for } |t| > NT/2$$

The consequence of reducing the ripples in the pass and stop bands is an increase in the width of the transition band.

The function which presents the smallest ripples for a given width of the principal lobe is the Dolf–Chebyshev function:

$$G(x) = \frac{\cos\left[K\cos^{-1}(Z_0\cos\pi x)\right]}{\operatorname{ch}\left[K\operatorname{ch}^{-1}(Z_0)\right]} \quad \text{for } x_0 \leqslant x \leqslant 1 - x_0$$
$$\frac{\operatorname{ch}\left[K\operatorname{ch}^{-1}(Z_0\cos\pi x)\right]}{\operatorname{ch}\left[K\operatorname{ch}^{-1}(Z_0)\right]} \quad \text{for } 0 \leqslant x \leqslant x_0 \text{ and } 1 - x_0 \leqslant x \leqslant 1 \qquad (5.20)$$

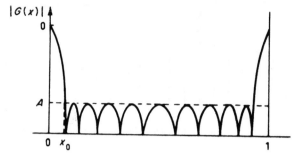

Fig. 5.10   The Dolf–Chebyshev function

with $x_0 = (1/\pi)\cos^{-1}(1/Z_0)$, K an integer and Z_0 a parameter. This function, as shown in Figure 5.10, presents a main lobe of width B, given by

$$B = 2x_0 = \frac{2}{\pi}\cos^{-1}\left(\frac{1}{Z_0}\right)$$

and secondary lobes of constant amplitude given by

$$A = \frac{1}{\text{ch}\,[K\,\text{ch}^{-1}(Z_0)]}$$

It is periodic and its inverse Fourier transform is formed of a set of $K+1$ discrete non-zero values, used to weight the coefficients of the Fourier series expansion of the filter function to be approximated.

Example: Let us calculate the coefficients of a low-pass filter with a sampling frequency $f_s = 1$, a cut-off frequency $f_c = 0.25$, transition bandwidth $\Delta f = 0.115$, and comprising $N = 17$ coefficients. The appropriate Dolf–Chebyshev function has parameters $K = 16$ and Z_0, where

$$2x_0 = \frac{2}{\pi}\cos^{-1}\left(\frac{1}{Z_0}\right) \simeq \Delta f$$

This corresponds to ripples of amplitude $A = 1/(\text{ch}\,[16\,\text{ch}^{-1}(Z_0)]$, whose value is taken as $A = 0.1$. The inverse Fourier transform $g(t)$ of this function (except for a scale factor) is composed of 17 discrete non-zero values which are given in Figure 5.11. These values form the weighted coefficients of the impulse response $h(t)$ of the ideal low-pass filter, given in Figure 5.12.

The resulting filter has the coefficients $a_i = g_i h_i$ given by

$$a_0 = 0.5 \qquad\qquad a_4 = 0$$
$$a_1 = 0.3141 \qquad\quad a_5 = 0.0451$$
$$a_2 = 0 \qquad\qquad\quad a_6 = 0$$

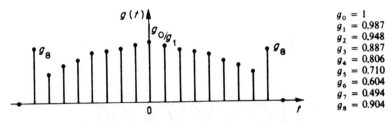

$$
\begin{aligned}
g_0 &= 1 \\
g_1 &= 0.987 \\
g_2 &= 0.948 \\
g_3 &= 0.887 \\
g_4 &= 0.806 \\
g_5 &= 0.710 \\
g_6 &= 0.604 \\
g_7 &= 0.494 \\
g_8 &= 0.904
\end{aligned}
$$

Fig. 5.11   *Weighting coefficients of a Dolf–Chebyshev window*

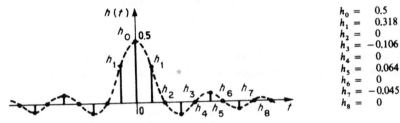

$$
\begin{aligned}
h_0 &= \quad 0.5 \\
h_1 &= \quad 0.318 \\
h_2 &= \quad 0 \\
h_3 &= -0.106 \\
h_4 &= \quad 0 \\
h_5 &= \quad 0.064 \\
h_6 &= \quad 0 \\
h_7 &= -0.045 \\
h_8 &= \quad 0
\end{aligned}
$$

Fig. 5.12   *Impulse response of an ideal filter*

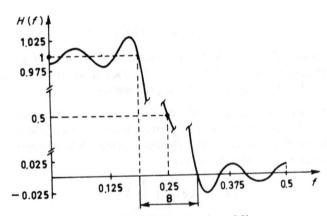

Fig. 5.13   *Transfer function of a real filter*

$$
a_3 = -0.0941 \qquad a_7 = -0.0224
$$
$$
a_8 = 0
$$

The transfer function is given in Figure 5.13.

It is worth noting that if the ripples in the function $G(x)$ are of constant amplitude those in the resulting filter decrease in amplitude, with distance from the pass band and stop band edges. The ripples in the pass and stop bands, however, are the same.

The technique of Fourier series expansion of the function to be approximated leads to a simple method of determining the coefficients of the filter but it involves two important restrictions:

(1) The ripples of the filter in the pass and stop bands are equal;
(2) The amplitude of the ripple is not constant.

As a consequence, this technique cannot provide a filter which is perfectly adapted to a particular characteristic, and the number of coefficients is generally greater than that which is strictly necessary, which adds complications to the equipment.

The first limitation can be eliminated by using a method which preserves the simplicity of direct calculation, the method of least squares. Further, this corresponds exactly to the objectives to be achieved in a number of applications.

5.4 CALCULATION OF COEFFICIENTS BY THE METHOD OF LEAST SQUARES

Let us calculate the N coefficients h_i of an FIR filter according to the criterion of least squares so that the transfer function approximates a given function.

The discrete Fourier transform applied to the set h_i, with $(0 \leqslant i \leqslant N - 1)$, produces a set H_k such that:

$$H_k = \frac{1}{N} \sum_{i=0}^{N-1} h_i e^{-j2\pi(ik/N)} \tag{5.21}$$

The set of H_k, $0 \leqslant k \leqslant N - 1$ forms a sampling of the frequency response of the filter with period f_s/N. Conversely, the coefficients h_i are related to the set H_k by the equation:

$$h_i = \sum_{k=0}^{N-1} H_k e^{j2\pi(ik/N)} \tag{5.22}$$

Consequently, the problem of calculating the N coefficients is equivalent to that of determining the frequency response of the filter at N points in the range $(0, f_s)$. The function $H(f)$ is then obtained by the interpolation formula which expresses the convolution product of the set of the samples $H_k \delta(f - (k/N)f_s)$ by the Fourier transform of the rectangular sampling window, calculated in Section 2.4:

$$H(f) = \sum_{k=0}^{N-1} H_k \frac{\sin\{\pi N[(f/f_s) - (k/N)]\}}{N \sin\{\pi[(f/f_s) - (k/N)]\}} \tag{5.23}$$

It should be noted that this expression simply forms a different type of series expansion of the function $H(f)$ to a limited number of terms.

Given the function to be approximated, $D(f)$, a first possibility is to choose

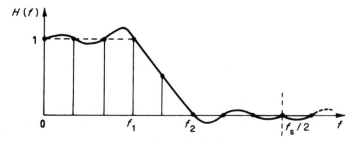

Fig. 5.14   Interpolated transfer function

the H_k such that:

$$H_k = D\left(\frac{k}{N} f_s\right) \quad \text{for} \quad 0 \leqslant k \leqslant N - 1$$

The transfer function of the filter $H(f)$, obtained by interpolation, shows ripples in the pass and stop bands, as shown in Figure 5.14.

The divergence between this function and that given represents an error $e(f) = H(f) - D(f)$ which can be minimized by the least squares criterion. The procedure begins by evaluating the squared error E which is the L_2 norm of the divergence function. To do this, the response $H(f)$ is sampled with a frequency step Δ less than f_s/N, in a way that produces interpolated values for example:

$$\Delta = \frac{f_s}{NL} \quad \text{with } L \text{ as an integer greater than 1}$$

The function $e(f)$ is calculated at frequencies which are multiples of Δ. Generally, when evaluating the squared error E only one part of the band $(0, f_s/2)$ is taken into account. For a low-pass filter this can be the pass band, the stop band or the set of both bands. In order to demonstrate the principle of the calculation it is assumed that the minimization relates to the pass band $(0, f_1)$ of a low-pass filter, whence

$$E = \sum_{n=0}^{N_0-1} e^2\left(n\frac{f}{NL}\right) \quad \text{with} \quad \frac{f_1}{f_s}NL < N_0 \leqslant \frac{f_1}{f_s}NL + 1$$

Further, it is often useful to find a weighting factor $P_0(n)$ for the error element of index n, so that the frequency response can be modelled. Thus:

$$E = \sum_{n=0}^{N_0-1} P_0^2(n)e^2\left(n\frac{f_s}{NL}\right) = \sum_{n=0}^{N_0-1} P_0^2(n)e^2(n) \tag{5.24}$$

With the error function obtained from the interpolation formula (5.23) the squared error E is a function of the set of H_k with $0 \leqslant k \leqslant N - 1$ and is expressed by $E(H)$. If these samples of the frequency response are given increases of ΔH_k a new squared error value is obtained:

$$E(H + \Delta H) = E(H) + \sum_{k=0}^{N-1} \frac{\partial E}{\partial H_k} \Delta H_k + \frac{1}{2} \sum_{k=0}^{N-1} \sum_{l=0}^{N-1} \frac{\partial^2 E}{\partial H_k \partial H_l} \Delta H_k \Delta H_l \quad (5.25)$$

From the defining equation for E and the interpolation equation (5.23) this becomes:

$$\frac{\partial E}{\partial H_k} = 2 \sum_{n=0}^{N_0-1} P_0^2(n) e(n) \frac{\partial e(n)}{\partial H_k}$$

$$\frac{\partial^2 E}{\partial H_k \partial H_l} = 2 \sum_{n=0}^{N_0-1} P_0^2(n) \frac{\partial e(n)}{\partial H_l} \frac{\partial e(n)}{\partial H_k}$$

These equations can be written in matrix form where A is the matrix with N rows and N_0 columns such that:

$$A = \begin{bmatrix} a_{00} & a_{01} & \cdots & a_{0(N_0-1)} \\ a_{10} & a_{11} & \cdots & a_{1(N_0-1)} \\ \vdots & \vdots & & \vdots \\ a_{0(N-1)} & a_{1(N-1)} & \cdots & a_{(N-1)(N_0-1)} \end{bmatrix} \quad \text{with} \quad a_{ij} = \frac{\partial e(j)}{\partial H_i}$$

Let P_0 be the diagonal matrix of order N_0 whose elements are the weighting factor $P_0(n)$. Then:

$$\left[\frac{\partial E}{\partial H_k} \right] = 2 A P_0^2 [e(n)] \quad (5.26)$$

The set of terms $\partial^2 E / \partial H_k \partial H_e$ form a square matrix of order N such that:

$$\frac{\partial^2 E}{\partial H_k \partial H_l} = 2 A P_0^2 A^t \quad (5.27)$$

The condition for $E(H + \Delta H)$ to be a minimum is that all its derivatives with respect to H_k $(0 \leqslant k \leqslant N - 1)$ are zero. Now,

$$\frac{\partial}{\partial H_k} E(H + \Delta H) = \frac{\partial E}{\partial H_k} + \sum_{l=0}^{N-1} \frac{\partial E}{\partial H_k} \frac{\partial E}{\partial H_l} \Delta H_l$$

The condition of least squares can be written:

$$A P_0^2 [e(n)] + A P_0^2 A^t [\Delta H] = 0 \quad (5.28)$$

Under these conditions the increments ΔH_k $(0 \leqslant k \leqslant N - 1)$ which transfer the initial values of the samples to the optimal values of the frequency response form a column vector which is written:

$$[\Delta H] = - [A P_0^2 A^t]^{-1} A P_0^2 [e(n)] \quad (5.29)$$

To summarize, the following operations are required for calculating the coefficients of the filter by the method of least squares:

(1) Sample the function to be approximated at N points in order to obtain N numbers H_k $(0 \leqslant k \leqslant N - 1)$.
(2) In the frequency band where the error should be minimized interpolate the response between the H_k in order to obtain N_0 numbers $e(n)$ $(0 \leqslant n \leqslant N_0 - 1)$ which represent the divergence between the response of the filter and the function to be approximated.
(3) Determine N_0 weighting coefficients $P_0(n)$ as a function of the constraints of the approximation.
(4) Using the interpolation equation, calculate the elements of the matrix A.
(5) Solve the matrix equation which gives the ΔH_k.
(6) Perform an inverse Fourier transformation on the set of numbers $(H_k + \Delta H_k)$ with $0 \leqslant k \leqslant N - 1$ to obtain the coefficients for the filter.

The weighting coefficients $P_0(n)$ allow certain constraints to be introduced — for example to obtain ripples in the pass and stop bands which are in a given ratio or to force the frequency response to a particular value. This latter condition can also be taken into account by a reduction by 1 in the number of degrees of freedom; this is more elegant but more complicated to program.

Implementation of this calculation does not present any particular difficulties and permits the calculation of a filter in a direct way. However, the filter obtained does not have constant amplitude. This is a commonly required feature, and to achieve it an iterative technique is employed.

5.5 CALCULATION OF COEFFICIENTS BY DISCRETE FOURIER TRANSFORM

A first iterative approach consists of using the discrete Fourier transform, which is calculated efficiently by a fast algorithm.

Consider calculation of a linear phase filter with N coefficients meeting the specification of Figure 5.7. A discrete Fourier transform of order N_0 with $N_0 \simeq 10N$ will be used.

The procedure consists of taking initial values for the coefficients, for example the terms h given by (5.18) for $-P \leq i \leq P$, if $N = 2P + 1$. This set of N values is completed symmetrically with zeros to obtain a set of N_0 real values, symmetrical with respect to the origin.

A DFT calculation then gives the response $H(f)$ at N_0 points on the frequency axis:

$$H(f) = H_{id}(f) + E(f)$$

where $H_{id}(f)$ is the ideal response and $E(f)$ is the deviation from this response.

A reduction of the deviation $E(f)$ is then performed, by replacing $H(f)$ with the function $G(f)$ such that:

$$G(f) = H_{id}(f) + E_L(f) \quad \text{if} \quad H(f) > H_{id}(f) + E_L(f)$$
$$G(f) = H_{id}(f) - E_L(f) \quad \text{if} \quad H(f) < H_{id}(f) - E_L(f)$$

where $E_L(f)$ represents the limit of the deviation given by the characteristic, for example δ_1 and δ_2 for the low-pass filter of Figure 5.7.

Calculation of the inverse DFT gives N_0 terms; the N values which encircle the origin are retained and the others are discarded. The procedure is repeated by taking the DFT of the N_0 values obtained in this way.

Denoting the sum of the squares of the $N_0 - N$ terms discarded in the time domain at iteration k by $J(k)$, a decreasing function is obtained if the specification of the filter is compatible with the number of coefficients N. The procedure is terminated when $J(k)$ falls below a fixed threshold.

By applying the method for different numbers of coefficients N, an optimum solution can be approached and even achieved in special cases. All types of linear phase filters can be designed in this way.

To obtain the optimum filter, a method based on the Chebyshev approximation is used.

5.6 CALCULATION OF COEFFICIENTS BY CHEBYSHEV APPROXIMATION

The objective is to produce a filter whose frequency response has constant amplitude ripple in such a way as to best approximate the desired characteristic. An example was given in Figure 5.8 for a low-pass filter whose ripples must not exceed amplitude δ_1 in the pass band and δ_2 in the stop band. This problem depends upon the approximation of a function by a polynomial in the Chebyshev sense, and L_∞ is the norm to be considered for the error function.

Using the expression for the transfer function of a linear phase FIR filter, the calculation of the coefficients corresponds to determining the function $H_R(f)$ which is written:

$$H_R(f) = \sum_{i=0}^{r-1} h_i \cos(2\pi f i T) \tag{5.30}$$

when the number of coefficients is $N = 2r - 1$. The technique which will be presented is valid in all cases, whether N is even or odd, or whether the coefficients are symmetric or antisymmetric. It is based on the following theorem [1].

A necessary and sufficient condition for $H_R(f)$ to be the unique and, in the Chebyshev sense, the best approximation to a given function $D(f)$ over a compact subset A for the range $[0, \frac{1}{2}]$ is that the error function $e(f) =$

$H_R(f) - D(f)$ presents at least $(r + 1)$ extremal frequencies on A. That is, there exist $(r + 1)$ extremal frequencies $(f_0, f_1, \ldots, f_r)$ such that $e(f_i) = -e(f_{i-1})$ with $1 \leqslant i \leqslant r$ and

$$|e(f_i)| = \max_{f \in A} |e(f)|$$

This result is still valid if a weighting function $P_0(f)$ for the error is introduced.

The problem is thus equivalent to the solution of the system of $(r + 1)$ equations:

$$P_0(f_i)[D(f_i) - H_R(f_i)] = (-1)^i \delta$$

The unknowns are the coefficients h_i ($0 \leqslant i \leqslant r - 1$) and the maximum of the error function: δ. In matrix form, by writing the unknowns as a column vector and by normalizing the frequencies so that $f_s = 1/T$ this becomes:

$$
\begin{bmatrix}
D(f_0) \\
D(f_1) \\
\vdots \\
D(f_r)
\end{bmatrix}
=
\begin{bmatrix}
1 & \cos(2\pi f_0)\ldots & \cos[2\pi f_0(r-1)] & \dfrac{1}{P_0(f_0)} & h_0 \\
1 & \cos(2\pi f_1)\ldots & \cos[2\pi f_1(r-1)] & \dfrac{-1}{P_0(f_1)} & h_1 \\
\vdots & \vdots & \vdots & \vdots & \vdots \\
 & & & & h_{r-1} \\
1 & \cos(2\pi f_r)\ldots & \cos(2\pi f_r(r-1)] & \dfrac{(-1)^r}{P_0(f_r)} & \delta
\end{bmatrix}
$$

This matrix equation results in the determination of the coefficients of the filter, under the condition that the $(r + 1)$ extremal frequencies f_i are known.

An iterative procedure based on an algorithm called the Remez algorithm is used to find the extremal frequencies. In this algorithm each stage has the following phases:

(1) Initial values are assigned or are available for the parameters $f_i (0 \leqslant i \leqslant r)$.
(2) The corresponding value δ is calculated by solving the system of equations, which leads to the following formula:

$$\delta = \frac{a_0 D(f_0) + a_1 D(f_1) + \cdots + a_r D(f_r)}{a_0/P_0(f_0) - a_1/P_0(f_1) + \cdots + (-1)^r a_r/P_0(f_r)}$$

with

$$a_k = \prod_{\substack{i=0 \\ i \neq k}}^{r} \frac{1}{\cos(2\pi f_k) - \cos(2\pi f_i)}$$

(1) The values of the function $H_R(f)$ are interpolated between the $f_i (0 \leqslant i \leqslant r)$ in order to calculate $e(f)$.

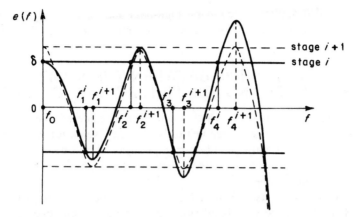

Fig. 5.15   Evolution of the error function in a stage of the Remez algorithm

(2) The extremal frequencies obtained in the previous phase are taken as the initial values for the next stage.

Figure 5.15 shows the evolution of the error function through one stage of the calculation. The procedure is halted when the difference between the value δ calculated with the new extremal frequencies and the previous value falls below a preset threshold. In the majority of cases this result is obtained after a few iterations.

The convergence of this procedure is related to the choice of initial values for the frequencies f_i. For the first iteration, the extremal frequencies obtained using a different method of calculation can be used as the filter coefficients. Even more simply, a uniform distribution of the extremal frequencies over the frequency range can be assumed.

As in the least squares method of the previous section, the values $H_R(f)$ have to be interpolated. Because of the non-uniform distribution of the extremal frequencies it is more convenient to use the Lagrange interpolation formulae:

$$H_R(f) = \frac{\sum_{k=0}^{r-1}\{\beta_k/(x - x_k)\}[D(f_k) - (-1)^k \delta/P_0(f_k)]}{\sum_{k=0}^{r-1}\beta_k(x - x_k)} \qquad (5.31)$$

with

$$\beta_k = \prod_{\substack{i=0 \\ i \neq k}}^{r-1} \frac{1}{x_k - x_i} \quad \text{and} \quad x = \cos(2\pi f)$$

At the end of the iteration, the extremal frequencies obtained are used to produce samples with a constant frequency increment which, by an inverse discrete Fourier transformation, produces the coefficients of the filter.

Filters having several hundreds of coefficients can be designed using this technique, and it applies to low-pass, high-pass and band-pass filters with or without fixed phase shift [2]. An example calculation is given in the appendix.

5.7 RELATIONSHIPS BETWEEN THE NUMBER OF COEFFICIENTS AND THE FILTER CHARACTERISTIC

In the calculation techniques which have been considered the number of coefficients N of the filter was assumed to have been given *a priori*. In practice, then, N is an important parameter (for example, in projects where the computational complexity needed to implement a digital filter satisfying given specifications must be determined).

For a low-pass filter as shown in Figure 5.7 the specifications concern the pass band ripple δ_1, the stop band ripple δ_2, the pass band edge f_1 and the transition band $\Delta f = f_2 - f_1$. An analysis of the results obtained in the design of a large number of filters with a wide range of specifications shows that the number of coefficients is proportional to the logarithms of $1/\delta_1$ and $1/\delta_2$ and also to the ratio of the sampling frequency f_s to the transition band Δf. The adjustment of parameters leads to the following estimate N_e for the filter order:

$$N_e = \tfrac{2}{3}\log\left(\frac{1}{10\delta_1\delta_2}\right)f_s/\Delta f \qquad (5.32)$$

This simple estimate is sufficient for most practical cases. It clearly points out the relative importance of the various parameters. The transition band Δf is the most sensitive, with the pass band and stop band ripples having less significant impact. For example, if $\delta_1 = \delta_2 = 0.01$, dividing one of these figures by 2 leads to an increase of only 10% in the filter order. Moreover, it is worth emphasizing that, according to the estimation, the filter complexity is independent of the pass band width.

Examples

(f_s is taken to be 1.0 unless otherwise stated.)
(1) Calculation for a filter with 39 coefficients ($N = 39$) leads to the following values:

$$\delta_1 = 0.017; \quad \delta_2 = 0.034; \quad f_1 = 0.10375; \quad f_2 = 0.14375; \quad \Delta f = 0.04$$

With these values for the parameters the estimate gives $N_e = 40$.
(2) A filter with 160 coefficients ($N = 160$) has the following parameters:

$$\delta_1 = 2.24 \times 10^{-2}; \quad \delta_2 = 1.12 \times 10^{-4}; \quad f_1 = 0.053125;$$
$$f_2 = 0.071875; \quad \Delta f = 0.01875$$

The estimate gives $N_e = 164$.

(3) A filter with 15 coefficients ($N = 15$) has the following parameters:

$$\delta_1 = 0.0411; \quad \delta_2 = 0.0137; \quad f_1 = 0.1725; \quad f_2 = 0.2875; \quad \Delta f = 0.115$$

The estimate gives $N_e = 13$.
The coefficients of the filter corresponding to the equation:

$$y(n) = \sum_{i=1}^{15} a_i x(n - i)$$

have values of:

$$a_1 = -0.00047 = a_{15} \quad a_2 = \quad 0.02799 = a_{14}$$
$$a_3 = \quad 0.02812 = a_{13} \quad a_4 = -0.03572 = a_{12}$$
$$a_5 = -0.07927 = a_{11} \quad a_6 = \quad 0.04720 = a_{10}$$
$$a_7 = \quad 0.30848 = a_9 \quad a_8 = \quad 0.44847$$

The ripples of the filter are shown in Figure 5.16.

It should be noted that non-negligible differences can occur between the value N which is necessary in practice and the estimated value N_e, when the transition band limits approach 0 and 0.5 or when N is a few units. A set of more complex formulae is given in Ref. [3]. As indicated later (Chapter 7), a high-pass filter can be derived from a low-pass one by inverting the sign of every second coefficient. Consequently, estimate (5.32) can also apply to high-pass filters. When the mask specifies different ripples for the pass and stop bands an over-estimation of the number of coefficients can be obtained by assuming the most rigid constraints on δ_1 and δ_2 in the pass and stop bands, respectively.

For band pass filters it is necessary to introduce several transition bands.

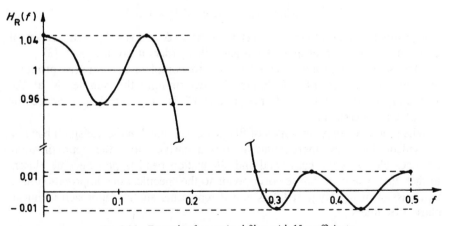

Fig. 5.16   Example of an optimal filter with 15 coefficients

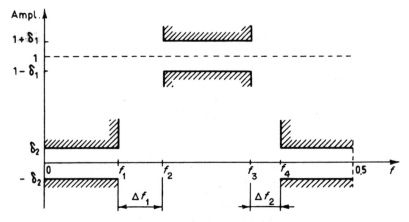

Fig. 5.17   Mask of a band-pass filter

Figure 5.17 gives the characteristic of such a filter, when has two transition bands Δf_1 and Δf_2. Experience shows that the number of coefficients N depends essentially on the smallest bandwidth, $\Delta f_m = \min(\Delta f_1, \Delta f_2)$. Then estimate (5.32) can be applied, with $\Delta f = \Delta f_m$. An upper bound is obtained by considering the band pass filter as a cascade of low-pass and high-pass filters and summing the estimates.

Example: A band-pass filter with 32 coefficients ($N = 32$) has the following characteristics:

$$\delta_1 = 0.015; \quad \delta_2 = 0.0015; \quad f_1 = 0.1; \quad f_2 = 0.2;$$
$$f_3 = 0.35; \quad f_4 = 0.425; \quad \Delta f_m = 0.075$$

The estimate using equation (5.32) with $\Delta f = \Delta f_m$ gives $N_e = 32$. A set of elaborate formulae to estimate band pass filter orders it given in Ref. [3].

The estimation formulae can be used to complete a computer program for the filter coefficient calculations by determining the number N at the beginning of the program. Expression (5.32) is very useful in practice for complexity assessment.

When the frequency responses of filters are examined in the transition band, it is evident that they approximate to raised cosine form; the approximation becomes closer as the pass band and attenuation band ripples become closer. In fact, this type of response corresponds to the specifications imposed on data transmission using Nyquist filters and it represents another approach to linear phase FIR filters.

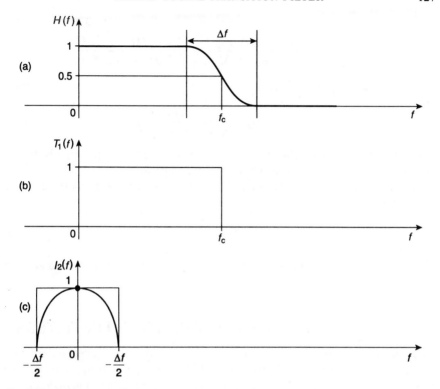

Fig. 5.18 (a) Response with raised cosine transition; (b) frequency impulse of width $2f_c$; (c) frequency impulse for the transition

5.8 RAISED COSINE TRANSITION FILTER

The frequency response $H(f)$ of a filter whose transition band has raised cosine form is represented in Figure 5.18(a).

It is expressed as the following convolution product:

$$H(f) = I_1(f) * \left[I_2(f) \frac{\pi}{2\Delta f} \cos\left(\frac{\pi f}{\Delta f} \right) \right] \tag{5.33}$$

where I_1 is an impulse of width $2f_c$ and $I_2(f)$ is an impulse of width Δf.

Under these conditions the impulse response $h(t)$ can be written as the product of two impulse responses $i_1(t)$ and $i_2(t)$ given by:

$$i_1(t) = 2f_c \frac{\sin \pi 2 f_c t}{\pi 2 f_c t}$$

and

$$i_2(t) = \frac{1}{2}\frac{\sin \pi\Delta f(t + 1/2\Delta f)}{2\Delta f(t + 1/2/\Delta f)} + \frac{1}{2}\frac{\sin \pi\Delta f(t - 1/2\Delta f)}{2\Delta f(t - 1/2/\Delta f)}$$

After simplification this gives:

$$h(t) = 2f_c \frac{\sin 2\pi f_c t}{2\pi f_c t} \frac{\cos \pi f t}{1 - 4\Delta f^2 t^2} \tag{5.34}$$

The total number of coefficients of the filter is determined principally by the function $i_2(t)$ and the width of its main lobe, equal to $3/\Delta f$.

Hence in a raised cosine transition band filter, the number of coefficients can be estimated using:

$$N \simeq \frac{3fe}{\Delta f} \tag{5.35}$$

This estimate can be considered as a first approach, when it is compared with equation (5.32).

5.9 STRUCTURES FOR IMPLEMENTING FIR FILTERS

FIR filters are composed of circuits which carry out the three fundamental operations of storage, multiplication and addition. They are arranged to produce from the set of data $x(n)$ an output set $y(n)$ according to the equation of definition of the filter. As no real operation can be instantaneous, the equation which is implemented in place of equation (5.5) in Section 5.1 is

$$y(n) = \sum_{i=0}^{N-1} a_i x(n - i - 1) \tag{5.36}$$

N data memories are required, and for each output number N multiplications and $N - 1$ additions have to be performed. Various arrangements of the circuits can be devised to put these operations into effect [4–6].

Figure 5.19(a) outlines the filter in what is called the 'direct' structure. Transposition of the diagram of this scheme produces the 'transposed' structure. This is represented in Figure 5.19(b), where the same operators are arranged differently. This structure allows multiplication of each piece of data $x(n)$ by all of the coefficients in succession. The memories store partial sums, and, in effect, at time n the first memory stores the number $a_{N-1}x(n)$, the next memory stores $a_{N-1}x(n - 1) + a_{N-2}x(n)$ and the last memory stores the sum $y(n)$.

In linear phase filters the symmetry of the coefficients can be exploited to halve the number of multiplications to be performed for each output number. This is very important for the complexity of the filter and justifies the almost general use of linear phase filters. The corresponding structure is shown in

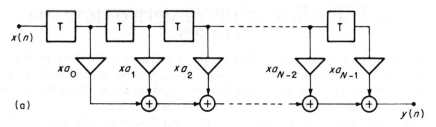

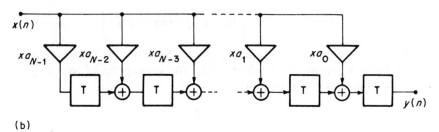

Fig. 5.19 Realization of FIR filters. (a) Direct structure; (b) transposed structure

Figure 5.20 for the direct form when the number of coefficients is odd: $N = 2P + 1$.

The complexity of the circuits depends on the number of operations to be performed and on their precision. Thus, the multiplication terms should have the minimum possible number of bits, which will reduce the amount of memory necessary both for the coefficients and for the data. These limitations modify the characteristics of the filter.

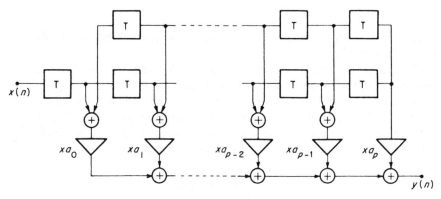

Fig. 5.20 Direct structure for a linear phase filter

5.10 LIMITATION OF THE NUMBER OF BITS FOR COEFFICIENTS

Limitation of the number of bits in the coefficients of a filter produces a change in the frequency response, which appears as the superposition of an error function. The consequences will be analysed for linear phase filters and extension of the results to any FIR filter is straightforward.

Assume a linear phase filter with $N = 2P + 1$ coefficients, for which the transfer function is written according to Section 5.2, equation (5.13), as

$$H(f) = e^{-j2\pi f PT} \left[h_0 + 2 \sum_{i=1}^{P} h_i \cos(2\pi f iT) \right]$$

The limitation of the number of bits in the coefficients introduces an error δh_i $(0 \leqslant i \leqslant P)$ in the coefficient h_i which, under the assumption of rounding with a quantization step q, is such that:

$$|\delta h_i| \leqslant \frac{q}{2}$$

This results in the superposition on the function $H(f)$ of an error function $e(f)$ such that:

$$e(f) = e^{-j2\pi f PT} \left[\delta h_0 + 2 \sum_{i=1}^{P} \delta h_i \cos(2\pi f iT) \right] \qquad (5.37)$$

The amplitude of this function must be limited so that the response of the real filter is within the required specification. An upper limit is obtained from

$$|e(f)| \leqslant |\delta h_0| + 2 \sum_{i=1}^{P} |\delta h_i| |\cos(2\pi f iT)|$$

$$|e(f)| \leqslant \frac{q}{2} N \qquad (5.38)$$

In general, this limit is much too large, and a more realistic estimate is obtained by statistical methods [7].

When a large number of filters with a wide range of specifications are considered and when general results are required the variables δh_i $(0 \leqslant i \leqslant P)$ can be viewed as random independent variables, uniformly distributed over the range $[-q/2, q/2]$. In these conditions they have variance $q^2/12$. The function $e(f)$ can equally be regarded as a random variable. Assume e_0 is the effective value of the function $e(f)$ over the frequency range $[0, f_s]$, that is,

$$e_0^2 = \frac{1}{f_s} \int_0^{f_s} |e(f)|^2 \, df \qquad (5.39)$$

In fact, the function $e(f)$ is periodic and defined by its Fourier series expansion. The Bessel–Parseval equation allows us to write, according to equation (5.8):

$$\frac{1}{f_s} \int_0^{f_s} |e(f)|^2 \, df = \sum_{i=0}^{N-1} (\delta h_i)^2$$

As a consequence, the variance σ^2 of the random variable e_0 can be written:

$$\sigma^2 = E[e_0^2] = N(q^2/12)$$

By assuming the second-order moment to be independent of frequency and by using equation (5.39) the variable $e(f)$ can be regarded as a random variable with variance such that:

$$\sigma = q/2\sqrt{(N/3)} \tag{5.40}$$

This relation provides an estimate of $|e(f)|$ which is much less than the bound equation (5.38). In fact, $e(f)$, resulting from equation (5.37) as a weighted sum of variables which are assumed to be independent, can be regarded as a Gaussian variable with a zero mean if quantization is performed by rounding and with variance σ^2. In order to determine the quantization step q, one can also argue from the confidence levels. For example, the probability that $|e(f)|$ exceeds the value 2σ is less than 5% using the table given in Appendix 2 of Chapter 1.

The above results will now be used to estimate the number of bits b_c needed to represent the coefficients of a filter specified by a mask.

Given a filter specification, let δ_m be the required value for the amplitude of the ripple and δ_0 the amplitude of the ripple before the number of bits in the coefficients is limited. The parasitic function $e(f)$ must be such that:

$$|e(f)| < \delta_m - \delta_0$$

The level of confidence in the estimate is deemed sufficient if q is chosen as

$$\frac{q}{2}\sqrt{\frac{N}{3}} < \frac{\delta_m - \delta_0}{2}$$

or

$$q < (\delta_m - \delta_0)\sqrt{(3/N)} \tag{5.41}$$

The number of bits b_c needed to represent the coefficients depends on the largest of the values h_i $(0 \leqslant i \leqslant P)$, and, allowing for the sign, the quantization step is given by

$$q = 2^{1-b_c}\left[\max_{0 \leqslant i \leqslant p} |h_i|\right] \tag{5.42}$$

If the filter is a low-pass one with a frequency response approaching unity in the pass band and corresponding to the mask in Figure 5.7 the values of the coefficients can have first approximations which are calculated by equation (5.18). Under these conditions, the maximum for h_0 is obtained with:

$$h_0 = (f_1 + f_2)/f_s \tag{5.43}$$

and equations (5.41) and (5.42) lead to the following estimate:

$$b_c \approx 1 + \log_2 [((f_1 + f_2)/f_s)(\sqrt{(N/3)})1/(\delta_m - \delta_0)] \tag{5.44}$$

where:

    b_c is the number of bits of the coefficients (including the sign),
    N is the number of coefficients of the filter,
    f_1 is the pass band edge,
    f_2 is the stop band edge,
    f_s is the sampling frequency,
    δ_m is the limit imposed on the amplitude of the ripple,
    δ_0 is the amplitude of the ripple of the filter before limitation of the number of bits in the coefficients.

Example: Consider the low-pass filter with 15 coefficients discussed in Section 5.7, which has the following parameters:

$$N = 15, \quad f_s = 1, \quad f_1 = 0.1725, \quad f_2 = 0.2875$$

The specification requires

$$\delta_1 = 0.05 \quad \delta_2 = 0.02$$

The ripples of the filter in the pass and stop bands, prior to limitation of the number of bits of the coefficients, have the values:

$$\delta_{10} = 0.0411 \quad \delta_{20} = 0.0137$$

Under these conditions:

$$\delta_m - \delta_0 = \min(\delta_1 - \delta_{10}, \delta_2 - \delta_{20}) = 0.0063$$

Hence,

$$b_c \simeq 1 + \log_2 [0.46(\sqrt{(5)}/0.0063)] = 8.3$$

Coefficients with 8 bits are chosen, i.e. $b_0 = 8$. The corresponding function $e(f)$ is represented in Figure 5.21.

In practice, equation (5.44) can be simplified. First, the tolerance provided by the filter mask is generally distributed equally between the ripples before limitation of the number of bits in the coefficients and before the supplementary error caused by this limitation (that is, $\delta_0 = \delta_m/2$). Further, the filters to be realized are generally such that:

$$0.5 f_s/\Delta f \leq N/3 \leq 1.5 f_s/\Delta f \tag{5.45}$$

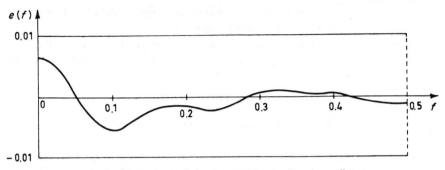

Fig. 5.21 *Frequency deviation caused by rounding the coefficients*

which, using equation (5.32), corresponds to a large range of values for the parameters δ_1 and δ_2. Under these conditions, a suitable estimate of the number of bits of the coefficients is given by

$$b_c \simeq 1 + \log_2 \left[((f_1 + f_2)/f_s)\sqrt{(f_s/\Delta f)}(2/\delta_m) \right]$$

By introducing the steepness of the cut-off of the filter (equation (5.17)) and the normalized transition band, one finally obtains:

$$b_c \simeq 3 + \log_2 \left[(f_1 + f_2)/2\Delta f \right] - \tfrac{1}{2}\log_2 (f_s/\Delta f) + \log_2 (1/\min\{\delta_1, \delta_2\}) \quad (5.46)$$

Thus, the number of bits for the coefficients is directly related to the filter specifications. It should be stated that filters with a narrow pass band require fewer bits than filters with a wide band. Equations (5.44) and (5.45) also apply to high-pass and band-pass filters.

The above analysis was made under the assumption that the two operations of the calculation of the coefficients and limitation of the number of bits are separate. It is equally possible to take an overall view of these two operations, and techniques exist for the appropriate calculations although they are somewhat complicated [8].

5.11 LIMITATION OF THE NUMBER OF BITS IN INTERNAL MEMORIES

Limitation of the number of bits of the memories in a filter represents a source of signal degradation across this filter. Strictly speaking, this degradation can be avoided in an FIR filter. In effect, if b_d denotes the number of bits of input data, with b_c being that of the coefficients, it is sufficient to be able to accumulate the products to $(b_d + b_c)$ bits in order to realize exactly the calculations defined by equation (5.36). Nevertheless, the products are generally rounded to b_m bits in order to simplify the multiplication and accumulation circuits. This operation produces a noise which is called the round-off noise. In the structures considered in the previous sections, this noise is added at the output of the filter. Figure 5.22 shows the arrangements of the numbers in the filter.

Allowing for the values of the coefficients h_i, the products are shifted by a number of bits b_0 which, for a low-pass filter, is written, according to equation (5.43), as:

$$b_0 = \log_2 (f_s/(f_1 + f_2))$$

The output of the filter is held in an accumulator having at least $b_a = (b_0 + b_m)$ bits.

In fact, the number of bits b_i inside the machine must be greater than this value b_a in order to avoid overflows, although with a 2's complement representation temporary overflows are acceptable, as shown in Section 13.1. This number of bits b_i will now be related to the filter specifications.

Noise of power N_1 is superposed on the signal which is presented at the input of the filter. This noise power is generally related to the number of bits b_d used to represent the signal. In fact, it is equal to k_0 times the noise power generated by the quantization to b_d bits (with $k_0 \geqslant 1$). If the round-off noise has a power N_c, and if the input signal S and the added noise N_1 have uniform spectral distributions, the signal-to-noise ratio SN at the output of the filter is expressed in decibels (dB) by:

$$SN = 10 \log \left[S\left(\frac{f_1 + f_2}{f_s}\right) \Big/ \left(N_1\left(\frac{f_1 + f_2}{f_s}\right) + N_c \right) \right] \qquad (5.47)$$

Then the reduction ΔSN of the signal-to-noise ratio across the filter is written:

$$\Delta SN = 10 \log \left[1 + \left(\frac{N_c}{N_1}\right) \frac{f_s}{f_1 + f_2} \right] \qquad (5.48)$$

If this degradation remains low, this becomes:

$$\Delta SN \simeq 4.3(N_c/N_1)f_s/(f_1 + f_2) \qquad (5.49)$$

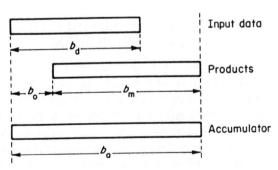

Fig. 5.22   Scaling numbers in an FIR filter

It is now necessary to determine the relation between the round-off noise power N_c and the number of bits b_i of the internal memories. By taking as unity the maximum value of the input numbers $x(n)$, i.e.

$$|x(n)| \leqslant 1$$

and using the defining equation (5.36) we obtain at the output

$$|y(n)| \leqslant \sum_{i=0}^{N-1} |a_i|$$

it can be shown that this sum is approximately proportional to $\log N$. For a low-pass filter with unit response at zero frequency we obtain:

$$\sum_{i=0}^{N-1} a_i = 1 \tag{5.50}$$

Under these conditions the sum of the absolute values of the coefficients generally remains less than several units and it can be considered that the following inequalities are valid:

$$1 \leqslant \sum_{i=0}^{N-1} |a_i| < 2 \tag{5.51}$$

Rounding to b_i bits in the accumulation process then leads to a round-off noise N_c of

$$N_c = N \frac{2^{2(2-b_i)}}{12}$$

when the noise at the input N_1 has a value of

$$N_1 = k_0 \frac{2^{2(1-b_d)}}{12} \quad \text{with} \quad k_0 \geqslant 1$$

Allowing for the inequality (5.45), the degradation of the signal-to-noise ratio across the filter is written

$$\Delta SN \simeq 4.3 \frac{3 f_s}{\Delta f} \frac{4}{k_0 2^{2(b_i - b_d)}} \frac{f_s}{(f_1 + f_2)} \tag{5.52}$$

In general, the filter specifications impose a limit on the value of ΔSN. By assuming $k_0 = 1$, an estimate of the number of memory bits in the machine is given by

$$b_i \simeq b_d + 3 + \tfrac{1}{2}[\log_2 (1/\Delta SN) + \log_2 (f_s/\Delta f) + \log_2 (f_s/(f_1 + f_2))] \tag{5.53}$$

It must be noted that the validity of this estimate is limited to low values of the term ΔSN, expressed in decibels. It would appear that filters with a narrow pass band require more bits. However, it is possible to use an internal rescaling of the numbers by allowing for the reduction of power of the signal after filtering, which corresponds to the number of bits b_R, with:

$$b_R = \tfrac{1}{2}\log_2\left(f_s/(f_1 + f_2)\right)$$

With rescaling, the number of memory bits in the machine is given by $b_{iR} = b_i - b_R$, that is,

$$b_{iR} \simeq b_d + 3 + \tfrac{1}{2}\log_2(1/\Delta SN) + \tfrac{1}{2}\log_2(f_s/\Delta f) \tag{5.54}$$

The estimates given in this section and in the preceding ones provide an evaluation of the complexity of the machines which are required for realizing FIR filtering functions.

5.12 Z-TRANSFER FUNCTION OF AN FIR FILTER

The Z-transfer function for an FIR filter with N coefficients is a polynomial of degree $N - 1$, which is written (equation (5.7))

$$H(Z) = \sum_{i=0}^{N-1} a_i Z^{-i}$$

This polynomial has $N - 1$ roots $Z_i (1 \leqslant i \leqslant N - 1)$ in the complex plane and can be written as the product

$$H(Z) = a_0 \prod_{i=1}^{N-1} (1 - Z_i Z^{-1}) \tag{5.55}$$

These roots have certain characteristics because of the properties of FIR filters.

First, if the coefficients are real, each complex root Z_i has a corresponding complex conjugate root $\overline{Z}_i$. Hence, $H(Z)$ can be written as a product of first- and second-degree terms with real coefficients. Each second-degree term is thus written as:

$$H_2(Z) = 1 - 2\,\mathrm{Re}\,(Z_i)Z^{-1} + |Z_i|^2 Z^{-2} \tag{5.56}$$

Second, the symmetry of the coefficients of a linear phase filter must appear in the decomposition into products of factors. For a second-degree term with real coefficients it is necessary that $|Z_i| = 1$ if the roots are complex, i.e. the root must lie on the unit circle. For a fourth-degree term with real coefficients the four complex roots have to be $Z_i, \overline{Z}_i, 1/Z_i, 1/\overline{Z}_i$. That is,

$$H_4(Z) = 1 - 2\,\mathrm{Re}\left(Z_i + \frac{1}{Z_i}\right)Z^{-1} + \left[|Z_i|^2 + \frac{1}{|Z_i|^2} + 4\,\mathrm{Re}\,(Z_i)\,\mathrm{Re}\left(\frac{1}{Z_i}\right)\right]Z^{-2}$$

$$- 2\,\mathrm{Re}\left(Z_i + \frac{1}{Z_i}\right)Z^{-3} + Z^{-4} \tag{5.57}$$

Under these conditions an FIR linear phase filter can be decomposed into a set of elementary filters of second or fourth degree having the symmetry properties of the coefficients.

The roots have been calculated for the low-pass filter with 15 coefficients given as an example in Section 5.7. The co-ordinates of the 14 roots are

$$Z_1 = -0.976 \pm j0.217 \quad Z_5 = 0.492 \pm j0.266$$
$$Z_2 = -0.797 \pm j0.603 \quad Z_6 = 1.573 \pm j0.851$$
$$Z_3 = -0.512 \pm j0.859 \quad Z_7 = 0.165$$
$$Z_4 = -0.271 \pm j0.962 \quad Z_8 = 6.052$$

Their positions in the complex plane are shown in Figure 5.23. This illustrates the characteristics of the frequency response of the filter and is related to Figure 5.16. The pairs of roots characteristic of the linear phase can be clearly seen. The configuration of the roots is modified if this constraint is no longer imposed.

5.13 MINIMUM PHASE FILTERS

The propagation delay through a linear phase filter can be excessive for some applications. Also it is not always possible or useful to use the symmetry of the coefficients of a linear phase filter to simplify the calculations [9]. If phase linearity is not an imposed characteristic, one can therefore hope to reduce the complexity of the filter by abandoning this constraint. A linear phase transfer function can be considered as the product of a minimum phase function and a pure phase shift. The condition for a Z-transfer function to be minimum phase is that its roots must be within or on the unit circle. This point is developed in Chapter 9.

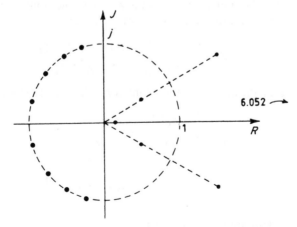

Fig. 5.23 Configuration of the zeros in an FIR filter

The calculation methods developed for linear phase filters can be adapted. However, the coefficients for a minimum phase filter can be obtained in a simple way using the coefficients of an optimal linear phase filter.

Let the frequency response of a linear phase filter with $N = 2P + 1$ coefficients be written:

$$H(f) = e^{-j2\pi fPT}\left[h_0 + 2\sum_{i=1}^{P} h_i \cos(2\pi f iT) \right]$$

The ripples in the pass and stop bands are δ_1 and δ_2, respectively. Let us consider the filter obtained by adding δ_2 to the above response and by rescaling it to approach unity in the pass band. Its response $H_2(f)$ will be

$$H_2(f) = e^{-j2\pi fPT}\frac{1}{1+\delta_2}\left[h_0 + \delta_2 + 2\sum_{i=1}^{P} h_i \cos(2\pi f iT) \right] \qquad (5.58)$$

In the pass band there are ripples with amplitude δ_1' such that:

$$\delta_1' = \frac{\delta_1}{(1+\delta_2)}$$

Its response in the stop band is represented in Figure 5.24; the ripples are limited to $\delta_2' = 2\delta_2/(1+\delta_2)$.

This is a linear phase filter because the symmetry of the coefficients is conserved. In contrast, it can be noticed that the roots of the Z-transfer function which are on the unit circle are double because $H_2(f)$ never becomes negative. Under these conditions the configuration of the roots is as shown in Figure 5.25.

The roots which are not on the unit circle are not double. However, the absolute value of the function $H_2(f)$ is not modified, except for a constant factor, if the roots Z_i outside the unit circle are replaced by the roots $1/Z_i$, which are inside the unit circle and thus also become double. This operation amounts simply to multiplication by $G(Z)$ such that:

$$G(Z) = \frac{[1-(Z^{-1}/Z_i)][1-(Z^{-1}/\overline{Z_i})]}{(1-Z_iZ^{-1})(1-\overline{Z_i}Z^{-1})}$$

Since $Z^{-1} = \overline{Z}$ on the unit circle, the symmetry with respect to the real axis yields:

$$\left|\frac{(Z^{-1}-Z_i)(Z^{-1}-\overline{Z_i})}{(Z-Z_i)(Z-\overline{Z_i})}\right|_{Z=e^{j\omega}} = 1 \qquad (5.59)$$

and thus:

$$|G(e^{j2\pi f})| = \frac{1}{|Z_i|^2}$$

Under these conditions we can write

$$H_2(f) = H_m^2(f)K \quad (K \text{ is a constant})$$

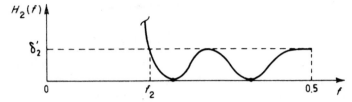

Fig. 5.24   Ripple in the stop band of the raised filter

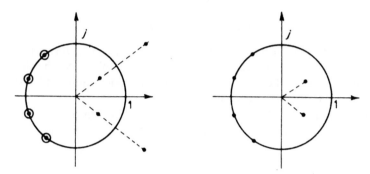

Fig. 5.25   Configuration of the zeros of $H_2(f)$ and of a minimum phase filter

where $H_m(f)$ is the response of a filter which has a Z-transfer function with P roots which are single and inside or on the unit circle. This filter satisfies the minimum phase condition, it has $P + 1$ coefficients and the amplitudes of the ripples in the pass and stop band are δ_{m_1} and δ_{m_2} such that:

$$\delta_{m_1} = \sqrt{\left(1 + \frac{\delta_1}{1 + \delta_2}\right) - 1} \simeq \frac{1}{2}\frac{\delta_1}{1 + \delta_2} \simeq \frac{\delta_1}{2} \qquad (5.60)$$

$$\delta_{m_2} = \sqrt{\left(\frac{2\delta_2}{1 + \delta_2}\right)} \simeq \sqrt{(2\delta_2)} \qquad (5.61)$$

To design this filter it is sufficient to start from the linear phase filter whose parameters δ_1 and δ_2 are determined from δ_{m_1} and δ_{m_2} and to follow the procedure described above. One drawback is that the extraction of the roots of a polynomial of degree $N - 1$ is required, which limits the values that can be envisaged for N. Other procedures can also be used [10, 11].

An estimate N'_e of the order of the FIR filter filter with minimum phase shift can be deduced from equation (5.32). According to the procedure described earlier for the specifications of δ_1 and δ_2, this becomes

$$N'_e \approx \frac{1}{2} \cdot \frac{2}{3} \log\left(\frac{1}{10\delta_1\delta_2^2}\right)\frac{f_s}{\Delta f}$$

or

$$N'_e \approx N_e - \tfrac{1}{3}\log\left(\frac{1}{10\delta_1}\right)\frac{f_s}{\Delta f} \qquad (5.62)$$

The validity of this formula is naturally limited to the case where $\delta_1 \ll 0.1$. The improvement in the order of the filter with minimum phase is a function of the ripple in the pass band. It generally remains relatively low.

Example: Assume the following specifications for a low-pass filter:

$$\delta_{m_1} = 0.0411; \quad \delta_{m_2} = 0.0137; \quad \Delta f = 0.115$$

whence

$$\delta_1 = 0.0822; \quad \delta_2 = 0.0000938$$

The number of coefficients needed for the corresponding linear phase filter is estimated at $N_e = 24$, which leads to $N'_e = 12$. Actually, it can be shown that the minimum phase filter satisfying the characteristic requires 11 coefficients instead of the 15 for the linear phase filter.

In conclusion, when the symmetries provided by the linear phase cannot be used it may be advantageous to resort to minimum phase filters.

5.14 DESIGN OF FILTERS WITH A LARGE NUMBER OF COEFFICIENTS

Optimization techniques become difficult to use or they no longer converge when the number of filter coefficients is very large, perhaps one thousand or more, corresponding to extremely narrow transition bands of the order of thousandths. One can then use sub-optimal techniques which require only the calculation of filters with a reduced number of coefficients. This is the case in the method described as frequency masking [12].

Consider a filter $H(Z)$ whose transition band Δf is centred on the cut-off frequency f_c. One starts by designing a low-pass filter $H_0(Z^M)$ with a reduced sampling frequency of f_s/M, where $M < f_s/4\Delta f$, such that the transition band of one of the alias frequencies of this filter coincides with the transition band of the required filter, as shown in Figure 5.26(b).

Two complementary filters are then constructed from $H_0(Z^M)$ as shown in Figure 5.26(c); this requires an odd number of coefficients, $2P + 1$, for $H_0(Z^M)$.

A diagram with two branches is obtained, to which are applied the filters $G_1(Z)$ and $G_2(Z)$; G_1 and G_2 are described, as interpolators and they have the responses given in Figure 5.26(c). It is then sufficient to sum the outputs to obtain the desired filter of Figure 5.26(a). The overall arrangement is shown in Figure 5.27.

The procedure thus requires three filters having transition bands of

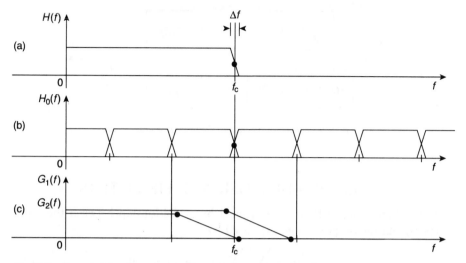

Fig. 5.26   *The principle of frequency masking: (a) desired filter, (b) subsampled filter, (c) interpolating filters*

$M\Delta f$, $f_c - kf_s/M$ and $(k + 1)f_s/M - f_c$, where k is the integer which permits the cut-off frequency f_c to be included.

The transfer function $H(Z)$ of the required filter takes the form:

$$H(Z) = H_0(Z^M)G_1(Z) + [Z^{-PM} - H_0(Z^M)]G_2(Z) \qquad (5.63)$$

which provides the coefficient values.

Note that the arrangement of Figure 5.27 provides an efficient realization of the overall filter since the filter $H_0(Z^M)$ has $M - 1$ zero coefficients among two non-zero coefficients. This arrangement can be simplified as shown in Figure 5.28. The interpolating filters can be taken as $F_1(Z) = G_1(Z) + G_2(Z)$ and $F_2(Z) = G_1(Z) - G_2(Z)$ but they can also be derived directly from their specifications, as deduced from Figure 5.26.

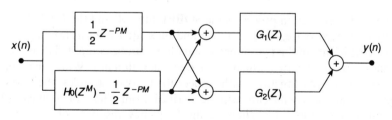

Fig. 5.27   *Diagram of filter using the frequency masking technique*

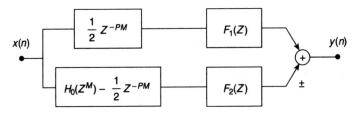

Fig. 5.28   *Simplified diagram of the frequency masking filter*

5.15 TWO-DIMENSIONAL FIR FILTERS

A two-dimensional FIR filter is defined by a relationship between the input $x(n, m)$ and the output $y(n, m)$ expressed by:

$$y(n, m) = \sum_{i=0}^{N_1-1} \sum_{j=0}^{N_2-1} a_{ij}\, x(n - i, m - j) \qquad (5.64)$$

The set of coefficients a_{ij} takes the form of a $N_1 \times N_2$ matrix denoted $A_{N_1 N_2}$. The corresponding two-variable transfer function

$$H(Z_1, Z_2) = \sum_{i=0}^{N_1-1} \sum_{j=0}^{N_2-1} a_{ij} Z_1^{-i} Z_2^{-j} \qquad (5.65)$$

is given in vector notations by:

$$H(Z_1, Z_2) = [1, Z_1^{-1}, \ldots, Z_1^{-(N_1-1)}] A_{N_1 N_2} \begin{bmatrix} 1 \\ Z_2^{-1} \\ \vdots \\ Z_2^{-(N_2-1)} \end{bmatrix} \qquad (5.66)$$

The coefficient matrix is also called the mask. As an illustration, the following high-pass filters are often used in image processing:

$$A' = \begin{bmatrix} -1 & 0 & 1 \\ -2 & 0 & 2 \\ -1 & 0 & 1 \end{bmatrix}; \qquad A'' = \begin{bmatrix} 1 & 1 & 1 \\ 1 & -2 & 1 \\ -1 & -1 & -1 \end{bmatrix}$$

In some edge extraction processes the SOBEL filter A' and the PREVITT filter A'' are employed twice, once as above and once after a 90° rotation.

The coefficients of the two-dimensional filters can be computed directly from specifications in the two-dimensional frequency domain. When the impulse response is an even function, with respect to the two variables, the frequency response and the coefficients can be derived from those of a linear phase one-dimensional filter. Let $H(\omega)$ be the frequency response of such a filter.

Equation (5.13), when the phase term is neglected, yields:

$$H(\omega) = h_0 + 2 \sum_{i=1}^{P} h_i \cos i\omega$$

A polynomial relation exists between $\cos i\omega$ and $\cos \omega$:

$$\cos i\omega = T_i(\cos \omega) \tag{5.67}$$

where $T_i(x)$ is the Chebyshev polynomial of degree i. Under these conditions $H(\omega)$ can also be written:

$$H(\omega) = \sum_{i=0}^{P} g_i (\cos \omega)^i \tag{5.68}$$

Changing variables as follows:

$$\cos \omega = H_1(\omega_1, \omega_2) = \sum_{k=0}^{K-1} \sum_{l=0}^{L-1} t(k, l) \cos k\omega_1 \cos l\omega_2 \tag{5.69}$$

yields a two-variable function:

$$H(e^{\omega_1}, e^{\omega_2}) = \sum_{i=0}^{P} g_i \left(\sum_{k=0}^{K-1} \sum_{l=0}^{L-1} t(k, l) \cos k\omega_1 \cos l\omega_2 \right)^i \tag{5.70}$$

which can be rewritten as:

$$H(\omega_1, \omega_2) = \sum_{i=0}^{N_1-1} \sum_{j=0}^{N_2-1} h_{ij} \cos i\omega_1 \cos j\omega_2 \tag{5.71}$$

where:

$$N_1 = 2KP + 1; \quad N_2 = 2LP + 1$$

The function $t(k, l)$ can be chosen so as to map the points in the frequency response of the one-dimensional filter into contours in the (ω_1, ω_2) plane. For example, the circular symmetry is approximately achieved by:

$$\cos \omega = \tfrac{1}{2}[\cos \omega_1 + \cos \omega_2 + \cos \omega_1 \cos \omega_2 - 1] \tag{5.72}$$

as can be seen from a series expansion of $\cos \omega_1$, with ω_1 small. The frequency response of a filter designed that way is shown in Figure 5.29.

The implementation of a two-dimensional filter can be obtained through straight application of equation (5.64). For filters derived from a one-dimensional function, expression (5.70) suggests an important simplification: in the one-dimensional FIR filter with $P + 1$ coefficients g_i, the delays are replaced by two-dimensional sections corresponding to the function $H_1(\omega_1, \omega_2)$ [13].

Separable filters are particularly simple to realize; in this case the coefficient

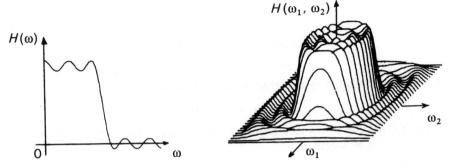

Fig. 5.29 *Two-dimensional FIR filter designed from a one-dimensional linear phase filter*

matrix is dyadic, i.e.

$$A_{N_1 N_2} = V_1 V_2^t$$

where V_1 and V_2 are the vectors. Then, in accordance with equation (5.66), the transfer function factorizes:

$$H(z_1, z_2) = H_1(z_1) H_2(z_2) \qquad (5.73)$$

The specifications of such filters are subject to limitations. Firstly, they must have quadrantal symmetry along the coordinate axes. As shown in Figure 5.30, the useful frequency domain is divided into four parts: low pass/low pass (LL), low pass/high pass (LH), high pass/low pass (HL) and high pass/high pass (HH).

Consequently, the ripple specifications must be defined. For a two-dimensional low-pass filter, the HH domain is subjected to the attenuation of two filters, horizontal and vertical. An illustration is given in Figure 5.31 which shows the frequency response of a two-dimensional separable filter based on the half-band filter of Figure 5.13.

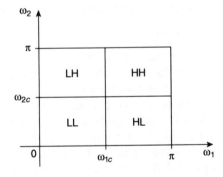

Fig. 5.30 *Frequency domains for a two-dimensional separable filter*

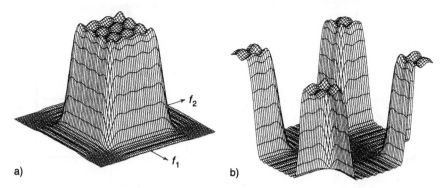

a)   b)

Fig. 5.31   Two dimensional half-band separable filter: (a) low pass/low pass, (b) high pass/high pass

Realization can be achieved by following the definition exactly; that is, a data table representing an image can be processed line by line with the horizontal filter and column by column with the vertical filter.

When the image is subjected to a horizontal scan as in television, the signal appears as one-dimensional and can be processed as such. If each line contains N points, the transfer function can be written:

$$H(z_1, z_2) = H_1(z)H_2(z^N) \tag{5.74}$$

For example, for the Sobel filter A', one has:

$$A' \begin{bmatrix} 1 \\ 2 \\ 1 \end{bmatrix} [-1 \quad 0 \quad 1]$$

and the corresponding circuit is given in Figure 5.32. Realization is particularly simple as the circuits do not contain multipliers.

5.16 COEFFICIENTS OF TWO-DIMENSIONAL FIR FILTERS BY THE METHOD OF LEAST SQUARES

The method will be developed for an important special case—filters with quadrantal symmetry. Two types of filter correspond to this category, rectangular and lozenge filters with the frequency domains of Figure 5.33.

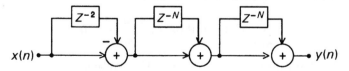

Fig. 5.32   Realization of a filter by contour extraction

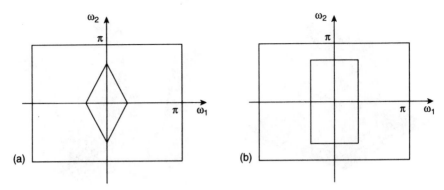

Fig. 5.33   Filters: (a) two-dimensional lozenge and (b) rectangle

The frequency response of a zero phase filter having $(2M + 1) \times (2N + 1)$ coefficients with quadrantal symmetry is expressed by:

$$H(\omega_1, \omega_2) = h_{00} + 2\sum_{i=1}^{M} h_{i0} \cos i\omega_1 + 2\sum_{j=1}^{N} h_{0j} \cos j\omega_2 + 4\sum_{i=1}^{M} \sum_{j=1}^{N} h_{ij} \cos i\omega_1 \cos j\omega_2$$

(5.75)

In total the filter has $(1 + M + N + MN)$ coefficients h_{ij} with different values.

The method of least squares with weighting will be applied directly, to approach the desired response, $D(\omega_1, \omega_2)$. With an oversampling factor of k, the quadratic deviation function, or cost function, to be minimized is:

$$J = \sum_{m=0}^{K_M} \sum_{n=0}^{K_N} \left| H\left(\frac{m\pi}{K_M}, \frac{n\pi}{K_N}\right) - D\left(\frac{m\pi}{K_M}, \frac{n\pi}{K_N}\right) \right|^2 W\left(\frac{m\pi}{K_M}, \frac{n\pi}{K_N}\right)$$

(5.76)

with $K_M = k(M + 0.5)$ and $K_N = k(N + 0.5)$ in order to cover all the useful frequency domains.

The weighting function $W(\omega_1, \omega_2)$ enables the approximation to be adjusted in accordance with the ripple specifications, for example.

With simplified notation, this gives:

$$J = \sum_{m=0}^{K_M} \sum_{n=0}^{K_N} e^2(m, n) W(m, n)$$

(5.77)

The minimum of the cost function is obtained for:

$$\sum_{m=0}^{K_M} \sum_{n=0}^{K_N} e(m, n) W(m, n) \frac{\partial e(m, n)}{\partial h_{ij}} = 0$$

(5.78)

Which yields a system with $(1 + M + N + MN)$ equations.

Designating the coefficient vector by $[h_{ij}]$ and the frequency vector by $V(m, n)$:

$$V^t(m, n) =$$
$$\left[1, \ldots, 2\cos\left(i\frac{m\pi}{K_M}\right), \ldots, 2\cos\left(j\frac{n\pi}{K_N}\right), \ldots, 4\cos\left((i\frac{n\pi}{K_M}\right)\cos\left(j\frac{n\pi}{K_N}\right), \ldots \right]$$

the solution can be written:

$$[h_{ij}] = \left[\sum_{m=0}^{K_M} \sum_{n=0}^{K_N} W(m, n)V(m, n)V^t(m, n) \right]^{-1} \left[\sum_{m=0}^{K_M} \sum_{n=0}^{K_N} W(m, n)V(m, n)D(m, n) \right]$$

$$(5.79)$$

If the number of coefficients is even, it is necessary to modify the parameters. For a filter with $(2M) \times (2N + 1)$ coefficients, it is necessary to take:

$$V^t(m, n) = \left[\ldots, 2\cos\left((i - 0.5)\frac{m\pi}{K_M}\right), \ldots, 4\cos\left((i - 0.5)\frac{n\pi}{K_M}\right)\cos\left(j\frac{n\pi}{K_N}\right), \ldots \right]$$

$$(5.80)$$

with $K_M = kM$ and $K_N = k(N + 0.5)$. The coefficient vector obtained in this case has $(M + MN)$ elements.

An important characteristic of filters used in image processing is the response to a unit step. Ringing at the transition can produce repetitions of contours and thus degrade the image. It is possible to reduce ringing by modifying the desired response $D(\omega_1, \omega_2)$ using a slope at the end of the pass band and the start of the attenuation band.

The method is illustrated by the design of a rectangular filter with $(2M + 1) \times (2N + 1) = 9 \times 9$ coefficients, with 0.125 and 0.25 as the end of the pass band and the start of the attenuation band on the horizontal frequency axis and 0.0625 and 0.125 on the vertical axis. The 25 different coefficients obtained are:

$$h_{ij} = \begin{bmatrix} 0.052427 & 0.0419028 & 0.0184534 & -0.0002861 & -0.006258 \\ 0.0491981 & 0.0393451 & 0.0173566 & -0.0002629 & -0.0059292 \\ 0.041534 & 0.0332908 & 0.0147612 & -0.000261 & -0.005282 \\ 0.0299102 & 0.0240605 & 0.107414 & -0.0002704 & -0.0041828 \\ 0.0180912 & 0.0146366 & 0.0065523 & -0.0003836 & -0.0031209 \end{bmatrix}$$

and the corresponding frequency response is given in Figure 5.34. Evidently this response is very close to that of a separable filter. Considering now a lozenge filter with $(2M + 1) \times (2N) = 9 \times 8$ coefficients, a pass band end at 0.125 and an attenuation band starting at 0.25 on the horizontal and vertical axes, the coefficients for one quadrant are:

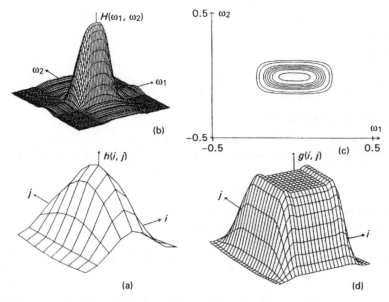

Fig. 5.34   *Rectangular filter with 9 × 9 coefficients: (a) impulse response, (b) frequency response, (c) horizontal section of the frequency response, (d) response to a unit step*

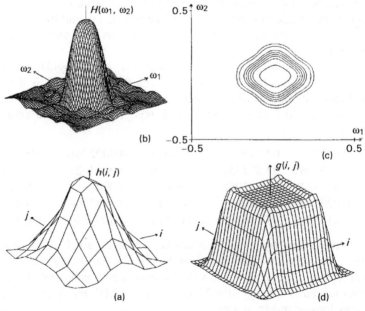

Fig. 5.35   *Lozenge filter with 9 × 8 coefficients: (a) impulse response, (b) frequency response, (c) horizontal section of the frequency response, (d) response to a unit step*

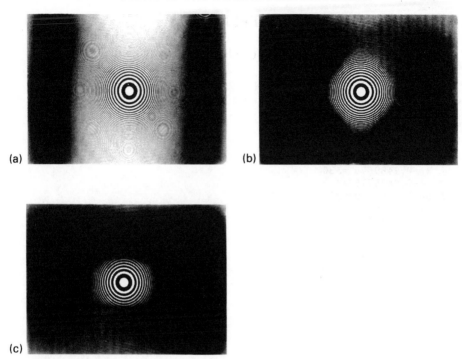

Fig. 5.36   Filtering of a test pattern: (a) original image, (b) lozenge filtering of the pattern, (c) rectangular filtering of the pattern

$$h_{ij} = \begin{bmatrix} 0.0763835 & 0.680674 & 0.0403862 & 0.0130039 & 0.000071 \\ 0.0642979 & 0.03951 & 0.0217936 & 0.0008111 & -0.002745 \\ 0.0276109 & 0.0195655 & 0.0068997 & -0.0050102 & -0.0110481 \\ 0.0065124 & 0.0011002 & 0.0085984 & -0.0099831 & -0.0073724 \end{bmatrix}$$

The frequency response is given in Figure 5.35. Calculation has been guided by attempting to reduce the unit step response $g(i,j)$ defined by:

$$g(i,j) = \sum_{i_1=-M}^{i} \sum_{j_1=-N}^{j} h(i_1, j_1) \tag{5.81}$$

This response is also shown in the figure, where it has been repeated in the four quadrants to provide a complete picture.

The two filters have been applied to a test card. Figure 5.36 shows the elimination of replicas obtained with the rectangular filter and the lozenge filter.

See Refs [14, 15] for complementary developments in the design techniques for two-dimensional FIR filters, including those with limited precision coefficients and constraints on the unit step response.

APPENDIX

Example of FIR filter design

FILTER WITH 2 FREQUENCY BANDS
NUMBER OF COEFF = 32

```
COEFFICIENTS
A  1= 0.01064416=A32
A  2=-0.00703074=A31
A  3=-0.2489676.=A30
A  4=-0.04257757=A29
A  5=-0.04550193=A28
A  6=-0.02596906=A27
A  7=-0.00801637=A26
A  8=-0.03449482=A25
A  9=-0.03164973=A24
A 10=-0.00394829=A23
A 11=-0.04885845=A22
A 12=-0.06340511=A21
A 13=-0.01710309=A20
A 14=0.08742009=A19
A 15=0.20962396=A18
A 16=0.29156661=A17
```

BAND NUMBER	1	2
LOW LIMIT	0.0000	0.1900
HIGH LIMIT	0.1400	0.5000
REQ. VALUE	1.0000	0.0000
WEIGHTING	1.00	10.00
DEVIATION	0.21175	0.02118
DEVIATION IN DB	1.668	-33.484

```
EXTREME FREQUENCIES
0.0000   0.0410   0.0801   0.1172   0.1400   0.1700   0.1798   0.2032
0.2305   0.2618   0.2930   0.3243   0.3555   0.3888   0.4200   0.4513
0.4845
```

REFERENCES

[1] T. W. PARKS and J. H. MACCLELLAN, Chebyshev approximation for non-recursive digital filters with linear phase. *IEEE Transactions on Circuit Theory*, **19**, March 1972.
[2] J. H. MACCLELLAN, T. W. PARKS and L. RABINER, A computer program for designing optimum FIR filters. *IEEE Transactions on Audio and Electroacoustics*, Dec 1973.
[3] F. MINTZER and B. LIU, Practical design rules for optimum FIR bandpass digital filters. *IEEE Transactions*, **ASSP27**(2), 1979.

[4] R. CROCHIÈRE and A. OPPENHEIM, Analysis of linear digital networks. *Proceedings of the IEEE*, April 1975.

[5] W. SCHUSSLER, On structures for non-recursive digital filters. *Arch. Elek. Ubertragung*, June 1972.

[6] M. BELLANGER and G. BONNEROT, Premultiplication scheme for digital FIR filters. *IEEE Transactions*, **ASSP26**, Feb 1978.

[7] D. CHAN and L. RABINER, Analysis of quantization errors in the direct form of FIR filters. *IEEE Transactions on Audio and Electroacoustics*, August 1973.

[8] F. GRENEZ, Synthèse des filtres numériques non récursifs à coefficients quantifiés. *Annales des Télécom.*, **34**(1/2), 1979.

[9] M. FELDMANN, J. HENAFF, B. LACROIX and J. C. REBOURG, Design of minimum phase charge-transfer transversal filters. *Electronics Letters*, **15**(8), 1979.

[10] R. BOITE and H. LEICH, A new procedure for the design of high order minimum phase FIR filters. *Signal Processing*, **3**(2), 101–8, 1981.

[11] Y. KAMP and C. J. WELLEKENS, Optimal design of minimum-phase FIR filters. *IEEE Transactions*, **ASSP31**(4), 1983.

[12] Y. C. LIM and Y. LIAN, The optimum design of one- and two-dimensional FIR filters using the frequency response masking technique. *IEEE Transactions on Circuits and Systems II*, **40**(2), 88–95, 1993.

[13] D. DUDGEON and R. MERSEREAU, *Multidimensional digital signal processing*, Prentice Hall, Englewood Cliffs NJ, 1984.

[14] P. SIOHAN, Contribution à l'étude des méthodes de conception des filtres numériques RIF: application au traitement d'images. Doctoral thesis, ENST, March 1989.

[15] V. OUVRARD and P. SIOHAN, Design of 2D video filters with spatial constraints. In Proceedings of EUSIPCO-92, North Holland, Brussels, 1992, pp. 1001–4.

EXERCISES

1 Consider the 17 coefficients of a low-pass filter with cut-off frequency $f_c = 0.25 f_s$ given in Figure 5.12. How many take on different values? Give the expression for the frequency response $H(f)$. Determine the frequencies for which it is zero and give the maximum ripple. Determine the zeros of the filter Z-transfer function.

2 Consider a filter for which the sampling frequency is taken as the reference ($f_s = 1$) and whose frequency response $H(f)$ is such that:

$$H(k \times 0.0625) = 1 \quad \text{for} \quad k = 0, 1, 2, 3$$
$$H(0.25) = 0.5$$
$$H(k \times 0.0625) = 0 \quad \text{for} \quad k = 5, 6, 7, 8$$

Using the discrete Fourier transform, calculate the 17 coefficients of this filter. Draw the frequency response and determine the zeros of the Z-transfer function.

3 Using the equations of Section 5.7, determine the ripple of a low-pass filter with 17 coefficients for which the upper frequency of the pass band is $f_1 = 0.2$ and the lower frequency of the stop band is $f_2 = 0.3$. Compare the results obtained with those in the preceding exercises.

4 Consider a filter with a transfer function $H(f)$ which, except for the phase shift, is given by the equation:

$$H(f) = h_0 + 2 \sum_{i=1}^{4} h_{2i-1} \cos\left[2\pi f(2i-1)T\right]$$

Give the direct and transposed structures which allow this filter to be achieved with a minimum number of elements. What simplifications are involved if the sampling frequency of the output is divided by two?

5 A narrow band low-pass filter is defined by the equation:

$$y(n) = \sum_{i=0}^{N-1} a_i x(n-i)$$

How is the frequency response modified if the coefficients a_i are replaced by $a_i(-1)^i$ and by $a_i \cos(i\pi/2)$? How are the Z-transfer function zeros affected?

6 Consider a low-pass filter which satisfies Figure 5.7 with the following values for the parameters:

$$f_1 = 0.05; \quad f_2 = 0.15; \quad \delta_1 = 0.01 \quad \text{and} \quad \delta_2 = 0.001$$

How many coefficients are needed and how many bits are required to represent them? If the input data have 12 bits and if the signal-to-noise ratio degradation is limited to $\Delta SN = 0.1\,\text{dB}$, what is the internal data wordlength?

7 Consider the filter given in Section 5.3. The coefficients are rounded to 6 bits (including the sign). Give the expression for the error function $e(f)$ which is introduced and calculate this expression in the neighbourhood of the point $f = 0.1925$ on the frequency axis. Following the steps outlined in Section 5.8, give an expression similar to equation (5.46) to estimate the number of bits required to represent the coefficients in this type of filter.

CHAPTER 6

INFINITE IMPULSE RESPONSE (IIR) FILTER SECTIONS

Digital filters with an infinite impulse response are discrete linear systems which are governed by a convolution equation based on an infinite number of terms. In principle, they have an infinite memory. This memory is achieved by feeding the output back to the input, which is why they are called recursive filters. Each element of the set of output numbers is calculated by weighted summation of a certain number of elements of the input set and of the previous output set.

In general, this infinite impulse response allows much more selective filtering functions to be obtained than with FIR filters of similar complexity. However, the feedback loop complicates the study of the properties and the design of these filters and leads to parasitic phenomena.

When examining IIR filters it is simpler initially to consider them in terms of first- and second-order sections. These simple structures are not only useful in introducing the properties of the IIR filters but also represent the most frequently used type of implementation. Indeed, even the most complex IIR filters appearing in practice are generally formed from a set of such sections.

6.1 FIRST-ORDER SECTION

Consider a system which, for the set of data $x(n)$, produces the set $y(n)$ such that:

$$y(n) = x(n) + by(n-1) \tag{6.1}$$

where b is a constant. This is a filter section of the first order.

The response of this system to the unit set $u_0(n)$ such that:

$$u_0(n) = 1 \quad \text{for } n = 0$$
$$0 \quad \text{for } n \neq 0$$

is the set $y_0(n)$ such that:

$$y_0(n) = 0 \quad \text{for } n < 0$$
$$b^n \quad \text{for } n \geq 0$$

This set constitutes the impulse response of the filter. It is infinite and the stability condition is written:

$$\sum_{n=0}^{\infty} |b|^n < \infty$$

Hence $|b| < 1$.

The response of the system for the set $x(n)$ such that:

$$x(n) = 0 \quad \text{for } n < 0$$
$$1 \quad \text{for } n \geq 0$$

is the set $y(n)$ such that

$$y(n) = 0 \qquad\qquad \text{for } n < 0$$
$$(1 - b^{n+1})/(1 - b) \quad \text{for } n \geq 0 \tag{6.2}$$

which tends towards $1/(1 - b)$ as n tends towards infinity if the system is stable. This response is shown in Figure 6.1.

By analogy with a continuous system with time constant τ, sampled with period T, and whose response $y_c(n)$ is written:

$$y_c(n) = [1 - e^{-(T/\tau)(n+1)}]$$

the time constant of the first-order digital filter is defined by:

$$e^{-T/\tau} = b$$

for $b > 0$. Then:

$$\tau = T/\log(1/b) \tag{6.3}$$

For b close to unity:

$$b = 1 - \delta \quad \text{with } 0 < \delta \ll 1$$

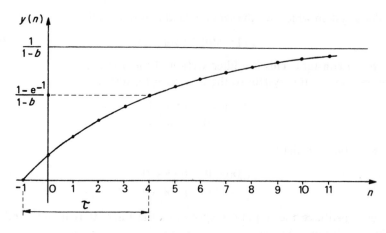

Fig. 6.1 *Response of the first-order section to a unit step*

that is, for systems defined by:

$$y(n) = x(n) + (1 - \delta)y(n - 1) \tag{6.4}$$

we get:

$$\tau \simeq T/\delta \tag{6.5}$$

This situation is encountered in adaptive systems, which are presented in Chapter 12.

If the input set $x(n)$ results for $n \geqslant 0$ from the sampling of the signal $x(t) = e^{j2\pi ft}$ or $x(t) = e^{j\omega t}$, with period $T = 1$, then

$$y(n) = \frac{e^{jn\omega}}{1 - be^{-j\omega}} - \frac{b^{n+1}e^{-j\omega}}{1 - be^{-j\omega}} \tag{6.6}$$

This expression exhibits a transient and a steady state term which corresponds to the frequency response $H(\omega)$ of the filter:

$$H(\omega) = 1/(1 - be^{-j\omega}) \tag{6.7}$$

The modulus and the phase of this function are

$$|H(\omega)|^2 = 1/(1 - 2b\cos\omega + b^2); \quad \phi(\omega) = \tan^{-1}\frac{b\sin\omega}{1 - b\cos\omega} \tag{6.8}$$

The phase can also be written

$$\phi(\omega) = \tan^{-1}(\sin\omega/(\cos\omega - b)) - \omega$$

and the group delay:

$$\tau_g(\omega) = \frac{d\phi}{d\omega} = \frac{b\cos\omega - b^2}{1 - 2b\cos\omega + b^2} \tag{6.9}$$

It can be noted that for very small ω:

$$|H(\omega)|^2 \simeq 1 \Big/ \left\{(1 - b^2)\left[1 + \frac{b}{(1 - b)^2}\omega^2\right]\right\} \tag{6.10a}$$

This expression approximates the response $H_{RC}(\omega)$ of an RC circuit, which is written

$$|H_{RC}(\omega)|^2 = \frac{1}{1 + R^2C^2\omega^2} \tag{6.10b}$$

It appears that for frequencies which are very small in comparison with the sampling frequency the digital circuit has a frequency response similar to that of an RC network. Figure 6.2(a) shows the form of the frequency response for a digital first-order circuit. Figure 6.2(b) gives the phase response and Figure 6.2(c) the group delay.

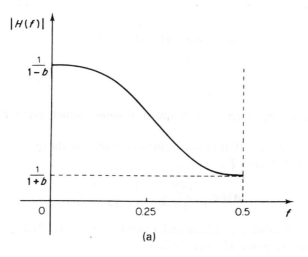

(a)

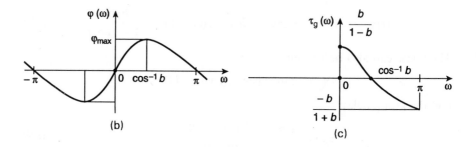

(b)   (c)

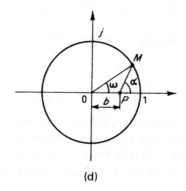

(d)

Fig. 6.2   First-order section: (a) frequency response, (b) phase response, (c) group delay, (d) pole

The phase can be written:

$$\varphi(\omega) = \tan^{-1}\left(\frac{\sin \omega}{\cos \omega - b}\right) - \omega \qquad \cos \omega > b$$

$$\varphi_{\max} = \frac{\pi}{2}\cos^{-1} b \qquad \cos \omega = b$$

$$\varphi(\omega) = \pi + \tan^{-1}\left(\frac{\sin \omega}{\cos \omega - b}\right) - \omega \qquad \cos \omega > b \qquad (6.11a)$$

It passes through a maximum for ω such that $\cos \omega = b$, which corresponds to cancellation of the group delay. The coefficient b thus directly controls the maximum phase of the Section.

The transfer function of the first-order section can also be obtained using the Z-transform. Assume $Y(Z)$ and $X(Z)$ are the transforms of the output and input sets, respectively. Then

$$Y(Z) = X(Z) + bZ^{-1}Y(Z)$$

and thus the Z-transfer function, $H(Z)$ is:

$$H(Z) = 1/(1 - bZ^{-1}) = Z/(Z - b)$$

The frequency response is obtained simply by replacing Z with $e^{j\omega}$, in the expression for $H(Z)$, with $\omega = 2\pi f$.

A graphical interpretation is given in Figure 6.2(d), which represents the pole P of this function in the complex plane. This is a point on the real axis with co-ordinate b.

Following this figure:

$$|H| = \frac{1}{MP} \quad \text{and} \quad \phi = \alpha - \omega$$

The stability condition implies that the pole P is inside the unit circle.

An interesting special case is the narrowband integrator, defined by the following transfer function:

$$H_{\text{int}}(\tau) = \frac{\varepsilon}{1 - (1 - \varepsilon)Z^{-1}} \qquad (6.11b)$$

with ε small such that $0 < \varepsilon \ll 1$.

It can also be shown that the 3 dB bandwidth is approximately equal to ε and the time constant equal to $1/\varepsilon$.

The one-sided Z-transform generates the transient responses and introduces the initial conditions. Indeed:

$$\sum_{n=0}^{\infty} y(n)Z^{-n} = \sum_{n=0}^{\infty} x(n)Z^{-n} + b\sum_{n=0}^{\infty} y(n-1)Z^{-n}$$

$$Y(Z) = X(Z) + by(-1) + bZ^{-1}Y(Z)$$

whence

$$Y(Z) = \frac{X(Z)}{1 - bZ^{-1}} + \frac{by(-1)}{1 - bZ^{-1}}$$ (6.12)

If $x(n) = e^{jn\omega}$, $X(Z)$ is written:

$$X(Z) = \sum_{n=0}^{\infty} e^{jn\omega} Z^{-n} = \frac{1}{1 - e^{j\omega}Z^{-1}}$$ (6.13)

The value $y(n)$ is obtained from the equation for the inverse Z-transform:

$$y(n) = \frac{1}{j2\pi} \int_{\Gamma} Z^{n-1} \left[\frac{1}{1 - e^{j\omega}Z^{-1}} \cdot \frac{1}{1 - bZ^{-1}} + \frac{by(-1)}{1 - bZ^{-1}} \right] dZ$$

By taking a circle with radius greater than unity as the contour of integration, Γ, the theory of residues gives:

$$y(n) = \frac{e^{jn\omega}}{1 - be^{-j\omega}} - \frac{b^{n+1}e^{-j\omega}}{1 - be^{-j\omega}} + y(-1)b^{n+1}$$ (6.14)

which can also be obtained directly from a series expansion of $Y(Z)$. This expression shows the steady state and transient responses and also the response due to the initial conditions. The last two items disappear when n increases if $|b| < 1$, i.e. if the system is stable.

This analysis shows that the first-order filter offers restricted possibilities because it has only one pole, which must be real if the filter has real coefficients. Further, its frequency response is a monotonic function. The second-order filter has a wider variety of possibilities. It is the structure most commonly used in digital filtering because of the modularity that it allows for even the most complex filters and because of its properties relating to limitations in the coefficient wordlengths and the round-off noise. We will first examine the purely recursive filter section.

6.2 PURELY RECURSIVE SECOND-ORDER SECTION

Consider a system which, for the data set $x(n)$, produces the corresponding set $y(n)$ such that:

$$y(n) = x(n) - b_1 y(n-1) - b_2 y(n-2)$$ (6.15)

In this expression the signs of the coefficients b_1 and b_2 have been changed compared to the previous section in order to facilitate the writing of the Z-transfer function of the system, $H(Z)$:

$$H(Z) = \frac{1}{1 + b_1 Z^{-1} + b_2 Z^{-2}} = \frac{Z^2}{Z^2 + b_1 Z + b_2}$$

This function has a double root at the origin and two poles P_1 and P_2 such that:

$$P_{1,2} = -\frac{b_1}{2} \pm \frac{1}{2}\sqrt{(b_1^2 - 4b_2)} \qquad (6.16)$$

Two cases occur, depending on the sign of $b_1^2 - 4b_2$:

(1) $b_1^2 \geqslant 4b_2$: both poles lie on the real axis of the complex plane. The transfer function is simply the product of two first-order functions with real coefficients. The corresponding filter is composed of two first-order sections in cascade and its properties are deduced accordingly. The amplitudes are multiplied and the phases are added. The step response at the output of the second section is

$$y_2(n) = \frac{1}{(1-b_1)(1-b_2)}\left[1 - b_2^{n+1} - (1-b_2)\frac{b_1^{n+1} - b_2^{n+1}}{b_1 - b_2} \right] \qquad (6.17)$$

where b_1 and b_2 are the coefficients. The corresponding time constant τ_{12} is given by:

$$\tau_{12} \approx \sqrt{2}\sqrt{\tau_1 \tau_2} \qquad (6.18)$$

for coefficients sufficiently close to unity. For N identical sections the time constant τ_N can be approximated by:

$$\tau_N \approx \sqrt{(N)}\tau_1 \qquad (6.19)$$

(2) $b_1^2 < 4b_2$: the two poles are complex conjugates written as P and $\bar{P}$, with

$$P = -\frac{b_1}{2} + j\frac{1}{2}\sqrt{(4b_2 - b_1^2)} \qquad (6.20)$$

Figure 6.3 illustrates this, the most interesting case, and the remainder of this section will concentrate on this.

The relation between the position of the poles and the filter coefficients is

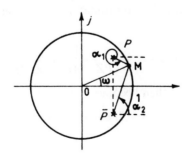

Fig. 6.3 Second-order section with complex poles

very simple:

$$b_1 = -2\operatorname{Re}(P) \tag{6.21}$$

that is, the coefficient of the Z^{-1} term in the expression for $H(Z)$ is equal in modulus to twice the real part of the pole and has the opposite sign. Then:

$$b_2 = |OP|^2 \tag{6.22}$$

The coefficient of the Z^{-2} term is equal to the square of the modulus of the pole or to the square of the distance from the pole to the origin. As will be seen later, both relations are very useful in determining filter coefficients.

If M denotes the co-ordinate $e^{j\omega}$ in the complex plane the modulus of the transfer function is

$$|H(\omega)| = \frac{1}{\mathrm{MP \cdot M\overline{P}}}$$

and the phase is

$$\phi(\omega) = \alpha_1 + \alpha_2 - 2\omega$$

where α_1 and α_2 denote the angles between the vectors $\overrightarrow{PM}$ and $\overrightarrow{\overline{P}M}$ and the real axis.

The analytical expressions are deduced from $H(Z)$ by letting $Z = e^{j\omega}$. By using

$$H(Z) = \frac{1}{1 + b_1 Z^{-1} + b_2 Z^{-2}}$$

we have

$$|H(\omega)|^2 = \frac{1}{1 + b_1^2 + b_2^2 + 2b_1(1 + b_2)\cos\omega + 2b_2\cos 2\omega} \tag{6.23a}$$

$$\phi(\omega) = -\arctan\left[\frac{b_1\sin\omega + b_2\sin 2\omega}{1 + b_1\cos\omega + b_2\cos 2\omega}\right] \tag{6.24a}$$

A very elegant form for the frequency response and the phase is obtained by representing the poles in polar co-ordinates, $P = re^{j\theta}$, and expressing $H(Z)$ as a factor product:

$$H(Z) = \frac{1}{(1 - PZ^{-1})(1 - \overline{P}Z^{-1})}$$

The coefficients b_1 and b_2 then become

$$b_1 = -2r\cos\theta; \quad b_2 = r^2$$

For $H(\omega)$ we obtain:

$$H(\omega) = \frac{1}{[1 - re^{j(\theta - \omega)}][1 - re^{-j(\theta + \omega)}]} \tag{6.25}$$

Hence:

$$|H(\omega)|^2 = \frac{1}{[1 + r^2 - 2r \cos(\theta - \omega)][1 + r^2 - 2r \cos(\theta + \omega)]} \tag{6.23b}$$

$$\phi(\omega) = \arctan\left[\frac{r \sin(\theta + \omega)}{1 - r \cos(\theta + \omega)}\right] - \arctan\left[\frac{r \sin(\theta - \omega)}{1 - r \cos(\theta - \omega)}\right] \tag{6.24b}$$

These expressions permit the curves for $|H(\omega)|$ and $\phi(\omega)$ to be plotted as a function of the frequency $\omega = 2\pi f$. It can be shown that $|H(\omega)|$ is an even function and that $\phi(\omega)$ is an odd function of ω.

The values corresponding to the extrema of $|H(\omega)|$ are the roots of the following equation, which is obtained by taking the derivative of equation (6.23) with respect to ω:

$$\sin\omega[b_1(1 + b_2) + 4b_2 \cos\omega] = 0$$

The extremum frequencies are 0 and 0.5 and another extremal frequency f_0 exists if

$$\left|\frac{b_1(1 + b_2)}{4b_2}\right| < 1 \tag{6.26a}$$

or, in polar co-ordinates:

$$\cos\theta < \frac{2r}{1 + r^2} \tag{6.26b}$$

In this case,

$$\cos(2\pi f_0) = \cos\omega_0 = -\frac{b_1(1 + b_2)}{4b_2} \tag{6.27}$$

The frequency f_0 is the resonance frequency of the filter section. The amplitude at the resonance is written as

$$H_m = \frac{1}{1 - b_2}\sqrt{\left(\frac{4b_2}{4b_2 - b_1^2}\right)} \tag{6.28}$$

or, in polar co-ordinates:

$$H_m = \frac{1}{1 - r}\frac{1}{(1 + r)\sin\theta} \tag{6.29}$$

It thus appears that the frequency response at resonance is inversely proportional to the distance from the pole to the unit circle. This is a fundamental expression which will be used frequently in the following chapters.

It is also important for the second-order section to determine the 3-decibel bandwidth, B_3, such that

$$B_3 = f_2 - f_1 = (\omega_2 - \omega_1)/2\pi$$

with:

$$|H(\omega_1)|^2 = |H(\omega_2)|^2 = H_m^2/2$$

For a strongly resonant filter section ($r \simeq 1$), using equations (6.22) and (6.23), the following approximation holds in the neighbourhood of the resonance frequency:

$$|H(\omega_1)^2| \simeq \frac{1}{4 \sin^2 \theta} \frac{1}{1 + r^2 - 2r \cos (\theta - \omega_1)}$$

$$= \frac{1}{2} \left[\frac{1}{(1 - r^2)^2 \sin^2 \theta)} \right]$$

whence:

$$\cos (\theta - \omega_1) = ((1 + r^2)/2r) - (1 - r^2)^2/4r$$

By expansion and limiting the number of terms we derive:

$$|\theta - \omega_1| \simeq 1 - r$$

Hence, the approximation for a strongly resonant filter section is:

$$B_3 = (1 - r)/\pi \tag{6.30a}$$

This result is used below for calculating the arithmetic complexity of filters.

Another characteristic is sometimes used for a purely recursive second-order section, the equivalent noise bandwidth B_2. This is the bandwidth of a noise source whose spectral density is assumed to be constant within this band and equal to H_m^2 and whose total noise power is equal to the power obtained at the output of the section when white noise of unit power is applied. By definition:

$$B_b.H_m^2 = \|H\|_2^2$$

Taking account of the expression for $\|H\|_2^2$ given below (6.36) and expression (6.29) above yields:

$$B_b = \frac{(1 - r^2) \sin^2 \theta}{1 + r^4 - 2r^2 \cos 2\theta} \tag{6.30b}$$

This expression is useful in spectral analysis for example.

The main characteristics of a purely recursive second-order filter section can be illustrated by an example.

Example: Assume a second-order filter section having poles with co-ordinates:

$$P = 0.6073 + j0.5355$$
$$\bar{P} = 0.6073 - j0.5355$$

The various parameters are

$$b_1 = -2\,\mathrm{Re}\,(P) = -1.2146$$
$$b_2 = |OP|^2 = 0.6556$$

$$H(Z) = \frac{1}{1 - 1.2146Z^{-1} + 0.6556Z^{-2}}$$

$$|H(\omega)|^2 = \frac{1}{2.905 - 4.02\cos\omega + 1.31\cos(2\omega)}$$

$$\theta = 2\pi \times 0.1156; \quad r = 0.81; \quad f_0 = 0.111; \quad H_m = 4.39; \quad B_3 = 0.06$$

The modulus of the response is shown in Figure 6.4 as a function of the frequency.

The phase response of the second-order section can be considered using equation (6.24) for the function $\phi(\omega)$. In order to describe the variations in this function it is useful first to calculate its derivative using equation (6.24b). Thence,

$$\frac{d\phi}{d\omega} = \frac{r\cos(\theta + \omega) - r^2}{1 - 2r\cos(\theta + \omega) + r^2} + \frac{r\cos(\theta - \omega) - r^2}{1 - 2r\cos(\theta - \omega) + r^2} \tag{6.31}$$

This derivative is interesting as it is the group delay of the filter. By definition (1.29):

$$\tau(\omega) = \frac{d\phi}{d\omega}$$

Thus, the group delay can be written as

$$\tau(\omega) = \frac{r[\cos(\theta + \omega) - r]}{1 - 2r\cos(\theta + \omega) + r^2} + \frac{r[\cos(\theta - \omega) - r]}{1 - 2r\cos(\theta - \omega) + r^2} \tag{6.32}$$

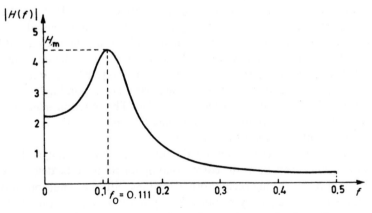

Fig. 6.4 *Frequency response of a purely recursive second-order section*

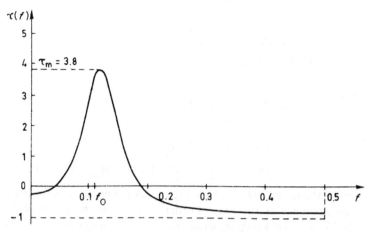

Fig. 6.5   Theoretical group delay of a purely recursive section

The function $\tau(\omega)$ has a maximum in the neighbourhood of the resonance frequency. At the frequency of $f = \theta/2\pi$ this becomes:

$$\tau(\theta) = \frac{r}{1-r}\left[1 + \frac{(1-r)[\cos 2\theta - r]}{1 - 2r\cos 2\theta + r^2}\right] \simeq \frac{r}{1-r} \qquad (6.33)$$

In physical systems this function is positive and $\phi(\omega)$ is an increasing function which has a value of 0 at the origin and a multiple of π at frequency 0.5.

Example: $r = 0.81; \quad \theta = 2\pi \times 0.1156$

Figure 6.5 shows the curve $\tau(f)$ as a function of frequency. This curve has a maximum of 3.8 in the neighbourhood of the resonance. The unit of time is the sampling period T. The values obtained should be multiplied by T if this period is different from unity.

The function $\tau(f)$ can be seen to have negative values. In fact, it is the theoretical group delay of the filter. The system as it has been presented, however, cannot actually be realized. Each output element $y(n)$ is calculated by an addition which involves an input number $x(n)$. This operation cannot be instantaneous. To enable the system to be realized $y(n)$ has to be delayed, for example, by one period. The group delay will then be increased correspondingly, and it is necessary to add the value ω to the phase $\phi(\omega)$. The function $\phi(\omega)$ obtained under these conditions is represented in Figure 6.6, and the curve has maximum slope in the neighbourhood of the resonance.

The equations which were given for the functions $|H(\omega)|$, $\phi(\omega)$, and $\tau(\omega)$ are important because they allow the corresponding functions to be determined for filters realized by cascading second-order sections, either by multiplication for the modulus of the frequency response or by addition for the phase and the group delay.

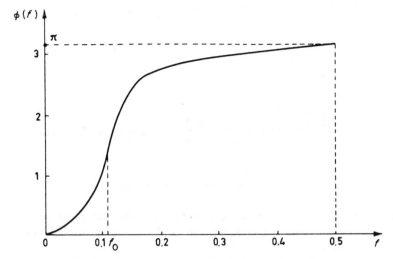

Fig. 6.6 Phase characteristic of a purely recursive section

To introduce the initial conditions and find the transient responses the one-sided Z-transform is used. From the equation of definition of the filter section the following relation can be found between the one-sided transforms $Y(Z)$ and $X(Z)$:

$$Y(Z) = \frac{X(Z)}{1 + b_1 Z^{-1} + b_2 Z^{-2}} - \frac{b_1 y(-1) + b_2 [y(-2) + y(-1)Z^{-1}]}{1 + b_1 Z^{-1} + b_2 Z^{-2}} \quad (6.34a)$$

For $x(n) = e^{jn\omega}$, $y(n)$ is given by the equation

$$y(n) = \frac{1}{j2\pi} \int_\Gamma Z^{n-1} Y(Z) dZ$$

with

$$X(Z) = \frac{1}{1 - e^{j\omega} Z^{-1}}$$

taking a circle with radius greater than unity as the integration contour Γ.

The study of the purely recursive section has been carried out in the frequency domain. In the time domain this section has an impulse response which is a set $h(n)$ determined by examining the response to the unit set, or by series expansion of the function $H(Z)$. For complex poles, one has

$$H(Z) = \frac{1}{1 - PZ^{-1}} \frac{P}{P - \bar{P}} + \frac{1}{1 - \bar{P}Z^{-1}} \frac{\bar{P}}{\bar{P} - P} = \sum_{n=0}^{\infty} h(n) Z^{-n}$$

Hence,

$$h(n) = r^n \frac{\sin(n+1)\theta}{\sin \theta} \quad (6.35)$$

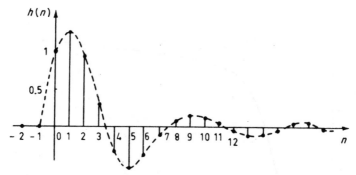

Fig. 6.7   Impulse response of a second-order section

Figure 6.7 shows the impulse response of the filter described in the above example. The response to a unit step can be derived from the definition, after several manipulations:

$$g(n) = \frac{1}{1 + b_1 + b_2}[1 + b_2 h(n) - h(n + 1)]$$

This then gives:

$$g(n) = \frac{1}{1 + r^2 - 2r \cos \theta}\left[1 + \frac{r^{i+1}}{\sin \theta}[r \sin(i + 1)\theta - \sin(i + 2)\theta]\right] \qquad (6.34b)$$

The norm $\|H\|_2$ of the function $H(\omega)$ is used, and can be calculated by two methods, as was shown in Section 4.3. By summation of the series:

$$\|H\|_2^2 = \sum_{n=0}^{\infty} |h(n)|^2 = \frac{1}{\sin^2 \theta} \sum_{n=0}^{\infty} r^{2n} \frac{1 - \cos[2(n + 1)\theta]}{2}$$

Using integration following the theorem of residues:

$$\|H\|_2^2 = \frac{1}{j2\pi} \int_{|Z|=1} \frac{Z dZ}{(Z - P)(Z - \bar{P})(1 - PZ)(1 - \bar{P}Z)}$$

Finally,

$$\|H\|_2^2 = \frac{1 + r^2}{1 - r^2} \frac{1}{1 + r^4 - 2r^2 \cos 2\theta} \qquad (6.36)$$

The value $\|H\|_1$ is also used:

$$\|H\|_1 = \sum_{n=0}^{\infty} |h(n)|$$

This value is bounded by the inequality:

$$\|H\|_1 = \frac{1}{\sin \theta} \sum_{n=0}^{\infty} r^n |\sin[(n + 1)\theta]| \leqslant \frac{1}{(1 - r)\sin \theta} \qquad (0 < \theta < \pi) \qquad (6.37)$$

Example: When the poles are located on the imaginary axis in the Z-plane $\theta = \pi/2$ and the impulse response is given by:

$$h(2p) = r^{2p}(-1)^p$$

Then:

$$\|H\|_2^2 = \sum_{p=0}^{\infty} |h(2p)|^2 = \frac{1}{1 - r^4}$$

$$\|H\|_1 = \sum_{p=0}^{\infty} h(2p) = \frac{1}{1 - r^2}$$

These results are in agreement with equations (6.36) and (6.37).

The results obtained for the purely recursive second-order section can be extended to the general second-order section.

6.3 GENERAL SECOND-ORDER SECTION

The most general second-order filter introduces the input data $x(n-1)$ and $x(n-2)$ to the calculation of an element $y(n)$ of the output set at time n. Its equation of definition is written as

$$y(n) = a_0 x(n) + a_1 x(n-1) + a_2 x(n-2) - b_1 y(n-1) - b_2 y(n-2) \quad (6.38)$$

and it leads to the Z-transfer function:

$$H_T(Z) = \frac{a_0 + a_1 Z^{-1} + a_2 Z^{-2}}{1 + b_1 Z^{-1} + b_2 Z^{-2}}$$

which has two zero's which, since the numerator coefficients are real, are either real or complex conjugates. The position of these zeros, written as Z_0 and $\overline{Z_0}$, is often special, and two cases are encountered in practice. The first corresponds to a filtering function. The zeros are then almost always on the unit circle, both to optimize the attenuation of the filter by introducing an infinite attenuation frequency and because, under these conditions, a symmetry in the coefficients appears and the calculations can be simplified. The second case corresponds to a pure phase shifter, when the zeros are harmonic conjugates of the poles.

The filter case will be considered first. The transfer function is written now:

$$H_T(Z) = a_0 \frac{1 + a_1 Z^{-1} + Z^{-2}}{1 + b_1 Z^{-1} + b_2 Z^{-2}} = a_0 \frac{(Z - Z_0)(Z - \overline{Z_0})}{(Z - P)(Z - \overline{P})} \quad (6.39)$$

or:

$$H_T(Z) = a_0 \frac{1 - 2\operatorname{Re}(Z_0)Z^{-1} + Z^{-2}}{1 - 2\operatorname{Re}(P)Z^{-1} + |P|^2 Z^{-2}}$$

The modulus of the frequency response of the general second-order section when the zeros are placed on the unit circle is expressed by:

$$|H_T(\omega)|^2 = \frac{(a_1 + 2a_0 \cos \omega)^2}{1 + b_1^2 + b_2^2 + 2b_1(1 + b_2)\cos \omega + 2b_2 \cos 2\omega} \qquad (6.40)$$

Such a filter can be regarded as the cascade of a purely recursive IIR filter section and a linear phase FIR filter section. Consequently, the phase characteristics and the group delay of the complete filter section are sums of the characteristics of the elementary components. That is,

$$\tau_T(\omega) = 1 + \frac{r \cos(\theta + \omega) - r^2}{1 - 2r \cos(\theta + \omega) + r^2} + \frac{r \cos(\theta - \omega) - r^2}{1 - 2r \cos(\theta - \omega) + r^2} \qquad (6.41)$$

$$\phi_T(\omega) = \omega + \arctan \left[\frac{r \sin(\theta + \omega)}{1 - r \cos(\theta + \omega)} \right] - \arctan \left[\frac{r \sin(\theta - \omega)}{1 - r \sin(\theta - \omega)} \right] \qquad (6.42)$$

These two expressions give the phase and the group delay of a second-order section when both zeros are on the unit circle. This filter section is usually called the elliptic section of second order, from the technique used for calculating the coefficients.

Example: To illustrate the properties of a general second-order filter section, let us use the example given earlier by completing the filter with two zeros:

$$Z_0 = 0.3325 + j0.943 \quad \text{and} \quad \overline{Z}_0 = 0.3325 - j0.943$$

The positions of the singularities in the complex plane are given by Figure 6.8(a). The transfer function $H_T(Z)$ is the quotient of the two second-order polynomials $N(Z)$ and $D(Z)$:

$$H_T(Z) = a_0 \frac{N(Z)}{D(Z)}$$

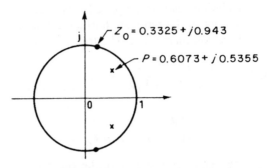

Fig. 6.8(a)   Poles and zeros of a general second-order section

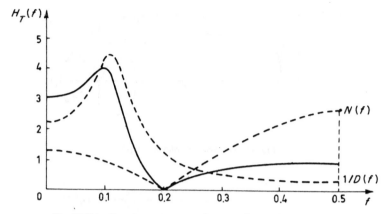

Fig. 6.8(b)   Frequency response of a general second-order section

with

$$N(Z) = (1 - 0.665Z^{-1} + Z^{-2})$$
$$D(Z) = 1 - 1.2146Z^{-1} + 0.6556Z^{-2}$$

Figure 6.8(b) shows the frequency response for the filter. The contributions of the numerator and the denominator of the transfer are also indicated. The factor a_0 corresponds to a scaling factor which is calculated so that the response of the filter has a specified value at a given frequency. For example:

$$H_T(0) = 1 \quad \text{results in} \quad a_0 = 0.33$$

The group delay and the phase are given by Figures 6.5 and 6.6, respectively. The norm $\| H_T \|_2$ of the function $H_T(\omega)$ is calculated as given earlier, and is

$$\| H_T \|_2^2 = a_0^2 \frac{2 + a_1^2 + a_1^2 b_2 - 4a_1 b_1 + 2b_1^2 - 2b_2^2}{(1 - b_2)[(1 + b_2)^2 - b_1^2]} \tag{6.43}$$

An important particular case is the notch filter, which is used to remove a line in a spectrum without disturbing the other components. The transfer function is

$$H_N(Z) = \frac{1 + a_1 Z^{-1} + Z^{-2}}{1 + a_1(1 - \varepsilon)Z^{-1} + (1 - \varepsilon)^2 Z^{-2}} \tag{6.44}$$

where ε is a small positive real value. As shown in Fig. 6.9(a), ε is the distance of the poles to the unit circle. For very small ε, the 3dB-attenuation Bandwidth can be approximated by

$$B_{3N} \approx \frac{\varepsilon}{\pi}$$

Outside the notch, the poles compensate for the zeros and the frequency response

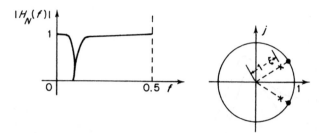

Fig. 6.9(a)   Second-order notch filter

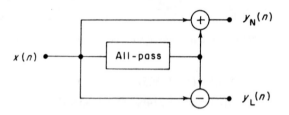

Fig. 6.9(b)   Implementation of the notch filter and its complement

is almost flat and close to unity. Moreover, such a filter achieves a very small amplification of the input white noise, since equation (6.43) yields:

$$\|H_N\|_2^2 \approx \frac{2-3\varepsilon}{2-5\varepsilon} \approx 1+\varepsilon$$

If the frequency of the signal to be removed is not precisely known then the zeros have to be moved inside the unit circle in order to increase the notch width.

Another class of general second-order sections is that of phase shifter circuits. The phase shifter circuit is characterized by the fact that the numerator and the denominator of the transfer function have the same coefficients but in the reverse order:

$$H_D(Z) = \frac{b_2 + b_1 Z^{-1} + Z^{-2}}{1 + b_1 Z^{-1} + b_2 Z^{-2}} = \frac{N(Z)}{D(Z)} \tag{6.45}$$

The polynomials $N(Z)$ and $D(Z)$ are image polynomials. As a result $|H_D(e^{j\omega})| = 1$, i.e. the circuit is a pure phase shifter.

The transfer function $H_D(Z)$ is written as a function of the poles and zeros as:

$$H_D(Z) = \frac{(P - Z^{-1})(\bar{P} - Z^{-1})}{(1 - PZ^{-1})(1 - \bar{P}Z^{-1})}$$

It is clear that the poles and zeros are harmonic conjugates and Figure 6.10 shows their position in the Z-plane.

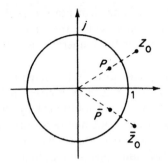

Fig. 6.10 Poles and zeros of the second-order phase shifter

The calculation of the phase and group delay for this circuit can be very simply deduced from equations (6.24) and (6.32) for the purely recursive element as

$$H_D(Z) = \frac{N(Z)}{D(Z)} = \frac{Z^{-2}D(Z^{-1})}{D(Z)}$$

However,

$$|D(\omega)| = |D(-\omega)|; \quad \phi(\omega) = -\phi(-\omega)$$

Hence,

$$\phi_D(\omega) = 2\phi(\omega) - 2\omega$$

After a few arithmetic manipulations the group delay $\tau_g(\omega)$ of the phase shifter becomes

$$\tau_g(\omega) = \frac{1 - r^2}{1 + 2r\cos(\theta - \omega) + r^2} + \frac{1 - r^2}{1 - 2r\cos(\theta + \omega) + r^2} \qquad (6.46)$$

It is not difficult to show that, as ω varies from 0 to π, the phase $\phi_D(\omega)$ varies by 2π:

$$\phi_D(\pi) = \int_0^\pi \tau_g(\omega)\,d\omega = 2\int_0^\pi \frac{1 - r^2}{1 + r^2 - 2r\cos\alpha}\,d\alpha = 2\pi$$

An interesting application of this result is that the above-mentioned notch filter can be implemented with the help of a phase shifter. In fact, two complementary filters can be obtained with a single all-pass section, as shown in Figure 6.9(b) [1].

6.4 STRUCTURES FOR IMPLEMENTATION

The elements can be implemented by circuits which directly produce the operations represented in the expression for the transfer functions. The term

Z^{-1} corresponds to a delay of one elementary period and is achieved by one memory element. The coefficients used in the circuits are those of the transfer function, with the same sign for the numerator and the opposite sign for the denominator.

The circuit which corresponds directly to the equation for the definition of the purely recursive second-order section is given in Figure 6.11. The output numbers $y(n)$ are delayed twice, and multiplied by the coefficients $-b_1$ and $-b_2$ before being added to the input numbers $x(n)$. The circuit includes two memory locations for data and two for the coefficients. For each output number, two multiplications and two additions are required.

The general second-order filter section can be realized to conform with the equation of definition. However, two data memory locations are required for the input numbers and two for the output numbers. The structure obtained is not canonic as it contains more than the minimum number of components. Indeed, only two data memories are necessary if the transfer function is factorized as

$$H_T(Z) = \frac{N(Z)}{D(Z)} = \frac{1}{D(Z)} N(Z)$$

i.e. the calculations involved in the denominator are performed first, followed by those for the numerator. This structure, called D–N, is shown in Figure 6.12(a). It corresponds to the introduction of two internal variables $u_1(n)$ and $u_2(n)$ forming a state vector $U(n)$ with $N = 2$ dimensions. The system is described by the following equations.

$$u_1(n + 1) = x(n) - b_1 u_1(n) - b_2 u_2(n)$$
$$u_2(n + 1) = u_1(n)$$
$$y(n) = a_0 x(n) - a_0 b_1 u_1(n) - a_0 b_2 u_2(n) + a_1 u_1(n) + a_2 u_2(n)$$

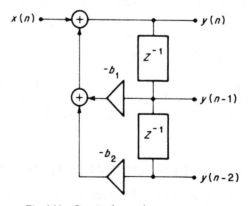

Fig. 6.11   *Circuit of a purely recursive section*

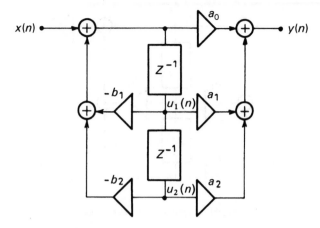

Fig. 6.12(a)   Second-order section in the D–N structure

or, in matrix form, conforming to equation (6.34a):

$$U(n+1) = \begin{bmatrix} -b_1 & -b_2 \\ 1 & 0 \end{bmatrix} U(n) + \begin{bmatrix} 1 \\ 0 \end{bmatrix} x(n)$$

$$y(n) = [-a_0b_1 + a_1, -a_0b_2 + a_2] U(n) + a_0 x(n) \tag{6.47}$$

This representation thus results in a canonic realization, which has the minimum number of internal variables and consequently the minimum number of memories.

Using the results of Section 6.6, there is a dual structure which corresponds to the internal variables $v_1(n)$ and $v_2(n)$ such that:

$$\begin{bmatrix} v_1(n+1) \\ v_2(n+1) \end{bmatrix} = \begin{bmatrix} -b_1 & 1 \\ -b_2 & 0 \end{bmatrix} \begin{bmatrix} v_1(n) \\ v_2(n) \end{bmatrix} + \begin{bmatrix} -a_0b_1 + a_1 \\ -a_0b_2 + a_2 \end{bmatrix} x(n)$$

$$y(n) = v_1(n) + a_0 x(n) \tag{6.48}$$

This alternative canonic structure is represented in Figure 6.13. It corresponds to performing the operations on the numerator of the Z-transfer function first, and is called the N–D structure.

The elliptic second-order section is usually achieved as shown in Figure 6.12(b). Four multiplications have to be performed; the one for the coefficient a_0, the scaling factor, is carried out either on the input numbers $x(n)$ or on the output as in the diagram. The computational savings compared with the earlier methods can be clearly seen.

Realization of an all-pass section is a special case. The canonical structure does not permit the characteristics of this function to be exploited—constant signal amplitude and the same coefficient values for the numerator and

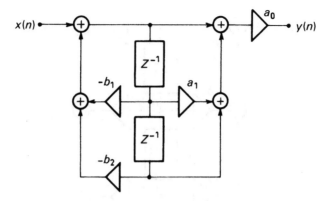

Fig. 6.12(b)   Elliptic second-order section

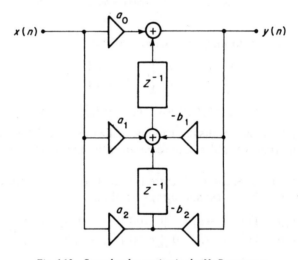

Fig. 6.13   Second-order section in the N–D structure

denominator. A structure with two multiplications adapted to this function is given in Figure 6.14. The corresponding input/output relation can be written:

$$y(n) = x(n-2) + b_1[x(n-1) - y(n-1)] + b_2[x(n) - y(n-2)]$$

In total, two multiplications, four additions and four memories are required. Note that an all-pass function, by definition, does not have a scale factor. Figure 6.14 allows only constant amplitude signals to be stored in memory; this minimizes memory length and simplifies estimates of calculation error. In order to optimize the circuit and to reduce the size of the multiplier it is important to minimize the number of bits in each of the factors of the multiplication. The coefficients will be considered first.

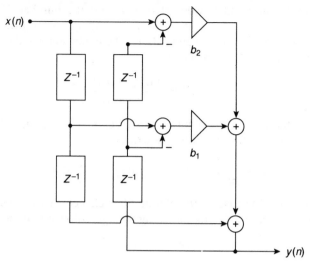

Fig. 6.14 Second order all pass section with two multipliers

6.5 COEFFICIENT WORDLENGTH LIMITATION

Limitation of the coefficient wordlength means that the coefficients can have only a limited number of values. It follows, therefore, that the poles have a limited number of possible positions inside the unit circle. The same effect occurs for zeros on the unit circle when the filter is elliptic. Thus, quantization of the absolute value of the coefficients to b bits limits the number of positions that the poles can take in a quadrant of the unit circle to 2^{2b}, and the number of frequencies of infinite attenuation to 2^b. Figure 6.15 shows these positions for $b = 3$.

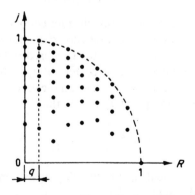

Fig. 6.15 Positions of the poles and zeros with quantization of the coefficient magnitude to 3 bits

If the transfer function is calculated first, and the number of bits in the coefficients is then limited, for example, by rounding, the transfer function is modified by perturbations $e_N(Z)$ and $e_D(Z)$ in the numerator and the denominator [2]. The function $H_R(Z)$ is obtained:

$$H_R(Z) = \frac{N(Z) + e_N(Z)}{D(Z) + e_D(Z)} \tag{6.49}$$

If the round-off errors in the coefficients are denoted by δa_i and δb_i $(0 \leqslant i \leqslant 2)$, the perturbation transfer functions are written:

$$e_N(f) = \sum_{i=0}^{2} \delta a_i e^{-j2\pi f i}; \qquad e^D(f) = \sum_{i=1}^{2} \delta b_i e^{-j2\pi f i}$$

Let us consider the case of the elliptic filter element whose coefficients are quantized by rounding to b_c bits, including the sign. Allowing for the inequalities:

$$|a_i| \leqslant 2; \qquad |b_i| < 2$$

the quantization step q is written:

$$q = 2 \times 2^{1-b_c} = 2^{2-b_c}$$

Thus

$$|e_D(f)| \leqslant 2(q/2) = 2^{2-b_c} \tag{6.50}$$

The modifications of the transfer function caused by quantization of the coefficients of the denominator are a maximum for frequencies near the poles, because the function $D(f)$ is then a minimum.

Neglecting the effect of rounding the numerator coefficients, we get:

$$H_R(f) = N(f)/[D(f) + e_D(f)] \simeq N(f)[1 - \{e_D(f)/D(f)\}]/D(f)$$

The relative error, $e(f) = (H_R(f) - H(f))/H(f)$, is bounded by:

$$|e_D(f)| \leq q(1/|D(f)|) \tag{6.51}$$

This expression allows us to determine the number of bits needed to represent the coefficients of the denominator as a function of the tolerance of the frequency response and the values of the coefficients. It is used in the next chapter.

In the numerator, quantization of the coefficient a_1 of the elliptic filter section leads to a displacement of the zeros which lie on the unit circle. The displacement df_i of the infinite attenuation point f_i is such that:

$$|df_i| \leqslant \frac{1}{2\pi} \frac{2^{-bc}}{|\sin 2\pi f i|} \tag{6.52}$$

Quantization of the coefficient a_0 for the elliptic section results simply in a change in the gain of the filter.

6.6 INTERNAL DATA WORDLENGTH LIMITATION

In the $D-N$ structure, which is the one most frequently used, the second multiplication factor is the number held in the data memory. Of necessity, this memory has a limited capacity; the feedback structure (Figure 6.12(a) implies that even if the input numbers $x(n)$ have a limited number of bits and the memories are empty when the operation begins, the number of bits of the data to be stored in the memory increases indefinitely. Limitation, usually by rounding, is required. On the other hand, the filter sections can introduce large gains and instabilities, so that logic saturation devices must be introduced to limit the amplitude of the data being stored in the memory. The elliptic second-order section with a logic saturation unit and quantization device is shown in Figure 6.16 for the $D-N$ structure. For simplification, this figure assumes a single wordlength limitation device sited immediately before the memory.

The quantization device involves a degradation in the signal, which passes through the filter. This is the round-off noise. Following the circuit in Figure 6.16, it can be seen that quantization has the effect of superimposing on the input signal $x(n)$, an error signal $e(n)$ which also passes through the filter. If the quantization step has the value q, this error signal can be regarded as having a spectrum with uniform distribution and a power $q^2/12$. Under these conditions, the round-off noise N_c at the output can be determined if $f_s = 1$ by using equation (4.25) in Section 4.4, as

$$N_c = \frac{q^2}{12} \int_0^1 \left| \frac{N(f)}{D(f)} \right|^2 \mathrm{d}f$$

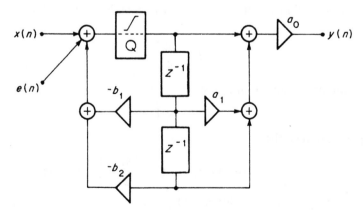

Fig. 6.16   Elliptic section with wordlength-limitation device

or, as a function of the set $h(n)$, the impulse response of the filter:

$$N_c = \frac{q^2}{12} \sum_{n=0}^{\infty} |h(n)|^2$$

By using the results in the earlier sections, for a purely recursive element with complex poles and polar co-ordinates (r, θ), this becomes:

$$N_c = \frac{q^2}{12} \frac{1+r^2}{1-r^2} \frac{1}{1+r^4-2r^2 \cos 2\theta} \tag{6.53}$$

and for the elliptic section:

$$N_c = \frac{q^2}{12} \frac{a_0^2(2 + a_1^2 + a_1^2 b_2 - 4a_1 b_1 + 2b_1^2 - 2b_2^2)}{(1-b_2)[(1+b_2)^2 - b_1^2]} \tag{6.54}$$

The quantization step q is related to the number of bits of the internal data memories. This relation involves the amplitude of the frequency response of the purely recursive part. It is studied in detail in the following chapter, for a cascade of second-order sections.

In this section only the D–N structure has been considered. The calculations can be readily adapted to the N–D structure [3]. The introduction of the quantization device also has consequences in the absence of a signal.

6.7 STABILITY AND LIMIT-CYCLES

Even if there is no signal at the input of an IIR filter, there can still be a signal at the output. This is particularly likely if the coefficients are such that the filter is unstable.

The condition for stability of the filter is that the poles must lie inside the unit circle. This condition defines a stability domain in the plane (b_1, b_2). From the results of Section 6.2, the domain of the complex poles is limited by the parabola:

$$b_2 = \frac{b_1^2}{4}$$

and the stability condition imposes:

$$0 \leqslant b_2 < 1$$

If the poles are real, then

$$-\frac{b_1}{2} + \frac{1}{2}\sqrt{(b_1^2 - 4b_2)} < 1; \quad -1 < -\frac{b_1}{2} - \frac{1}{2}\sqrt{(b_1^2 - 4b_2)}$$

Now, besides $b_2 < 1$, the stability condition is:

$$b_2 > -b_1 - 1; \quad b_2 > -1 + b_1; \quad \text{or} \quad |b_1| < 1 + b_2 \tag{6.55}$$

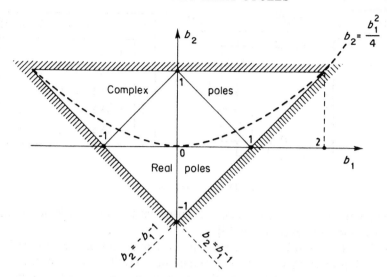

Fig. 6.17 Stability domain of the second-order section

The domain of stability is, therefore, a triangle defined by the three straight lines:

$$b_2 = 1; \quad b_2 = -b_1 - 1; \quad b_2 = b_1 - 1$$

as shown in Figure 6.17.

Nevertheless, even if the stability condition is fulfilled, there may still be a signal at the output in the absence of an input. This is usually a constant or periodic signal which corresponds to an auto-oscillation of the filter, and which is often called a limit-cycle. Such auto-oscillations can be produced with large amplitudes if overflow occurs when the capacity of the memories is exceeded in the absence of a logic saturation device. The equation for the system when there is no input signal is

$$y(n) + b_1 y(n-1) + b_2 y(n-2) = 0$$

The natural condition for the absence of oscillations is given by the inequality:

$$|b_1 y(n-1) + b_2 y(n-2)| < 1$$

and thus the condition which is necessary and sufficient for the absence of large amplitude auto-oscillations is:

$$|b_1| + |b_2| < 1 \tag{6.56}$$

This inequality determines a square in the plane (b_1, b_2), within the triangle of stability of the filter element.

To eliminate all possibility of large amplitude oscillations caused by overflow

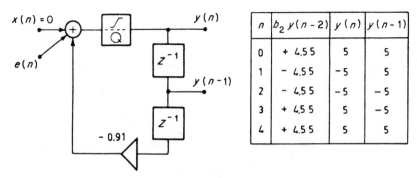

n	$b_2\,y(n-2)$	$y(n)$	$y(n-1)$
0	+ 4.55	5	5
1	− 4.55	−5	5
2	− 4.55	−5	−5
3	+ 4.55	5	−5
4	+ 4.55	5	5

Fig. 6.18   Low-level limit cycles

of the memory one can show that it is sufficient to employ a logic saturation device as shown in Section 6.6. [4].

Limit-cycles are also produced through quantization before storage in the memory. However, these have small amplitudes in well-designed systems. They arise through the fact that in practice the input signal is never zero, because, even in the absence of data $x(n)$, the error signal $e(n)$ caused by quantization of the internal data before storage in the memory is still applied to the filter. An example is shown in Figure 6.18.

An upper limit for such signals can be obtained quite simply by bearing in mind that the error signal $e(n)$ is itself bounded in the rounding process:

$$|e(n)| \leqslant \frac{q}{2}$$

If the filter has an impulse response represented by the set $h(i)$, the following inequality holds:

$$|y(n)| \leqslant (q/2)\sum_i |h(i)|$$

This bound is in fact very large, and a more realistic estimate of the amplitude A_a of the limit cycle is given by the expression:

$$A_a = \frac{q}{2}\max|H(\omega)|$$

where $H(\omega)$ is the transfer function for the filter section.

Application to a purely recursive second-order section with complex poles produces, according to equations (6.37) and (6.29):

$$|y(n)| \leqslant \frac{q}{2}\frac{1}{(1-r)\sin\theta} \tag{6.57}$$

$$A_a = \frac{q}{2}\frac{1}{(1-r^2)\sin\theta} \tag{6.58}$$

These signals often have a spectrum formed of lines with frequencies close to those at which $H(\omega)$ is a maximum, which are either factors of, or in simple ratios to, the sampling frequency.

When designing filters the number of bits in the internal data memories must be chosen to be sufficiently large, and the quantization step q has to be chosen to be sufficiently small to prevent the limit-cycles from being troublesome. It should be noted also that they can be eliminated by using a type of quantization other than rounding (for example, truncation of the absolute value) [5]. However, this is only achieved with an increase in the power of the round-off noise in the presence of a signal.

The results obtained in this chapter will be used in the following chapter where second-order sections in cascade will be discussed.

REFERENCES

[1] P. A. REGALIA, S. K. MITRA and P. P. VAIDYANATHAN, All-pass filter: a versatile signal processing building block. *Proceedings of the IEEE*, **76**(1), 19–37, 1988.
[2] J. B. KNOWLES and E. M. OLCAYTO, Coefficient accuracy and digital filter response. *IEEE Transactions on Circuit Theory*, Mar 1968.
[3] L. B. JACKSON, On the interaction of round-off noise and dynamic range in digital filters. *BSTJ*, Feb 1970.
[4] P. EBERT, J. MAZO and M. TAYLOR, Overflow oscillations in digital filters. *BSTJ*, Nov 1969.
[5] T. CLAASEN, W. MECKLENBRAUKER and J. PEEK, Effects of quantizations and overflow in recursive digital filters. *IEEE Transactions*, **ASSP24**(6), 1976.
[6] S. K. MITRA and J. F. KAISER, *Handbook for Digital Signal Processing*, John Wiley, New York, 1993.

EXERCISES

1 Study the first-order section

$$y(n) = x(n) + by(n-1)$$

in the following conditions:

$$x(n) = 0; \qquad n < 0$$
$$x(n) = \cos n\omega; \quad n \geqslant 0; \qquad b = -0.8; \quad \omega = \frac{\pi}{2}; \quad y(-1) = 0$$

Give the expression of $y(n)$ with transient and steady state terms. From $y(n)$, compute the amplitude and phase responses and check with the results of Section 6.1. Considering the steady state term in $y(n)$ give the filter delay. Compare with the value obtained for the group delay. Justify the difference between the two values.

2 Calculate the response of the system which is defined by the following equation:

$$y(n) = x(n) + x(n-1) - 0.8y(n-1)$$

to the unit set $u_0(n)$ and the set $x(n)$ such that:

$$x(n) = 0 \quad \text{for } n < 0$$
$$1 \quad \text{for } n \geqslant 0$$

Give the steady state frequency response and the transient response.

3 Assume a purely recursive second-order section which has the following coefficients:

$$b_1 = -1.56; \quad b_2 = 0.8$$

State the position of the poles. Calculate the frequency response, the phase response and the group delay. How are the functions modified if two zeros are added at j and $-j$? For this case, show the circuit diagram using the $D-N$ form and count the number of multiplications required for each output number.

4 Give the expression for the impulse response of a purely recursive second-order filter section which has the following coefficients:

$$b_1 = 1.60; \quad b_2 = 0.98$$

Calculate the frequency and amplitude of the response. Give the response $H(\omega)$ and calculate the norm $\|H\|_2$.

The zeros are added on the unit circle to produce an infinite attenuation at frequency $3f_s/8$. What are the coefficients of the filter? Calculate the new expression for $H(\omega)$ and the new value of $\|H\|_2$.

For this filter, find the amplitude of the limit cycles, using the $D-N$ form and then the $N-D$ form. Give an example of a limit cycle.

Does this filter produce large-amplitude oscillations when there is no logic saturation device? Give an example.

5 How many bits are needed to represent the coefficients of the filter with the following Z-transfer function

$$H(Z) = \frac{1 - 0.952Z^{-1} + Z^{-2}}{1 - 1.406Z^{-1} + 0.917Z^{-2}}$$

in order that the frequency response is not modified by more than 1% in the neighbourhood of the poles? Calculate the displacement of the point of infinite attenuation.

6 Assume the realization of a phase shifter of second order having poles $P_{1,2}$ such that:

$$P_{1,2} = 0.71 \pm j0.54$$

Calculate the coefficients and give the expression for the function $\tau_g(\omega)$. Show that an implementation scheme exists which produces a reduced number of multiplications. When there is no logic saturation device, can this element exhibit large-amplitude oscillations? Can it produce low-amplitude limit cycles?

CHAPTER 7

INFINITE IMPULSE RESPONSE FILTERS

Digital filters with an infinite impulse response (IIR), or recursive filters, have properties similar to those of analogue filters, and consequently their coefficients can be determined by similar techniques [1–3].

Before discussing the method for calculating the coefficients it is useful to give some general expressions for the properties of these filters.

7.1 GENERAL EXPRESSIONS FOR THE PROPERTIES OF IIR FILTERS

The general IIR filter is a system which produces from the set of data $x(n)$ the set $y(n)$ such that:

$$y(n) = \sum_{l=0}^{L} a_l x(n-l) - \sum_{k=1}^{K} b_k y(n-k) \tag{7.1}$$

The Z-transfer function for this system is written:

$$H(Z) = \frac{\sum_{l=0}^{L} a_l Z^{-1}}{1 + \sum_{k=1}^{K} b_k Z^{-k}} \tag{7.2}$$

This is the quotient of two polynomials in Z, which are often of the same degree.

As the coefficients a_l and b_k are real numbers, $H(Z)$ is a complex number such that:

$$\overline{H(Z)} = H(\bar{Z})$$

and the frequency response of the filter can be written with the same conventions as in the earlier chapters:

$$H(\omega) = |H(\omega)| e^{-j\phi(\omega)}$$

The modulus and the phase are expressed in terms of $H(Z)$ by the following equations:

$$|H(\omega)|^2 = [H(Z)H(Z^{-1})]_{Z=e^{j\omega}} \tag{7.3}$$

By squaring $H(\omega)$ and using equation (7.3),

$$\phi(\omega) = -\tfrac{1}{2}j \log \left[\frac{H(Z)}{H(Z^{-1})} \right]_{Z=e^{j\omega}} \tag{7.4}$$

and, by taking the derivative of $\phi(Z)$ with respect to the complex variable Z, we obtain:

$$\frac{d\phi}{dZ} = -\tfrac{1}{2}j\left[\frac{H'(Z)}{H(Z)} + \frac{1}{Z^2}\frac{H'(Z^{-1})}{H(Z^{-1})}\right]$$

For $Z = e^{j\omega}$ this becomes:

$$\frac{d\phi}{dZ} = -\frac{1}{jZ}\,\text{Re}\left[Z\frac{d}{dZ}\log(H(Z))\right]$$

Thus, the equation for the group delay becomes:

$$\tau(\omega) = \frac{d\phi}{d\omega} = \frac{d\phi}{dZ}jZ = -\text{Re}\left[Z\frac{d}{dZ}\log(H(Z))\right]_{Z=e^{j\omega}} \tag{7.5}$$

Equations (7.3)–(7.5) allow the analysis of IIR filters of any order.

Example: Assume:

$$H(Z) = \frac{1}{D(Z)} = \frac{1}{1 + b_1 Z^{-1} + b_2 Z^{-2}}$$

Since

$$\tau(\omega) = \text{Re}\left[Z\frac{D'(Z)}{D(Z)}\right]_{Z=e^{j\omega}} = -\text{Re}\left[\frac{b_1 Z^{-1} + 2b_2 Z^{-2}}{1 + b_1 Z^{-1} + b_2 Z^{-2}}\right]_{Z=e^{j\omega}}$$

one has

$$\tau(\omega) = 1 - \frac{1 - b_2^2 + b_1(1 - b_2)\cos\omega}{1 + b_1^2 + b_2^2 + 2b_1(1 + b_2)\cos\omega + 2b_2\cos(2\omega)} \tag{7.6}$$

which is equivalent to the expression given in the previous chapter when the poles are complex with $b_1 = -2r\cos\theta$ and $b_2 = r^2$.

Others expressions for IIR filters can be obtained in terms of the poles and zeros of $H(Z)$. If the numerator and the denominator have the same degree N, and if N is even, then

$$H(Z) = a_0 \prod_{i=1}^{N/2} \frac{1 + a_1^i Z^{-1} + a_2^i Z^{-2}}{1 + b_1^i Z^{-1} + b_2^i Z^{-2}}$$

The square of the modulus of the transfer function is equal to the product of the squares of the moduli of the elementary functions. The phase and the group delay are the sums of the contributions of the sections:

$$|H(\omega)|^2 = a_0 \prod_{i=1}^{N/2} |H_i(\omega)|^2$$

$$\tau(\omega) = \sum_{i=1}^{N/2} \tau_i(\omega)$$

The general equations for IIR filters given above are used to calculate the coefficients.

7.2 DIRECT CALCULATION OF THE COEFFICIENTS USING MODEL FUNCTIONS

A direct method for calculating the coefficients of an IIR filter consists of using a model function, which is a real function defined on the frequency axis. The model functions which will be considered are those of Butterworth. Bessel and Chebyshev, and the elliptic functions, all of which have known selectivity properties. They are also used to calculate analogue filters and form a model for the square of the transfer function to be derived. However, one drawback to their use for calculating digital filters is that they are not periodic when the desired function has period f_s. It is then necessary to establish a mapping between the real axis and the range $[0, f_s]$. Such a mapping is supplied by a conformal transformation in the complex plane with the following properties:

(1) It transforms the imaginary axis onto the unit circle;
(2) It transforms a rational fraction of the complex variable s into a rational fraction of the complex variable Z;
(3) It conserves stability.

7.2.1 Impulse invariance

Consider the analogue filter defined by the equation

$$y'_a(t) = by_a(t) + x(t) \tag{7.7}$$

It has a transfer function

$$H(s) = \frac{1}{s - b} \tag{7.8}$$

and an impulse response:

$$h(t) = e^{bt} \tag{7.9}$$

Sampling of this response with period T provides the sequence:

$$h(nT) = e^{bTn} \tag{7.10}$$

which has a Z-transform:

$$H(Z) = \frac{1}{1 - e^{bT}Z^{-1}} \tag{7.11}$$

The pole b of the analogue filter has become e^{bT} for the digital filter. The method generalizes to any number of poles.

This simple method is used in the simulation of analogue systems by digital computers. It has a serious disadvantage for the design of filters due to aliasing of the frequency response. During the sampling operation, the frequency response of the analogue filter is replicated in the useful band of the digital filter and the amplitude specification cannot be conserved.

Another approach consists of establishing a direct correspondence between the operation of the analogue filter and the digital filter.

Reconsidering the differential equation of the analogue filter, at time nT one can write:

$$y_a(nT) = y_a(nT - T) + \int_0^T y_a'(nT - T + \tau)\, d\tau \qquad (7.12)$$

Evaluating the integral using the trapezium rule leads to:

$$y_a(nT) - y_a(nT - T) = \frac{T}{2}[y_a'(nT) + y_a'(nT - T)] \qquad (7.13)$$

Hence:

$$y_a(nT) - y_a(nT - T) = \frac{T}{2}[by_a(nT) + x(nT) + by_a(nT - T) + x(nT - T)]$$

Taking the Z-transform of both sides leads to the following Z-transfer function:

$$H(Z) = \frac{1}{(2/T)(1 - Z^{-1})/(1 + Z^{-1}) - b} \qquad (7.14)$$

which yields a relation between s and Z, called the bilinear transformation.

7.2.2 Bilinear transformation

Assume a transformation which transforms a point Z on the complex plane into the point s where

$$s = \frac{2}{T}\frac{1 - Z^{-1}}{1 + Z^{-1}} \qquad (7.15)$$

To each point on the unit circle $Z = e^{j\omega T}$ there is a corresponding point s such that:

$$s = \frac{2}{T}\frac{1 - e^{-j\omega T}}{1 + e^{-j\omega T}} = j\frac{2}{T}\tan\frac{\omega T}{2} \qquad (7.16)$$

Consequently, the imaginary axis corresponds to the unit circle. The equation which gives Z as a function of s is

$$Z = \frac{(2/T) + s}{(2/T) - s} \qquad (7.17)$$

A rational fraction involving s is transformed to a rational fraction involving Z in which the numerator and the denominator have the same degree.

If the real part of s is negative, the modulus of Z is less than 1, i.e. the part of the complex plane of the variable s to the left of the imaginary axis is transformed inside the unit circle. This property permits the stability characteristics of the system to be conserved.

In the definition of the transformation, the factor $2/T$ is a scale factor. $T = 1/f_s$ is the sampling period of the digital system. This factor controls the warping of the frequency axis which is introduced when the bilinear transformation is used to obtain the Z-transfer function of a digital system using a complex function of s. Indeed, the digital system obtained has a frequency response which is a function of the variable f_N, which in turn is related by equation (7.16) to the values jf_A of the initial function on the imaginary axis. Thus,

$$\pi f_A T = \tan(\pi f_N T) \tag{7.18}$$

Figure 7.1 illustrates this equation. It will be noted that the warping is negligible for very low frequencies, which justifies the choice of the scale factor $2/T$.

Although other transformations can also be applied to the calculation of digital filter coefficients the bilinear transformation is the one most commonly used. It permits the calculation of digital filters from the transfer functions for analogue filters or by using computer programs developed for analogue filters.

It should be remembered, however, that the frequency response is warped, as shown above, since the analogue and digital frequencies ω_A and ω_N are

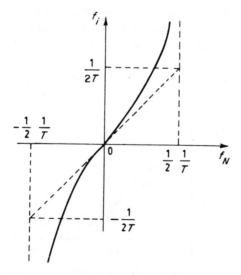

Fig. 7.1   Frequency warping introduced by the bilinear transformation

related by the equation:

$$\omega_A = \frac{2}{T} \tan \left[\frac{\omega_N T}{2} \right] \tag{7.19}$$

The group delay is also modified:

$$\tau_N = \tau_A \left[1 + \left(\frac{\omega_A T}{2} \right)^2 \right] \tag{7.20}$$

i.e. an analogue filter with a constant group delay is transformed into a digital filter which does not have this property.

In order to design a digital filter from a mask using this method, the mask must first be modified to take account of the subsequent frequency warping. The analogue filter satisfying the new mask can then be calculated, and finally the bilinear transform can be applied.

7.2.3 Butterworth filters

Two examples will be used to illustrate the calculation of the coefficients by a model function. The first is the case of Butterworth filter functions, because of their simplicity, and the second is the case of elliptic functions, because they are the ones most frequently used.

A Butterworth function of order n is defined by

$$|F(\omega)|^2 = \frac{1}{1 + (\omega/\omega_c)^{2n}} \tag{7.21}$$

The parameter ω_c gives the value of the variable for which the function has the value $\frac{1}{2}$. Figure 7.2 represents this function for various values of n.

By analytic extension, taking $\omega_c = 1$, this can be written as:

$$|F(\omega)|^2 = |H(j\omega)|^2 = |H(s)H(-s)|_{s=j\omega} = \frac{1}{1 + \omega^{2n}}$$

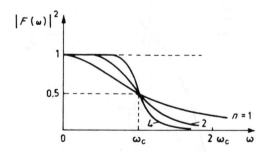

Fig. 7.2   Butterworth functions

$$H(s)H(-s) = \frac{1}{1 + (s/j)^{2n}} = \frac{1}{1 + (-s^2)^n}$$

The poles of this function lie on the unit circle. For example, when n is odd, one can write:

$$H(s)H(-s) = \frac{1}{\prod_{k=1}^{2n} (s - e^{j\pi(k/n)})}$$

By setting to $H(s)$ the poles which are to the left of the imaginary axis, to obtain a stable filter, and after proper factorization to obtain first- and second-order sections with real coefficients, one has

$$H(s) = \frac{1}{1+s} \prod_{k=1}^{(n-1)/2} \frac{1}{s^2 + 2\cos(\pi(k/n))s + 1}$$

Similarly, for even n, one obtains

$$H(s) = \prod_{k=1}^{n/2} \frac{1}{s^2 + 2\cos[\pi(2k-1)/2n]s + 1}$$

The corresponding digital filter is produced by change of variable according to equation (7.15):

$$s = \frac{2}{\tan(\omega_c T/2)} \frac{1 - Z^{-1}}{1 + Z^{-1}}$$

The point on the frequency axis where the response of the digital filter has a value of $2^{-1/2}$ is f_c with

$$\omega_c = 2\pi f_c$$

By setting $u = 1/\tan(\pi f_0 T)$ and $\alpha_k = 2\cos(\pi k/n)$ the Z-transfer function for the digital filter of odd order n is obtained:

$$H(Z) = \frac{1 + Z^{-1}}{(1 + u) + (1 - u)Z^{-1}} \prod_{k=1}^{(n-1)/2} a_0^k \frac{(1 + Z^{-1})^2}{1 + b_1^k Z^{-1} + b_2^k Z^{-2}}$$

with:

$$a_0^k = \frac{1}{1 + u\alpha_k + u^2}; \quad b_1^k = 2a_0^k(1 - u^2); \quad b_2^k = a_0^k(1 - u\alpha_k + u^2)$$

For even n, with $\alpha_k = 2\cos(\pi(2k-1)/2n)$, this becomes

$$H(Z) = \prod_{k=1}^{n/2} a_0^k \frac{(1 + Z^{-1})^2}{1 + b_1^k Z^{-1} + b_2^k Z^{-2}} \tag{7.22}$$

It would thus appear that the zeros of the Z-transfer function are all found at the point $Z = -1$, which can simplify the realization of the filter. On the other hand, the function is completely determined by the data for the parameters n and u.

The order n is calculated from the specification of the filter. Assume that a filter is to be produced with a frequency response greater than or equal to $1 - \delta_1$ in the band $[0, f_1]$ and less than or equal to δ_2 in the band $[f_2, f_s/2]$. In terms of the model function $F(\omega)$, these constraints imply the following inequalities:

$$\frac{1}{1 + (\omega_1/\omega_c)^{2n}} \geq (1 - \delta_1)^2 \quad \text{and} \quad \frac{1}{1 + (\omega_2/\omega_c)^{2n}} \leq \delta_2^2$$

For small δ_1 and δ_2, the following equation for n results:

$$n \geq \frac{\frac{1}{2}\log(2\delta_1) + \log(\delta_2)}{\log(\omega_1) - \log(\omega_2)}$$

The order N of the digital filter is obtained using equation (7.18):

$$N \geq \frac{\log(1/\delta_2\sqrt{(2\delta_1)})}{\log\tan(\pi f_2 T) - \log\tan(\pi f_1 T)} \tag{7.23}$$

Once n is chosen, the parameter u must lie in the interval:

$$\frac{1}{\tan(\pi f_2 T)}\left(\frac{1}{\delta_2}\right)^{1/n} \leq u \leq \frac{(2\delta_1)^{(1/2)n}}{\tan(\pi f_1 T)}$$

These parameters allow $H(Z)$ to be calculated.

Example: Assume the following characteristic, which was studied earlier for FIR filters:

$$\delta_1 = 0.045; \quad \delta_2 = 0.015; \quad f_s = 1; \quad f_1 = 0.1725; \quad f_2 = 0.2875$$

It is found that $N \simeq 7.3$ so the value adopted is $N = 8$:

$$H(Z) = 0.00185 \frac{(1 + Z^{-1})^2}{1 - 0.36Z^{-1} + 0.04Z^{-2}} \times \frac{(1 + Z^{-1})^2}{1 - 0.39Z^{-1} + 0.12Z^{-2}}$$

$$\times \frac{(1 + Z^{-1})^2}{1 - 0.45Z^{-1} + 0.31Z^{-2}} \times \frac{(1 + Z^{-1})^2}{1 - 0.58Z^{-1} + 0.69Z^{-2}}$$

The frequency response obtained is shown by Figure 7.3 and the group delay is given in Figure 7.4.

When the filter transition band, $\Delta f = f_2 - f_1$, is sufficiently small, equation (7.23) can be simplified:

$$N \simeq \log\frac{1}{\delta_2\sqrt{(2\delta_1)}} \frac{2.3}{2\pi} \frac{f_s}{\Delta f}\sin\left(2\pi\frac{f_1}{f_s}\right) \tag{7.24}$$

Thus, the filter order is proportional to the inverse of the transition band, as for FIR filters. It follows that the selectivity of this kind of filter is rather limited in practice.

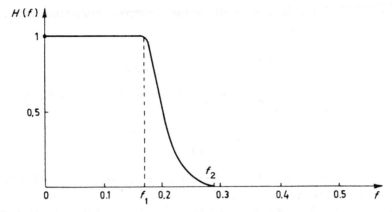

Fig. 7.3   Frequency response of a Butterworth filter of order 8

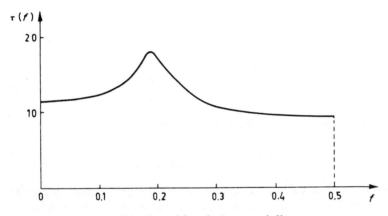

Fig. 7.4   Group delay of a Butterworth filter

In conclusion, Butterworth filters are straightforward to calculate. Significant simplications can occur in their realization because of the arrangement of the roots and, in certain cases, of the poles, but they are much less selective than elliptic filters.

7.2.4 Elliptic filters

The elliptic filter displays ripples in the pass band and in the stop band. It is optimal in that for a given order n and for fixed ripple amplitudes it has the smallest transition band. The model function involves elliptic functions and is written as

$$T^2(u) = \frac{1}{1 + \varepsilon^2 sn^2(u, k_1)} \tag{7.25}$$

where $y = sn(u, k)$ is defined implicitly by the incomplete elliptic function of the first type:

$$u = \int_0^{\arcsin y} \frac{d\theta}{(1 - k^2 \sin^2 \theta)^{1/2}} \qquad (7.26)$$

Figure 7.5 represents the function $T^2(u)$ for $u = j\omega$, and shows the parameters corresponding to k_1 such that:

$$k_1 = \frac{\varepsilon}{\sqrt{(A^2 - 1)}}$$

The function $sn^2(\omega, k)$ oscillates between 0 and 1 for $\omega < \omega_1$ and between $\sqrt{(A^2 - 1)}/\varepsilon$ and infinity for $\omega \geqslant \omega_2$.

One can show that the order n of the filter is determined from the parameters k_1 and k, the selectivity factor:

$$k = \frac{\omega_1}{\omega_2}$$

by the expression:

$$n = \frac{K(k)K(\sqrt{(1 - k_1^2)})}{K(k_1)K(\sqrt{(1 - k^2)})} \qquad (7.27a)$$

where $K(k)$ is the complete elliptic integral of the first type:

$$K(k) = \int_0^{\pi/2} \frac{d\theta}{(1 - k^2 \sin^2 \theta)^{1/2}} \qquad (7.27b)$$

This integral is calculated by the Chebyshev polynomial approximation method which results in an error of the order of 10^{-8} with a polynomial of degree 4. The inverse function to the incomplete integral of the first type is calculated as the quotient of two rapidly converging series.

A simplified equation for the order n of the filter can be obtained from the

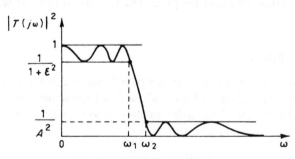

Fig. 7.5 *Elliptic filtering function*

general specification given in Figure 5.7. With the assumption of a ripple in the pass band of between 1 and $1 - 2\delta_1$, and with the following parameters (Figure 7.5):

$$\delta_2 = 1/A; \quad 2\delta_1 = 1 - 1/\sqrt{(1 + \varepsilon^2)}; \quad f_s = 1$$

one has

$$n \simeq (2/\pi^2) \ln(2/\delta_2\sqrt{\delta_1}) \ln[8\omega_1/(\omega_2 - \omega_1)] \qquad (7.28)$$

The order N of the digital filter satisfying the mask in Figure 5.7 is then given by:

$$N \simeq (2/\pi^2) \ln(2/\delta_2\sqrt{\delta_1}) \times \ln[8 \tan(\pi f_1/f_s)/(\tan(\pi f_2/f_s) - \tan(\pi f_1/f_s))]$$

The transition band $\Delta f = f_2 - f_1$ is generally narrow, and thus (using logarithms to base 10),

$$N \simeq 1.076 \log(2/\delta_2\sqrt{\delta_1}) \log[(f_s/\Delta f)(4/\pi)\sin(2\pi f_1/f_s)] \qquad (7.29)$$

This relation should be compared with equation (5.32) for finite impulse response filters. It shows that, for elliptic IIR filters, the order is proportional to the logarithm of the inverse of the normalized transition band. This leads to much lower values than for FIR filters. Further, equation (7.29) shows that the width of the band is also involved. The maximum value of N is found for f_1 close to $f_s/4$, i.e. for a pass band approximating half of the useful band. A further simplification can be obtained for filters with a narrow pass band. In this case, the order N' of the filter is given by:

$$N' \simeq 1.076 \log[2/(\delta_2\sqrt{\delta_1})] \log(8f_1/\Delta f) \qquad (7.30)$$

and, as in analogue filters, it is the steepness of the cut-off which is important here.

Once the filter order has been determined, the calculation procedure involves determining the poles and zeros of $T^2(u)$, which show a double periodicity in the complex plane. By changing the variable and then applying the bilinear transform the configuration of the poles and zeros of the digital filter in the Z-plane is obtained [4]. A program to design a filter in accordance with this method is given in an appendix. The form of the filter is specified by:

(1) The peak-to-peak amplitude of the pass band ripples, expressed in dB:

$$BP = -20\log(1 - 2\delta_1)$$

(2) The amplitude of the ripples in the attenuation band, expressed in dB:

$$AT = 20\log(1/\delta_2)$$

(3) The frequency of the end of the pass band, FB
(4) The frequency of the start of the attenuation band, FA
(5) The sampling frequency, FS

The program calculates the poles and zeros of the filter in the Z-plane.

Example: Consider the specification given in the previous section: $BP = 0.4$, $AT = 36.5$, $FS = 1$, $FB = 0.1725$, $FA = 0.2875$.

It is found that $N = 3.37$, and the adopted values in $N = 4$. The zeros and poles have co-ordinates (Figure 7.6):

$$Z_1 = -0.816 + j0.578 \quad Z_2 = -0.2987 + j0.954$$
$$P_1 = 0.407 + j0.313 \quad P_2 = 0.335 + j0.776$$

To demonstrate the point of infinite attenuation the curve $1/|H(f)|$ giving the attenuation of the filter as a function of frequency is shown in Figure 7.7.

Figure 7.8 shows the group delay of the filter obtained. The curves can be

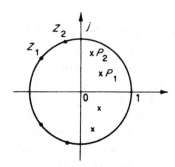

Fig. 7.6   Poles and zeros of an elliptic filter of order 4

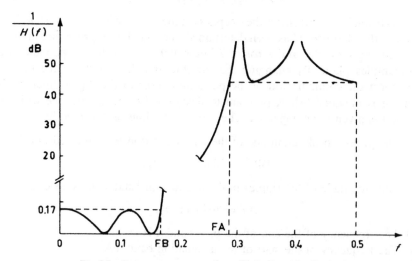

Fig. 7.7   Frequency response of an elliptic filter of order 4

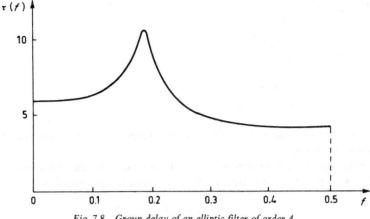

Fig. 7.8 *Group delay of an elliptic filter of order 4*

compared with the results obtained for the same characteristic using the Butterworth filter. They demonstrate the advantage of the elliptic filter, which requires an order a factor of 2 lower, and produces a corresponding reduction in the complexity of the circuits.

The methods which have been described permit calculation of low-pass filters, from which suitable frequency transformations make it possible to obtain high-pass and band-pass filters.

7.2.5 Calculating any filter by transformation of a low-pass filter

The calculation procedures which were presented in the previous sections result in a function $H(s)$ which provides, by bilinear tranformations, the Z-transfer function for the digital filter. For $s = j\omega$ the function $H(\omega)$ is a low-pass filter function in the frequency domain which extends from zero to infinity. It is possible to apply transformations to it which result in other types of filter [5]. For example, in order to obtain a low-pass filter whose pass band extends from 0 to ω'_1 using a function whose pass band covers the domain $[0, \omega_1]$, the following transformation can be made:

$$s \to s\frac{\omega_1}{\omega'_1}$$

By starting from a low-pass filter with a pass band limited to 1 the following filters can be obtained, by denoting the lower and upper limits of the pass band by ω_B and ω_H:

(1) Another low pass: $s \to \dfrac{s}{\omega_H}$

(2) High pass: $s \to \dfrac{\omega_B}{s}$

(3) Band pass: $s \to \dfrac{s^2 + \omega_H \omega_B}{s(\omega_H - \omega_B)}$

(4) Band stop: $s \to \dfrac{s(\omega_H - \omega_B)}{s^2 + \omega_H \omega_B}$

These transforms conserve the ripple in the response of the filter but result in frequency warping.

A more direct method consists of using transforms other than the bilinear transform to reach the function $H(Z)$. For example, the transform

$$s = \frac{1}{T} \frac{1 - 2\cos(\omega_0 T)Z^{-1} + Z^{-2}}{1 - Z^{-2}} \tag{7.31}$$

allows a band-pass digital filter to be obtained from a low-pass filter function in s.

For $Z = e^{j\omega T}$ this becomes:

$$s = j\frac{1}{T} \frac{\cos(\omega_0 T) - \cos(\omega T)}{\sin(\omega T)} \tag{7.32}$$

If the pass band of the digital filter extends from ω_B to ω_H, ω_0 must be chosen so that the abscissae of the transformed points are equal in absolute value but are of opposite sign:

$$\frac{\cos(\omega_0 T) - \cos(\omega_B T)}{\sin(\omega_B T)} = - \frac{\cos(\omega_0 T) - \cos(\omega_H T)}{\sin(\omega_H T)}$$

and thus:

$$\cos(\omega_0 T) = \frac{\cos[(\omega_B + \omega_H)T/2]}{\cos[(\omega_B - \omega_H)T/2]}$$

This approach avoids adding a stage to the calculation procedure for a band-pass filter.

It is also possible to use transformations in the Z-plane which conserve the unit circle. The simplest is to transform from Z to $- Z$, which changes a low-pass filter into a high-pass one.

The transformation

$$Z^{-1} \to (Z^{-1} - \alpha)/(1 - \alpha Z^{-1}) \tag{7.33}$$

where α is a real number, changes a low-pass filter into another low pass one. In fact, it can be shown that the most general transformation is expressed by [5]:

$$Z^{-1} \rightarrow \pm \prod_{k=1}^{K} \frac{Z^{-1} - \alpha_k}{1 - \alpha_k Z^{-1}} \tag{7.34}$$

with $|\alpha_k| < 1$ to ensure stability.

For example, a low-pass filter is transformed into a band-pass filter by:

$$Z^{-1} \rightarrow \frac{Z^{-2} - \alpha_1 Z^1 + \alpha_2}{\alpha_2 Z^{-2} - \alpha_1 Z^{-1} + 1} \tag{7.35}$$

It would thus appear that every type of filter can be calculated in a direct way using model functions. However, there are important limitations. First, for example, for elliptic filters the ripples must be constant in the pass and stop bands. Second, the methods described do not allow for possible constraints on the impulse response. To overcome these limitations optimization techniques have to be employed.

7.3 ITERATIVE TECHNIQUES FOR CALCULATING IIR FILTERS WITH FREQUENCY SPECIFICATIONS

As with FIR filters, optimization methods permit the calculation of IIR filters with any specification. Nevertheless, the calculation is somewhat more sensitive than for FIR filters as precautions have to be taken to avoid producing an unstable system.

Two methods will be presented. These correspond to two different optimization criteria, the first of which is minimization of the mean square error [6].

7.3.1 Minimizing the mean square error

The transfer function of a filter is given in a factorized form by the equation which was introduced earlier:

$$H(Z) = a_0 \prod_{i=1}^{N/2} \frac{1 + a_1^i Z^{-1} + a_2^i Z^{-2}}{1 + b_1^i Z^{-1} + b_2^i Z^{-2}}; \quad a_0 > 0 \tag{7.36}$$

by regarding the numerator and denominator as having the same degree N (even).

Assume $D(f)$ is the function which approximates the frequency response of the filter $H(f)$. The difference between these functions represents an error which can be minimized by least squares for a number of points, N_0 on the frequency axis. Thus,

$$E = \sum_{n=0}^{N_0 - 1} (|H(f_n)| - |D(f_n)|)^2$$

The value E is a function of the set of $2N + 1$ parameters, which are the coefficients of the filter:

$$E = E(a_0, a_1^i, a_2^i, b_1^i, b_2^i) \quad \text{with} \quad 1 \leqslant i \leqslant \frac{N}{2}$$

The minimum corresponds to the set of $2N + 1$ parameters x_k such that:

$$\frac{\partial E}{\partial x_k} = 0; \quad 1 \leqslant k \leqslant 2N + 1$$

For the parameter a_0 one can set $H(Z) = a_0 H_1(Z)$, whence

$$\frac{\partial E}{\partial a_0} = 0 = 2 \sum_{n=0}^{N_0-1} (a_0 |H_1(f_n)| - |D(f_n)|)|H_1(f_n)|$$

Thus, the value of a_0 is

$$a_0 = \frac{\sum_{n=0}^{N_0-1} |D(f_n)| \, |H_1(f_n)|}{\sum_{n=0}^{N_0-1} H_1(f_n)|^2} \tag{7.37}$$

The optimization is restricted to $2N$ variables.

The procedure consists of taking an initial function $H_1^0(Z)$, which is found, for example, by the direct calculation method given in the preceding section for elliptic filters, and then assuming that it is sufficiently close to the optimum for the function E to be represented by a quadratic function with $2N$ parameters x_k. The desired optimum is then obtained through an increment in the parameters represented by the vector ΔX with $2N$ element such that:

$$E(X + \Delta X) \simeq E(X) + \sum_{k=1}^{2N} \frac{\partial E}{\partial x_k} \Delta x_k + \frac{1}{2} \sum_{k=1}^{2N} \sum_{l=1}^{2N} \frac{\partial^2 E}{\partial x_k \partial x_l} \Delta x_k \Delta x_l$$

By using A to denote the matrix with $2N$ rows and N_0 columns which has elements:

$$a_{ij} = 2 \frac{\partial}{\partial x_i} [a_0 |H_1(f_j)|]$$

and by using Δ to denote the column vector e_n with N_0 terms such that:

$$e_n = a_0 |H_1(f_n)| - |D(f_n)|$$

The condition of least squares is obtained by requiring $E(X + \Delta X)$ to be an extremum. As in Section 5.4, when calculating the coefficients of FIR filters we have

$$\Delta X = -[AA^t]^{-1}A\Delta$$

The calculation is then repeated with the new values for the parameters, which should ultimately lead to the required optimum. The chances of achieving this and the rate of convergence depend on the increments given to the parameters, and one of the best strategies is offered by the Fletcher and Powell algorithm [7].

To ensure stability in the resulting system either the stability can be controlled at each stage or the final system can be modified by replacing the poles P_i outside the unit circle by $1/P_i$, which does not modify the modulus of the frequency response except for a constant factor. In the latter case it is generally necessary to return to the optimization procedure to achieve the optimum.

Mean square error minimization can be applied to other functions as well as the frequency response: for example, the group delay [8].

7.3.2 Chebyshev approximation

This criterion corresponds to a limitation in the amplitude of the ripple in the frequency response of the filter in some frequency bands. One elegant approach consists of applying the Remez algorithm, already used to calculate the coefficients of linear phase FIR filters. The technique consists of using an initial filtering function $H_0(Z)$ which approximates the required $H(Z)$. This function can, for example, be of the elliptic type calculated by the method of Section 7.2.4 using adequate specifications. It is written as:

$$H_0(Z) = \frac{N_0(Z)}{D_0(Z)}$$

As the zeros of the filter functions are generally on the unit circle the numerator $N(Z)$ can be regarded as the transfer function of a linear phase FIR filter.

The first stage of the iterative technique is to calculate a new value for the numerator, $N_1(Z)$, using the algorithm for FIR filters. In this calculation, $D_0(f)$ is the function to be approximated in the pass band and $1/|D_0(f)|$ is used as a weighting factor.

One then looks for a new value for the denominator $D_1(Z)$. A function which approximates $|N_1(f)|$ in the pass band can be sought directly by again using the algorithm for FIR filters. It is more satisfactory, however, to use an adaptation of the calculation techniques employed for analogue filters.

By assuming that the required function $H(f)$, written as

$$H(f) = \frac{N(f)}{D(f)}$$

is such that:

$$|H(f)| \leqslant 1$$

one can write

$$|G(f)|^2 = |D(f)|^2 - |N(f)|^2$$

and hence

$$|H(f)|^2 = \frac{1}{1 + \left|\dfrac{G(f)}{N(f)}\right|^2}$$

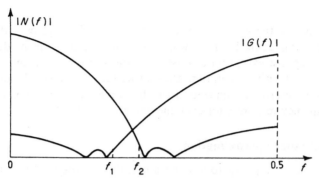

Fig. 7.9   The N(f) and G(f) functions for a low-pass filter

Figure 7.9 shows the functions $|G(f)|$ and $|N(f)|$ for a low-pass filter. The zeros of the function $G(Z)$ lie on the unit circle and they can be calculated using the algorithm for linear phase FIR filters. The weighting function is determined from $1/|N(f)|$.

By optimizing the stop and pass bands alternately the required filter function is obtained after several iterations and the filter coefficients are then obtained. The stability of the filter requires that only those poles of $H(z)$ which are inside the unit circle are conserved. Figure 7.10 shows a telephone channel filter which was calculated using this method.

More general optimization techniques, notably linear programming [3], can also produce the required filter.

7.4 FILTERS BASED ON SPHEROIDAL SEQUENCES

The filter design criterion can be the maximization of the energy concentration in a given frequency band, instead of a set of specifications on the frequency response.

Let λ be a scalar intended to represent the energy concentration and defined by:

$$\lambda = \int_{-f_c}^{f_c} H(f)\overline{H(f)}\,\mathrm{d}f \left/ \int_{-1/2}^{1/2} H(f)H(f)\,\mathrm{d}f \right. \tag{7.38}$$

where $[-f_c, f_c]$ is the band in which the energy has to be concentrated.

For the response

$$H(f) = \sum_{n=-P}^{P} a_n e^{-j2\pi f n}$$

a direct calculation yields:

$$\lambda = \frac{\sum_{n=-P}^{P} \sum_{m=-P}^{P} a_n a_m [\sin(n-m)2\pi f_c/(n-m)\pi]}{\sum_{n=-P}^{P} a_n^2} \tag{7.39}$$

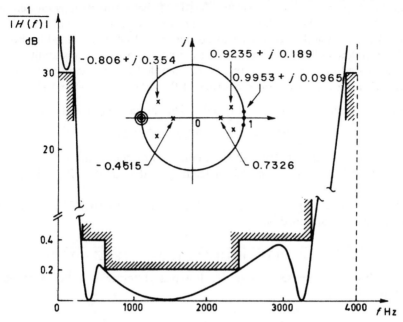

Fig. 7.10   Filter calculated by the iterative technique

and in matrix form:

$$A^t R A = \lambda A^t A$$

which is an eigenvalue equation. The filter coefficients are the elements of the eigenvector corresponding to the largest eigenvalue of the matrix R, whose elements are the terms

$$\frac{\sin (n - m) 2\pi f_c}{(n - m)\pi}$$

The elements of the eigenvectors of the matrix R are called the discrete Prolate Spheroidal sequences [9].

An FIR filter has been obtained; it is also possible to derive an IIR filter. To that end, consider the following purely recursive function:

$$|H(f)|^2 = \frac{1}{1 + \left| \sum_{n=1}^{N} b_n e^{j2\pi f n} \right|^2}$$

The coefficients can be calculated to minimize the energy of the denominator in the band $[-f_c, f_c]$, under the condition $|H(f_c)|^2 = 0.5$. Then the same method as above can be applied and the coefficients b_n $(1 \leqslant n \leqslant N)$ can be taken as the elements of the eigenvector associated with the smallest eigenvalue of the spheroidal matrix. First, the scaling factor of the eigenvector is chosen such that $|H(f_c)|^2 = \frac{1}{2}$. Then, the poles of the analytic expansion of $|H(f)|^2$ are calculated and the desired filter transfer function $H(Z)$ is obtained by keeping only those few poles which are inside the unit circle to ensure stability.

The procedure can be made reasonably simple by using iterative techniques and exploiting the structural properties of the spheroidal matrix [9].

Example: Assume: $N = 4$; $f_s = 1$; $f_c = 0.1$. The minimal eigenvector V_{min} is

$$V^t_{min} = [1.0, -2.773, -2.773, 1.0]$$

If T designates the matrix whose elements are the terms $e^{j2\pi f_c(n-m)}$ with $1 \leqslant n, m \leqslant N$, then the scaling factor leading to the equality $V^t_{min} T V_{min} = 1$ is 10.46.

After factorization of the analytical expansion $H(Z)H(Z^{-1})$, the transfer function finally obtained is:

$$H(Z) = \frac{0.0704}{(Z - 0.73 + j0.446)(Z - 0.73 - j0.446)(Z - 0.741)}$$

The technique presented above for low-pass filtering can be extended to high-pass filtering.

7.5 STRUCTURES REPRESENTING THE TRANSFER FUNCTION

IIR filters can be produced using circuits which directly perform the operations represented by the expression for their transfer function. The Z^{-1} term corresponds to a delay of one sampling period and is achieved by storage in the memory. The coefficients to be created by the circuits are those of the transfer function, with the same sign for the numerator and the opposite sign for the denominator.

Only canonic structures, i.e. those which require the minimum number of elementary operators, computing circuits and memories, will be examined.

7.5.1 Direct structures

These correspond to a direct realization of the Z-transfer function. Let us consider a purely recursive filter with the transfer function:

$$H(Z) = \frac{1}{1 + \sum_{i=1}^{N} b_i Z^{-i}}$$

A number $y(n)$ of the output set is obtained from the numbers $x(n)$ of the input set from the equation:

$$y(n) = x(n) - \sum_{i=1}^{N} b_i y(n - i)$$

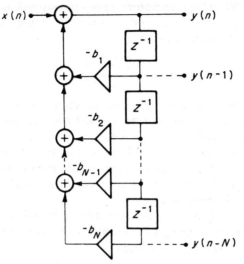

Fig. 7.11   Circuit of a purely recursive filter

which defines the operations to be performed in realizing the filter. The corresponding circuit is given in Figure 7.11. The circuit has N data memories to store the elements $y(n - i)$ $(i = 1, \ldots, N)$. The calculation of each element of the output set requires N multiplications and N additions.

A general IIR filter can be regarded as the cascade of a purely recursive filter

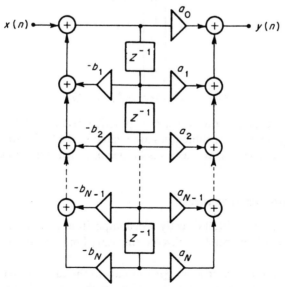

Fig. 7.12   IIR filter in the direct D–N structure

and an FIR filter. When the numerator and denominator of the transfer function are of the same degree the circuit obtained is as in Figure 7.12. It corresponds to the following Z-transfer function:

$$H(Z) = \frac{\sum_{i=0}^{N} a_i Z^{-i}}{1 + \sum_{i=1}^{N} b_i Z^{-i}}$$

As the denominator is calculated first, the structure is called D–N.

The order of the operations can be reversed and the numerator calculated first. This structure, called N–D, is deduced from the previous one by a transposition and is shown in Figure 7.13. The numbers stored in the memories are the partial sums. One interesting point is that each number $y(n)$ or $x(n)$ is multiplied by all of the coefficients in succession, which can simplify the organization of the multiplication.

As with second-order sections, these structures can be described by equations (4.34) and (4.37), by introducing the variables $u_i(n)$ and $v_i(n)$ with $1 \leqslant i \leqslant N$.

The matrix of the system A is written:

$$A = \begin{bmatrix} -b_1 & -b_2 & \cdots & -b_N \\ 1 & 0 & \cdots & 0 \\ 0 & 1 & \cdots & 0 \\ \vdots & \vdots & \vdots & \vdots \\ 0 & 0 & \cdots & 0 \end{bmatrix} \tag{7.40}$$

and

$$B = \begin{bmatrix} 1 \\ 0 \\ 0 \\ 0 \\ \vdots \\ 0 \end{bmatrix} : \quad d = a_0$$

$$C^t = (a_1 - a_0 b_1, a_2 - a_0 b_2, \ldots, a_N - a_0 b_N)$$

Direct structures are rarely used in practice because of the difficulties in implementing them. These difficulties are related to the limitation of the number of bits of the coefficients, and as a result, decomposed structures are preferred.

7.5.2 Decomposed structures

Instead of developing $H(Z)$ directly a decomposition can be carried out into a sum or product of separately realized elementary functions of the first or second order.

Decomposition into a product corresponds to the cascade structure in which the filter is generated from a set of sections of the first and second order:

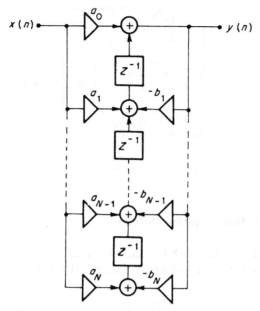

Fig. 7.13   *IIR filter in the direct N–D structure*

$$H(Z) = a_0 \frac{\prod_{i=1}^{N}(1 - Z_i Z^{-1})}{\prod_{i=1}^{N}(1 - P_i Z^{-1})}$$

$$= a_0 \cdots \frac{1 - Z_i Z^{-1}}{1 - P_i Z^{-1}} \cdots \frac{1 - 2\,\mathrm{Re}(Z_j)Z^{-1} + |Z_j|^2 Z^{-2}}{1 - 2\,\mathrm{Re}(P_j)Z^{-1} + |P_j|^2 Z^{-2}} \cdots \quad (7.41)$$

This structure is the one most frequently used because it presents, in addition to its modularity, the useful properties of low sensitivity to coefficient wordlength limitation and to round-off noise.

The function $H(Z)$ can also be decomposed into rational fractions:

$$H(Z) = a_0 + \cdots + \frac{\alpha_i}{1 - P_i Z^{-1}} + \cdots + \frac{\alpha_j + \beta_j Z^{-1}}{1 - 2\,\mathrm{Re}(P_j)Z^{-1} + |P_j|^2 Z^{-2}} + \cdots$$

$$(7.42)$$

The approach corresponds to connecting the M basic elements in parallel as shown in Figure 7.14. The numbers $y(n)$ are obtained by summing the outputs from the different elements to which the input numbers $x(n)$ are applied.

The choice between these different forms of realization is dictated by the facilities for their implementation, by the effect that the limitation of the number of bits in the representation of the coefficients has upon the properties of the resulting filter and by the power of the round-off noise produced in the arithmetic operations.

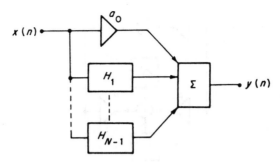

Fig. 7.14 *The parallel structure*

7.5.3 Basic structure of phase shifters

Transfer functions of the Butterworth, Chebyshev and elliptic type can be decomposed into a sum of two phase shifters [10]. For such a function, this gives:

$$H(Z) = \frac{N(Z)}{D(Z)} = \frac{1}{2}[A_1(Z) + A_2(Z)] \qquad (7.43)$$

where $A_1(Z)$ and $A_2(Z)$ are the transfer functions of the phase shifters.

Calculation of $A_1(Z)$ and $A_2(Z)$ from $H(Z)$ involves the complementary function $G(Z) = M(Z)/D(Z)$ such that:

$$|G(f)|^2 = 1 - |H(f)|^2 \qquad (7.44)$$

It is assumed that the initial function $H(Z)$ is such that $N(Z)$ is a symmetric polynomial and $M(Z)$ is asymmetric, that is:

$$\bar{N}(Z) = Z^N N(Z) \qquad \bar{M}(Z) = Z^N M(Z) \qquad (7.45)$$

Under these conditions, combining (7.44) and (7.45) yields:

$$\bar{N}(Z)N(Z) + \bar{M}(Z)M(Z) = \bar{D}(Z)D(Z) \qquad (7.46)$$

and

$$[N(Z) + M(Z)][N(Z) - M(Z)] = Z^{-N}D(Z^{-1})D(Z) \qquad (7.47)$$

Note that the zeros of $N(Z) + M(Z)$ and $N(Z) - M(Z)$ are harmonic conjugates and that the zeros of $D(Z)$ are their inverses. Designating the poles of the filters by $P_i(i = 1, \ldots, N)$, and hence the zeros of $D(Z)$, one can write, to within a constant:

$$N(Z) + M(Z) = \prod_{i=1}^{r}(1 - Z^{-1}P_i) \prod_{i=r+1}^{N} (Z^{-1} - P_i) \qquad (7.48)$$

and

$$N(Z) - M(Z) = \prod_{i=1}^{r}(Z^{-1} - P_i) \prod_{i=r+1}^{N} (1 - Z^{-1}P_i)$$

where r is the number of zeros of the polynomial $N(Z) + M(Z)$ within the unit circle. Dividing by $D(Z)$, one obtains:

$$H(Z) + G(Z) = \frac{\displaystyle\prod_{i=r+1}^{N} (Z^{-1} - P_i)}{\displaystyle\prod_{i=r+1}^{N} (1 - Z^{-1}P_i)} \tag{7.49}$$

and, similarly

$$H(Z) - G(Z) = \frac{\displaystyle\prod_{i=1}^{r}(Z^{-1} - P_i)}{\displaystyle\prod_{i=1}^{r}(1 - Z^{-1}P_i)} \tag{7.50}$$

The phase shifters $A_1(Z)$ and $A_2(Z)$ have the following expressions:

$$A_1(Z) = \prod_{i=r+1}^{N} \frac{Z^{-1}-P_i}{1-Z^{-1}P_i} \qquad A_2(Z) = \prod_{i=1}^{r} \frac{Z^{-1} - P_i}{1 - Z^{-1}P_i} \tag{7.51}$$

Finally, the filter $H(Z)$ and its complement $G(Z)$ are obtained by the arrangement of Figure 7.15.

The general procedure for designing the phase shifters from an elliptical filter is as follows:

(1) Calculate the transfer function $H(Z) = N(Z)/D(Z)$ of an elliptic filter of odd order N;
(2) Calculate the coefficients of the antisymmetric polynomial $M(Z)$ from $N(Z)$ and $D(Z)$ by using (7.46);
(3) Determine the inverses of the poles of $H(Z)$ which are the roots of the polynomial $N(Z) + M(Z)$;
(4) Calculate $A_1(Z)$ and $A_2(Z)$ using expression (7.51).

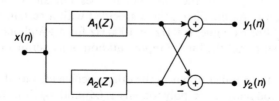

Fig. 7.15   Realization of an IIR filter and the complementary filter using two phase shifters

A simplified approach, when the order N is not very high, consists of finding $A_1(Z)$ and $A_2(Z)$ directly by combining poles. Hence for:

$$H(Z) = 0.0546 \left[\frac{1 + 1.8601Z^{-1} + 2.9148Z^{-2} + 2.9148Z^{-3} + 1.8601Z^{-4} + Z^{-5}}{(1 - 0.4099Z^{-1})(1 - 0.06611Z^{-1} + 0.4555Z^{-2})(1 - 0.4993Z^{-1} + 0.8448Z^{-2})} \right]$$

this gives:

$$A_1(Z) = \frac{0.4555 - 0.6611Z^{-1} + Z^{-2}}{1 - 0.6611Z^{-1} + 0.4555Z^{-2}}$$

$$A_2(Z) = \frac{(-0.4099 + Z^{-1})(0.8448 - 0.4993Z^{-1} + Z^{-2})}{(1 - 0.4099Z^{-1})(1 - 0.4993Z^{-1} + 0.8448Z^{-2})}$$

The basic structure of phase shifters is useful since it provides two complementary filters with the same calculations, which is useful in filter banks, as shown in Chapter 11. Furthermore, it is less sensitive than other structures to rounding of the coefficients.

Note that filters which can be decomposed as the sum of phase shifters are entirely defined by their poles.

A sum of phase shifters as in Figure 7.15 is the most efficient realization of an elliptic filter, since it requires a number of multiplications equal to the order of the filter.

7.6 LIMITING THE COEFFICIENT WORDLENGTH

Practical implementation of a filter implies limitation of the number of bits in the representation of filter coefficients, which form one of the multiplication terms. The effect of this on the complexity is important because multiplication is often the most critical factor. It is, therefore, necessary to find the minimum number of bits which allows the constraints imposed on the filtering function to be met.

Limitation of the number of bits of the scale factor a_0 appears as a modification in the gain of the filter but it does not affect the form of the frequency response. As the filter gain is specified with a certain tolerance at a given frequency (for example, 800 Hz or 1000 Hz for a telephone channel) it is necessary to ensure that the binary representation of a_0 permits this constraint to be met.

Limitation of the number of bits in the other coefficients modifies the transfer function by introducing error polynomials $e_N(Z)$ and $e_D(Z)$ in the numerator and the denominator. The actual transfer function is $H_R(Z)$:

$$H_R(Z) = (N(Z) + e_N(Z))/(D(Z) + e_D(Z)) \tag{7.52}$$

If the rounding errors in the coefficients are denoted by δa_i and δb_i these error functions can be written as a function of the normalized frequency ($f_s = 1$):

$$e_N(f) = \sum_{i=1}^{N} \delta a_i e^{-j2\pi f i}; \quad e_D(f) = \sum_{i=1}^{N} \delta b_i e^{-j2\pi f i}$$

In fact, these expressions form the Fourier series expansion of functions which are periodic in frequency. The Bessel–Parseval equation (1.7) relating the power of a signal to the power of its components allows the following equation to be written:

$$\int_0^1 |e_N(f)|^2 \, df = \sum_{i=1}^{N} |\delta a_i|^2$$

If q denotes the quantization step,

$$|\delta a_i| \leqslant \frac{q}{2}$$

and an upper limit is obtained for $|e_N(f)|$ by:

$$|e_N(f)| \leqslant N\frac{q}{2} \tag{7.53}$$

A statistical estimate σ of $|e_N(f)|$ can be obtained by regarding the δa_i as random variables uniformly distributed over the range $[-q/2, q/2]$. It is evaluated from the effective value of the function $e_N(f)$. Thus,

$$\sigma^2 = \int_0^1 |e_n(f)|^2 \, df = N\frac{q^2}{12}$$

and hence

$$\sigma = \frac{q}{2}\sqrt{\frac{N}{3}} \tag{7.54}$$

This estimate is valid for both $|e_N(f)|$ and $|e_D(f)|$. It is clearly less than the bound (7.53) given above and it is, in fact, more realistic when N exceeds several units.

The consequences of rounding the coefficients can be analysed separately for the numerator and denominator of the transfer function by considering the stop band for one and the pass band for the other. Examination of the configuration of the poles and zeros in the Z-plane shows that the poles determine the response of the filter in the pass band and the zeros in the stop band.

In the stop band, the denominator coefficient wordlength limitation can be neglected and, in terms of the variable, $\omega = 2\pi f$, one has:

$$H_R(\omega) = (N(\omega) + e_N(\omega))/D(\omega)$$

The error on the response is then estimated by:

$$|H_R(\omega) - H(\omega)| \simeq \sigma/|D(\omega)|$$

If the specification requires that the ripples in the stop band are less in modulus than δ_2, then, by separating the tolerance into two equal parts, one for the ripple when there is no rounding error on the coefficients and the other to allow for the error caused by this rounding, one has

$$\frac{\sigma}{|D(\omega)|} < \frac{\delta_2}{2} \tag{7.55}$$

In the pass band the numerator coefficient wordlength limitation can be neglected:

$$H_R(\omega) \simeq \frac{N(\omega)}{D(\omega) + e_D(\omega)} \simeq \frac{N(\omega)}{D(\omega)}\left[1 - \frac{e_D(\omega)}{D(\omega)}\right] \tag{7.56}$$

If the ripple in the pass band is less in modulus than δ_1, then by again dividing the tolerance into two equal parts,

$$\frac{\sigma}{|D(\omega)|} < \frac{\delta_1}{2} \tag{7.57}$$

This condition is generally much more restrictive than the earlier one because the function $|D(\omega)|$ has very low values in the pass band, and is even more restrictive with a more selective filter. Further, when the pass band is narrow, the coefficients can have large values. For a low-pass filter such as that in Figure 7.16 the following equation can be written:

$$D(Z) \simeq [1 - Z^{-1}]^N$$

and thus

$$b_i \simeq \frac{N!}{i\,!(N-i)!} \tag{7.58}$$

Under these conditions a very large number of bits is required if both large values of the coefficients are to be represented and a very low quantization

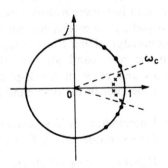

Fig. 7.16   *The poles and zeros of a narrow low-pass filter*

error is required. It is for this reason that decomposed structures are used almost exclusively with sections of first and second order.

Let us first consider the cascade structure which corresponds to the decomposition (7.41) of the transfer function. If the order N of the filter is even, this is written:

$$H(\omega) = N(\omega)/D(\omega) = \prod_{i=1}^{N/2} N_i(\omega)/D_i(\omega)$$

The polynomials $N_i(\omega)$ and $D_i(\omega)$ are of the second degree.

In the pass band, if rounding of the coefficients of the polynomials $N_i(\omega)$ is neglected, we obtain

$$H_R(\omega) \simeq \prod_{i=1}^{N/2} \frac{N_i(\omega)}{D_i(\omega) + e_i(\omega)}$$

or

$$H_R(\omega) \simeq \frac{N(\omega)}{D(\omega)} \left[1 - \sum_{i=1}^{N/2} \frac{e_i(\omega)}{D_i(\omega)} \right] \tag{7.59}$$

Then, using equation (7.53), we obtain

$$|e_i(\omega)| \leqslant q$$

and the overall relative error $e(\omega)$ in the frequency response is found to be bounded by:

$$|e(\omega)| \leqslant q \sum_{i=1}^{N/2} \frac{1}{|D_i(\omega)|} \tag{7.60}$$

This expression demonstrates the benefit of the decomposed structure as the bound of the error is proportional to

$$\sum_{i=1}^{N/2} \frac{1}{|D_i(\omega)|}$$

instead of (after equation (7.56)):

$$\prod_{i=1}^{N/2} \frac{1}{|D_i(\omega)|}$$

Further, the absolute value of the coefficients of the denominator cannot be greater than 2 if the filter is stable.

With the parallel structure, also in the pass band, the following equation is obtained:

$$H_R(\omega) \simeq \sum_{i=1}^{N/2} \frac{N_i'(\omega)}{D_i(\omega) + e_i(\omega)} \simeq \sum_{i=1}^{N/2} \frac{N_i'(\omega)}{D_i(\omega)} \left[1 - \frac{e_i(\omega)}{D_i(\omega)} \right]$$

Bearing in mind that the terms $|N'_2(\omega)/D_i(\omega)|$ are close to unity, the error in this case is approximately the same as that with the cascade structure.

In the stop band, away from the infinite attenuation frequencies, the following is obtained for the cascade structure:

$$H_R(\omega) \simeq \prod_{i=1}^{N/2} \frac{N_i(\omega) + e_i(\omega)}{D_i(\omega)} \simeq \frac{N(\omega)}{D(\omega)}\left[1 + \sum_{i=1}^{N/2} \frac{e_i(\omega)}{N_i(\omega)}\right] \tag{7.61}$$

For the parallel structure, the following estimate can be made:

$$H_R(\omega) \simeq \sum_{i=1}^{N/2} \frac{N'_i(\omega) + e_i(\omega)}{D_i(\omega)} = \sum_{i=1}^{N/2} \frac{N'_i(\omega)}{D_i(\omega)} + \sum_{i=1}^{N/2} \frac{e_i(\omega)}{D_i(\omega)}$$

or,

$$H_R(\omega) \simeq \frac{N(\omega)}{D(\omega)}\left[1 + \sum_{i=1}^{N/2} \frac{e_i(\omega)}{N_i(\omega)} \sum_{\substack{i=1 \\ j \neq i}}^{N/2} \frac{D_j(\omega)}{N_j(\omega)}\right] \tag{7.62}$$

By comparing equations (7.61) and (7.62) it appears that the terms

$$\alpha_j = \prod_{j=1}^{N/2} \frac{D_j(\omega)}{N_j(\omega)}$$

in the stop band can have values much greater than unity for example, in the neighbourhood of the zeros of the filter. The result is that in the stop band the parallel structure is more sensitive to rounding errors than the cascade structure.

Finally, the cascade structure allows the coefficients of the IIR filters to be represented using fewer bits. Thus, this structure is the one most frequently used.

7.7 COEFFICIENT WORDLENGTH FOR THE CASCADE STRUCTURE

In order to develop simple formulae for the number of bits needed to represent the coefficients of a low-pass filter realized by a cascade structure it is first necessary to restate some observations which have already been made earlier.

First, the transfer function $H(z)$ of an elliptic filter is

$$H(Z) = \sum_{i=1}^{N/2} a_0^i \frac{1 + a_1^i Z^{-1} + Z^{-2}}{1 + b_1^i Z^{-1} + b_2^i Z^{-2}} \tag{7.63}$$

From the results of Section 6.5, rounding of the coefficients a_1^i results only in a frequency displacement of the point of infinite attenuation of the filter. It is possible to allow for this in the calculation and so minimize the effect of rounding the coefficients of the numerator of the Z-transform function. Because of this, it is generally the denominator of the function $H(z)$ which determines the number of bits in the coefficients. The poles of this function should be inside the unit circle in the complex plane and, as shown in Section 6.5, limitation to

b_c bits corresponds to quantization of the coefficients with a quantum q such that:

$$q = 2^{2-b_c} \tag{7.64}$$

The number of bits b_c includes the sign. The resulting error in the frequency response is bounded by equation (7.60). In general, as shown at the beginning of this section, the error $e(\omega)$ is well below that bound and a statistical evaluation is needed to obtain a more realistic estimate. By analogy with the Gaussian distribution where the ratio of the peak value and the standard variation is about 4, we can write:

$$|e(\omega)| \simeq (q/4) \sum_{i=1}^{N/2} \frac{1}{|D_i(\omega)|} \tag{7.65}$$

The lowest values of the $|D_i(\omega)|$ terms are found in the pass band where the limit to be considered is δ_1. If δ_{10} denotes the amplitude of the filter ripples before limitation of the number of bits of the coefficients it is necessary that:

$$|e(\omega)| \leqslant \delta_1 - \delta_{10}$$

and consequently, allowing for equations (7.64) and (7.65), b_c has to be selected so that:

$$b_c \simeq \log_2 (1/(\delta_1 - \delta_{10})) + \log_2 \left(\max_{0 \leqslant \omega \leqslant \pi} \sum_{i=1}^{N/2} \frac{1}{|D_i(\omega)|} \right) \tag{7.66}$$

Of the $N/2$ second-order sections, the most important for this particular estimate is the one whose poles are closest to the unit circle and for which the gain at resonance has the highest value. Assuming r and θ are the polar co-ordinates of the poles of this section, using equation (6.29) we obtain:

$$H_m = \max_{0 \leqslant \omega \leqslant \pi} \frac{1}{|D_i(\omega)|} = \frac{1}{(1-r)} \frac{1}{(1+r)} \frac{1}{\sin \theta}$$

Moreover, the bandwidth at 3 decibels (B_3) of this section is given by equation (6.30a):

$$B_3 = ((1-r)/\pi) f_s$$

For high-selectivity filters in particular, there is a direct relationship between the bandwidth B_3 of this section and the transition band Δf of the filter, and so the following approximation can be made:

$$\Delta f \simeq 3B_3$$

and consequently

$$1 - r \simeq \Delta f / f_s \tag{7.67}$$

Allowing for the fact that it is generally possible to approximate θ by $2\pi f_1/f_s$, we obtain

$$H_m \simeq (f_s/\Delta f)1/(2\sin(2\pi f_1/f_s)) \qquad (7.68)$$

For the other sections, the amplitude of the resonance is clearly lower and the following supplementary approximation can be made:

$$\max_{0 \leqslant \omega \leqslant \pi} \sum_{i=1}^{N/2} \frac{1}{|D_i(\omega)|} \simeq \frac{f_s}{\Delta f} \frac{1}{2\sin(2\pi f_1/f_s)}$$

so that, finally,

$$b_c \simeq \log_2(1/(\delta_1 - \delta_{10})) + \log_2\left[\frac{f_s}{\Delta f} \frac{1}{2\sin(2\pi f_1/f_s)}\right] \qquad (7.69)$$

If the tolerance δ_1 is divided between the ripples of the filter before rounding the coefficients and the supplementary ripples caused by this rounding, $\delta_{10} = \delta_1/2$, we obtain:

$$b_c \simeq \log_2(1/\delta_1) + \log_2(f_s/\Delta f) + \log_2(1/\sin(2\pi f_1/f_s)) \qquad (7.70)$$

This equation should be compared with equation (5.46) for FIR filters. The normalized transition band contributes to the increase in b_c and IIR filters generally require a larger number of bits than FIR filters for representing the coefficients. Further, this number of bits increases as the width of the pass band decreases.

These estimates have been made under the assumption of limitation by rounding. The most important feature for the frequency response $H_R(\omega)$ of a filter is that of the parasitic polynomials $e_i(\omega)$ in the immediate neighbourhood of the frequencies which minimize the $D_i(\omega)$. This is shown by equation (7.59). In practice, configurations of quantized coefficients may be found which give better results than rounding. These configurations can be obtained, for example, by a systematic search in the neighbourhood of the rounding and can produce values of b_c which are significantly less than the estimate of equation (7.69).

7.8 ROUND-OFF NOISE

Another limitation occurs in the practical realization of IIR filters because of the limits on the capacity of the data memories. This limit is the origin of the round-off noise. It will be analysed for the $D-N$ structure, but the same arguments apply to the $N-D$ one. While this structure has specific advantages, mainly related to the calculation of partial sums and to the sequencing of the multiplications, it is the $D-N$ structure which is most often used because it is in general easier to design, construct and test.

In the presence of a signal, i.e. for non-zero values of $x(n)$, rounding before storage in the memory with quantization step q is equivalent to superposition

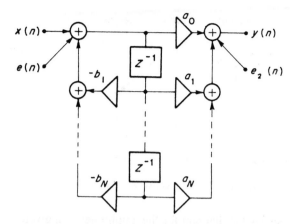

Fig. 7.17   Round-off errors in the D–N structure

on the input signal of an error signal $e(n)$ such that $|e(n)| < q/2$, assumed to have uniform spectrum and power so that $\sigma^2 = q^2/12$.

If other roundings are involved (for example, in the multiplications) it is apparent that the error signals produced are added either to the input or to the output signal depending upon whether they correspond to the coefficients of the recursive or non-recursive part, as shown in Figure 7.17. Consequently, in order to simplify the analysis only the case of single quantization is considered, since it is always possible to reduce to this case simply by modifying the power of the injected noise.

The error signal applied to the input of the filter undergoes the filtering function and, by applying equation (4.25), the power of the round-off noise at the output is

$$N_c = \frac{q^2}{12} \int_0^1 \left| \frac{N(f)}{D(f)} \right|^2 df \qquad (7.71)$$

or, as a function of the set $h(k)$, the impulse response of the filter,

$$N_c = \frac{q^2}{12} \sum_{k=0}^{\infty} |h(k)|^2 \qquad (7.72)$$

The implementation of the cascade structure presents some possibilities for reducing this noise power [11].

When the filter is realized as a cascade of $N/2$ second-order sections the round-off noise produced in each section undergoes the filtering function of that section and of the following ones. In this case it should be noted that the amplitude, or level, at the input of each section varies with the rank of the section and the frequency of the signal being considered.

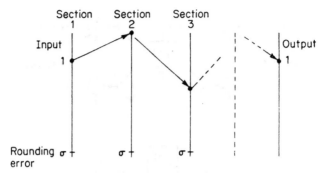

Fig. 7.18   The levels at one frequency in the pass band

Figure 7.18 shows the diagram for the amplitudes of a given frequency in the pass band such that the amplitude is 1 at the input and output of the filter. If the rounding procedure is the same for all sections the noise produced is the same and the contributions are added to each other. The total noise at the output of the filter under these conditions has a power of:

$$N_c = \frac{q^2}{12} \sum_{j=1}^{N/2} \left(\int_0^1 \prod_{l=j}^{N/2} \left| \frac{N_l(f)}{D_l(f)} \right|^2 \, df \right) \tag{7.73}$$

It is important to arrange the cascade of sections in such a way that the total round-off noise is minimized, and the parameters are available:

(1) The pairing of the poles and zeros to form a section;
(2) The order in which the sections are arranged;
(3) The scaling factor applied to each section.

These three parameters will be examined in turn.

(1) *The pairing of the poles and zeros.* The products

$$P_j(f) = \prod_{l=j}^{N/2} \left| \frac{N_l(f)}{D_l(f)} \right|^2$$

have to be minimized, which means that each of the factors is minimized and, in particular, the lowest maximal value for each factor must be obtained. This condition is approximately fulfilled by the very simple procedure of associating the pole nearest to the unit circle with its closest zero, then the next pole with the zero which is then closest, and so on, as shown in Figure 7.19.

(2) *Determining the order of the sections.* The factor making the largest contribution to the total noise is often the one which has the highest maximal value. It can be worthwhile to place it at the beginning of the chain so that its contribution appears only once in the total sum following equation (7.73) and to connect the sections in decreasing order of their maxima.

It should be noted, however, that equation (7.73) assumes that rounding is

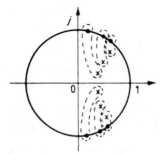

Fig. 7.19 Pairing of the poles and zeros in a cascade structure

performed with the same step in each section. This is not the case in practice because of the rescaling of the numbers in the internal memories. This involves the introduction of scale factors in order to avoid peak-limitation of the signals in each section. Strictly, determination of the order of the sections requires knowledge of these scale factors and the evaluation of the total round-off noise power in each particular case. The complexity of the problem for higher-order filters can be imagined. Also, the assumption that the sections are connected in decreasing order of their maxima is retained because it results in simple and generally realistic evaluations. It also involves the superposition of noise with a constant spectral distribution on the signal at the input of the filter. This can be important in practice, especially for narrow band filtering.

Thus, these two very simple rules of pairing and interconnection of the sections form an approximation which is sufficient for most practical applications.

(3) *Calculating the scale factors.* These are the parameters which control the scaling of the numbers in the internal data memories, and this is the final element needed for evaluating the round-off noise and for determining the capacity of the memories needed to satisfy the specifications.

The limitation of the capacity of the data memories means that the amplitude of the signal inside the filter is limited to a value A_m, which is a peak amplitude. That is, each number greater than A_m is limited to this value so that it can be stored in the memory. This results in harmonic distortion of the signal, which often is not tolerable, so methods of avoiding it have to be found.

The signal applied to the first section should have an amplitude such that, when multiplied by the gain of this element, it does not result in an unacceptable peak limitation of the signals. Consequently, the input signal must be multiplied by a preliminary factor a_0^0. The transfer function of the filter of order N realized in $N/2$ elements of the second order can be written, for an elliptic filter, in the form:

$$H(\omega) = a_0^0 \prod_{i=1}^{N/2} a_0^i \frac{N_i(\omega)}{D_i(\omega)}$$

with:

$$\frac{N_i(\omega)}{D_i(\omega)} = \frac{1 + a_1^i e^{-j\omega} + e^{-2j\omega}}{1 + b_1^i e^{-j\omega} + b_2^i e^{-2j\omega}}$$

Taking into account the hypothesis made earlier of single rounding in each section, equation (7.73) for the total round-off noise becomes:

$$N_c = \left(\frac{q^2}{12}\right) a_0^0 \sum_{j=1}^{N/2} \int_0^1 \sum_{i=j}^{N/2} a_0^i \left|\frac{N_i(f)}{D_i(f)}\right|^2 df \qquad (7.74)$$

Using the *D–N* structure, the arrangement of the filter is as shown in Figure 7.20, which also shows the points where the different powers of round-off noise e_i with $0 \leqslant i \leqslant N/2$ are introduced. The terms a_0^i $(0 \leqslant i \leqslant N/2)$ are scaling factors which must be chosen so as to avoid peak limitation of the signal in the sections.

To calculate the preliminary factor a_0^0 the response $1/D_1(\omega)$ of the first section must be considered. Also, the input numbers of the filter $x(n)$ are represented by a limited number of bits and have an assumed amplitude which is itself limited to A_m, i.e.:

$$|x(n)| \leqslant A_m$$

If $y_1(n)$ denotes the output numbers of the element with the transfer function $1/D_1(\omega)$ and if the set $h(k)$ is the impluse response, then

$$|y_1(n)| \leqslant \sum_{k=0}^{\infty} |h(k)| \, |a_0^0 x(n-k)| \leqslant a_0^0 A_m \sum_{k=0}^{\infty} |h(k)|$$

Using the results given in Section 6.2, for a section with complex poles of polar co-ordinates (r, θ), equation (6.37) holds:

$$\sum_{k=0}^{\infty} |h(k)| \leqslant \frac{1}{(1-r)\sin\theta}$$

By taking

$$a_0^0 = (1-r)\sin\theta \qquad (7.75)$$

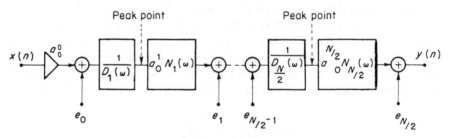

Fig. 7.20   *Arrangement of the cascade structure*

the following inequality is found:

$$|y(n)| \leqslant A_m$$

A less restrictive choice can be made for the factor a_0^0 by imposing the condition of absence of peak limitation to sinusoidal signals. In this case it is the value:

$$H_m^1 = \max \left| \frac{1}{D_1(\omega)} \right| = \frac{1}{(1-r^2)\sin\theta}$$

which has to be introduced and, under these conditions:

$$a_0^0 = (1-r^2)\sin\theta \qquad (7.76)$$

This value is less than twice the earlier one. A third scaling system is also used, which is based on the energy of the signals [12].

Using the results of Section 4.3, the energy of the signal $y_1(n)$ is written:

$$\sum_{n=-\infty}^{\infty} y_1^2(n) = \frac{1}{2\pi} \int_{-\pi}^{\pi} H_1(e^{j\omega})H_1(e^{-j\omega})(a_0^0)^2 X(e^{j\omega})X(e^{-j\omega})\,d\omega$$

$$\leqslant \left[\sum_{k=-\infty}^{\infty} x^2(k) \right](a_0^0)^2 \frac{1}{2\pi} \int_{-\pi}^{\pi} \frac{d\omega}{|D_1(\omega)|^2}$$

The condition:

$$(a_0^0)^2 \frac{1}{2\pi} \int_{-\pi}^{\pi} \frac{d\omega}{|D_1(\omega)|^2} = 1 \qquad (7.77)$$

results in the following equation for the factor a_0^0, based on equation (6.36) in Section 6.2:

$$(a_0^0) = \frac{(1-r^2)(1+r^4-2r^2\cos 2\theta)}{(1+r^2)}$$

Example: Assume $r = 0.845$; $\cos\theta = 0.396$ (Figure 7.6)

$$(1-r)\sin\theta = 0.142$$
$$(1-r^2)\sin\theta = 0.262$$
$$(\text{3rd case})\, a_0^0 = 0.644$$

Of the three scaling systems, which correspond to the L_1, L_∞ and L_2 norms of the signals, the second forms a good compromise. This is the method which will be used throughout the rest of this section. It should also be noted that it facilitates experimental verification and development using sinusoidal signals.

The quantities H^i_m and K^i_m can be defined for the section of rank i by

$$H^i_m = \max \left| \frac{1}{D_i(\omega)} \right|;$$

$$K^i_m = \max \left| \frac{a^i_0 N_i(\omega)}{D_i(\omega)} \right|; \quad \text{with} \quad 1 \leqslant i \leqslant \frac{N}{2}$$

With the chosen scaling system:

$$a^0_0 = \frac{1}{H^1_m}$$

and for sections with complex poles, using equation (6.28):

$$a^0_0 = (1 - b^1_2) \bigg/ \sqrt{\left(1 - \frac{(b^1_1)^2}{4b^1_2} \right)}$$

There is no peak limitation of the sinusoidal wave in the second section if the following inequality is satisfied:

$$\frac{1}{H^1_m} K^1_m H^2_m \leqslant 1$$

which leads to choosing:

$$K^1_m = \frac{H^1_m}{H^2_m}$$

Using equations (6.27) and (6.40), K^1_m can often be approximated by the expression:

$$K^1_m \simeq a^1_0 \left[a^1_1 - \frac{b^1_1(1 + b^1_2)}{2b^1_2} \right] H^1_m$$

and it follows that:

$$a^1_0 \simeq \frac{(1 - b^2_2)\sqrt{(1 - (b^2_1)^2/4b^2_2)}}{a^1_1 - (b^1_1(1 + b^1_2)/2b^1_2)}$$

A simpler estimate of a^1_0 can be obtained by considering the inequality:

$$\left| \frac{1}{H^1_m} \frac{1}{D_1(\omega)} N_1(\omega) \frac{1}{D_2(\omega)} \right| \leq \left| \frac{N_1(\omega)}{D_2(\omega)} \right|$$

For frequencies close to the value which minimizes $D_2(\omega)$, one often has $|N_1(\omega) < 1|$; this can easily be verified by the position of the poles and zeros. One can then take $a^1_0 = 1/H^2_m$.

The scale factors of the subsequent sections are determined in the same way, so that finally, the process is as follows:

(1) Apply the preliminary factor $a_0^0 = 1/H_m^1$ to the input set.
(2) Assign to the element of rank i $(1 \leqslant i \leqslant N/2 - 1)$ the scale factor a_0^i such that:

$$K_m^i = \frac{H_m^i}{H_m^{i+1}}$$

The approximation can often be made:

$$a_0^i = \frac{1}{H_m^{i+1}} \tag{7.78}$$

With this approximation it is necessary to verify that the values obtained for the later scaling factors are not too small, in order not to increase calculation errors unnecessarily.

(3) Calculate the scaling factor $a_0^{N/2}$ of the last section so that the filter has a gain of exactly 1 at a reference frequency in the pass band.

In practice, of the $(N/2 + 1)$ values of a_0^i, $N/2$ are taken as powers of two, so that multiplication is a simple shift operation.

The simple procedure described above is sufficient in most practical cases. A more precise method consists of allowing for all preceding sections when determining a scaling factor. Finally, it is found that the scaling factor of the section with rank i is a function of the sections which precede it, while those which follow it filter the noise that it produces.

Once the scale factors are known, all the elements involved in the implementation are available and the power of the rounding noise at the output of the filter can be determined for each value of the number of bits in the internal data memories.

7.9 DETERMINING INTERNAL DATA WORDLENGTH

At the filter input the signal is composed of $x(n)$ numbers which are represented by a limited number of bits b_d. This signal can be regarded as consisting of the ideal signal with a superimposed noise of power N_1. The numbers $x(n)$ are assumed to take positive and negative values and are limited in absolute value to 1, i.e.:

$$|x(n)| \leqslant 1$$

The representation of signals with b_d bits corresponds to the quantization step q in such a way that

$$q = 2^{-b_d + 1}$$

The noise power N_1 superimposed on the input signal is assumed to be k_0 ($k_0 \geqslant 1$) times the noise power generated by the quantization with step q:

$$N_1 = k_0 \frac{2^{-2b_d}}{3} \tag{7.79}$$

In general, the round-off noise, that is, the noise added by the filter, should be compared with the noise already present at the input, which means that it represents a degradation in the signal-to-noise ratio while crossing the filter.

As explained above, whatever the precision of the arithmetic operations, a rounding operation must be performed on the internal data before they are stored in the memory. This is represented by the superposition of an error $e(n)$ on the set representing the signal. As a consequence it appears as a degradation of the signal-to-noise ratio across the filter.

In a cascade-structure filter the noise caused by the limitation of internal data to b_i bits is assumed to be applied at the input of each second-order section. In order to allow for the method of obtaining this limitation it is assumed that the noise power created is equal to k times ($k \geqslant 1$) the noise power caused by quantization to b_i bits. In effect, this limitation can be a unique rounding, as in Figure 7.17, or a rounding after each multiplication, as is often the case in practice.

The determination of the capacity of the internal memories, and hence of the number of bits of the data b_i, should allow for the range of variation in the amplitudes. If, in the cascade of second-order sections, the first section is the one with the highest gain at the resonance H_m^1, then the signal should be divided by H_m^1 at the input of the filter in order to guarantee the absence of peak limitation. For sinusoidal signals at least, this factor H_m^1 is given by equation (7.47). As a result, the first section introduces a degradation SN in the signal-to-noise ratio which is expressed in decibels by:

$$\Delta SN = 10 \log \left[1 + (H_m^1)^2 (k/k_0) 2^{-2(b_i - b_d)} \right] \tag{7.80}$$

The degradation introduced by the other sections, which have a lower maximum amplitude, is generally less important and can be neglected. It follows that the capacity of the memories can be determined from equation (7.80). If the degradation ΔSN is to be low, then:

$$b_i \simeq b_d + \tfrac{1}{2} \log_2 \left[4.3 (k/k_0)(H_m^1)^2 / \Delta SN \right] \tag{7.81}$$

Allowing for equation (7.68), by assuming $k = k_0$, we obtain:

$$b_i \simeq b_d + \tfrac{1}{2} \log_2 (1/\Delta SN) + \log_2 \left[\frac{f_s}{\Delta f} \frac{1}{\sin(2\pi f_1/f_s)} \right] \tag{7.82}$$

In this equation the degradation of the signal-to-noise ratio ΔSN is expressed in decibels.

This expression should be compared with equation (5.53) for FIR filters. Very selective narrow band filters require more bits for the internal memories.

It is important to emphasize that the evaluations above, particularly equations (7.80) and (7.81), are based upon a simplified approach. In a more detailed study the degradation ΔSN of the signal-to-noise ratio across the filter must be calculated by accurately identifying and allowing for all the noise sources, and, as a consequence, by modifying equation (7.74) so as to determine the round-off noise of the whole calculation. Further, the signal power at the exit of the filter must be allowed for as it is at this point that the noise is calculated. Extension of the design of the filter to include overall optimization of the pairing of poles and zeros, the order of the sections and the scale factors can only rarely be justified.

The two following examples are used to illustrate the results which have been given in this and the preceding section.

Example 1: Consider the filter of the fourth order given in Section 7.2.4 (Figure 7.6):

$$H_m^1 = 3.8; \quad H_m^2 = 2.2$$

Select:

$$a_0^0 = 0.25; \quad a_0^1 = 0.5 \quad \text{and} \quad a_0^2 = 0.394$$

The transfer function eventually derived is:

$$H(Z) = 0.25 \frac{1 + 0.5974Z^{-1} + Z^{-2}}{1 - 0.670Z^{-1} + 0.7144Z^{-2}}$$

$$\times 0.5 \frac{1 + 1.632Z^{-1} + Z^{-2}}{1 - 0.814Z^{-1} + 0.2636Z^{-2}} \times 0.394$$

The diagram of this filter is given in Figure 7.21.

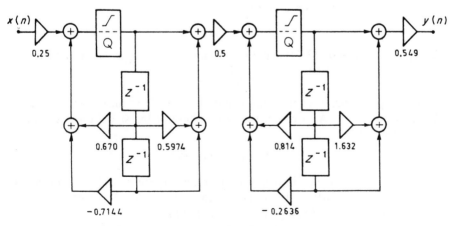

Fig. 7.21   *Arrangement of an elliptic filter of order 4*

If the input numbers are supplied by a 12-bit analogue-to-digital converter, and if the internal memories have a capacity of 16 bits, the degradation of the signal-to-noise ratio with $k = 2$ and $k_0 = 1$ is estimated using equation (7.79) as:

$$\Delta SN = 10 \log (1 + \tfrac{1}{8}) \simeq 0.5 \, \text{dB}$$

Example 2: Assume a very selective low-pass filter of order 10 with the following coefficients:

$$a_1^1 = -1.41956 \quad b_1^1 = -1.50269 \quad b_2^1 = 0.98242$$
$$a_1^2 = -1.37231 \quad b_1^2 = -1.49805 \quad b_2^2 = 0.93652$$
$$a_1^3 = -1.22241 \quad b_1^3 = -1.50915 \quad b_2^3 = 0.85767$$
$$a_1^4 = -0.73120 \quad b_1^4 = -1.53308 \quad b_2^4 = 0.73730$$
$$a_1^5 = 1.07660 \quad b_1^5 = -1.55640 \quad b_2^5 = 0.62646$$

and for which the mask has the following corresponding parameters (Figure 5.7):

$$\delta_1 = 0.01; \quad \delta_2 = 0.0002; \quad f_1 = 0.112; \quad f_2 = 0.117$$

Then:

$$H_m^1 = 87; \quad H_m^2 = 23; \quad H_m^3 = 12; \quad H_m^4 = 9; \quad H_m^5 = 15$$

Rounding after multiplication by a_0^0, b_1^1, and b_2^1 leads to taking $k = 3$ in estimate (7.80), which, for $k_0 = 1$ and $b_i = b_d$, results in $\Delta SN = 43.5 \, \text{dB}$.

This estimate can be compared with the optimum solution obtained by dynamic programming [13] which gives the following scale factors:

$$a_0^0 = 0.11838; \quad a_0^1 = 0.45112; \quad a_0^2 = 0.14724$$
$$a_0^3 = 0.70252; \quad a_0^4 = 0.43834; \quad a_0^5 = 0.38293$$

By using Z_i to denote the zero which corresponds to the coefficient a_1^i and by using P_i to denote the pole which corresponds to b_2^i ($1 \leq i \leq 5$), the pairings and the order of the sections are as follows:

$$P_4 - Z_2; \quad P_2 - Z_4; \quad P_5 - Z_1; \quad P_1 - Z_3; \quad P_3 - Z_5$$

By applying equation (7.74) and allowing for all noise sources, we obtain:

$$N_c = (q^2/12) \times 731$$

If the input signal has a uniform spectrum, its power at the output is reduced by the factor f_1/f_s. As a result, the degradation of the signal-to-noise power across the filter is

$$(\Delta SN)_{\text{opt}} = 34.7 \, \text{dB}$$

With respect to the simplified approach for this very selective filter, optimization brings a gain of 8.8 dB to the round-off noise, or, expressed as the number of bits of internal memories, a gain of less than 2 bits.

7.10 AUTO-OSCILLATIONS

Even when there is no input signal to the IIR filter, limiting the number of data memory bits can still result in the appearance of low- and high-amplitude auto-oscillations.

Auto-oscillations can be produced at large amplitudes by exceeding the capacity of the memories. The equation for the system can then be written:

$$y(n) + \sum_{i=1}^{N} b_i y(n-i) = 0 \tag{7.83}$$

The condition for the natural absence of such phenomena is expressed by the inequality:

$$\left| \sum_{i=1}^{N} b_i y(n-i) \right| < 1$$

For filters of order greater than 2 it can be shown that the presence of a logic saturation device is not sufficient to ensure the absence of large-amplitude auto-oscillations [14]. In contrast, a filter composed of a cascade of second-order elements with a logic saturation device does not allow for this possibility.

The low-amplitude auto-oscillations or limit cycles produced by a section are filtered by the subsequent sections, for which the input signal is not zero.

If the strategy of interconnecting the sections in order of decreasing maxima is applied, when there is no input signal to the filter, the first section can produce a limit cycle at a frequency near the resonance frequency, that is, at the limit of the pass band for a very selective filter. In fact, the auto-oscillation corresponds to the insertion into the chain in Figure 7.9 of a parasitic signal e_0 whose amplitude is limited by the quantization as indicated in Section 6.7. The amplitude A_a of the auto-oscillation at the output of a filter of unit gain and in the $D-N$ cascade structure can also be estimated by

$$A_a \simeq H_m^1 2^{-b_i}$$

where, as before, b_i denotes the number of bits of the internal memories.

Using equation (7.68), we obtain:

$$A_a \simeq 2^{b_i-1} (f_s/\Delta f)(1/(\sin(2\pi f_1/f_s))) \tag{7.84}$$

This expression provides a relation between the amplitude of the auto-oscillations and the filter characteristics for this particular realization.

Other methods of realization for example, the interconnection of sections in a different order can result in lower values for the amplitude and these parasitic signals.

7.11 COMPARISON OF IIR AND FIR FILTERS

Both IIR and FIR filters allow any given specification to be satisfied and a systems designer often has to choose between the two approaches. The criterion is the complexity of the circuits to be employed. In practice, as will be seen in a later chapter, the comparison is reduced mainly to evaluating a single parameter, the number of multiplications to be carried out.

Equations (5.32) and (7.29) give estimates for FIR and IRR filters of the order N, which is needed to satisfy the specifications of low-pass filtering as expressed by the ripple in the pass and stop bands and by the widths of the pass band and the transition band.

For an FIR linear phase filter the coefficients are symmetric and, if N is even, $N/2$ have different values. Such a filter, therefore, requires $N/2$ memories for the coefficients and N memories for the internal data. For each output number $N/2$ multiplications and N additions have to be made. If phase linearity is not imposed the order N can be reduced with minimum phase filters, as shown by equation (5.62). This reduction depends on the ripple in the pass band and is less than 50%. There is a consequent increase in the number of multiplications as the symmetry of the coefficients disappears. Amongst the advantages of FIR filters it should be stressed that they are always stable and they are not difficult to implement.

IIR filters are more difficult to achieve except for the particular case of the elliptic type, which is the most efficient and the most often used. The estimate of equation (7.29) shows that the order of the filter is a maximum near the frequency $f_s/4$, as shown in Figure 7.22.

Let n be the filter order. By assuming that the filter is formed of second-order sections, each comprising four coefficients and two data memories (Section 6.4), realization of a filter with n data memories requires $2n$ memories for the coefficients, and each output number involves $2n$ additions and $2n$ multiplications.

If comparison between FIR and IIR filters is limited to the number of multiplications to be carried out to produce an output number, then, in the case of the low-pass filter, the IIR type is more advantageous than the FIR one for values of the parameters such that:

$$N > 4n \qquad (7.85)$$

Following equations (5.32) and (7.29) it can be seen that the transition band is the most crucial parameter in the comparison, and similar numbers of multiplications are obtained in the most adverse conditions for the IIR filter if the transition band is given by:

$$\tfrac{1}{3}f_s/\Delta f \simeq 2\log(f_s/\Delta f)$$

that is, $\Delta f \simeq f_s/3$. It follows that the inequality (7.85) is satisfied as soon as Δf is smaller than $f_s/3$. This is the case in the great majority of applications. For example, for parameter values corresponding to Figure 7.22 this inequality is always valid.

Linearity in the phase can be approximated by an IIR filter over a limited frequency band by completing, for example, the basic elliptic filter with a group

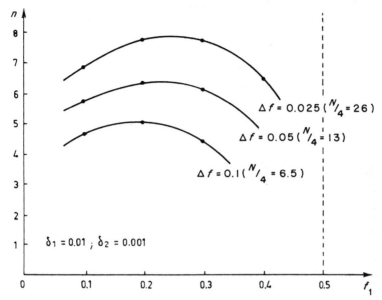

Fig. 7.22 Order of an elliptic low-pass filter as a function of the width of the pass band

delay equalizer, made of phase shifter sections whose properties have been given in Section 6.3. Experience has shown that the FIR filter, which exhibits perfectly linear phase, always requires fewer calculations [15]. It is also easy to implement

Finally, it is recommended that FIR filters are used when linear phase is required and that IIR filters are used in other cases.

Nevertheless, the comparison above has been made with the implicit hypothesis that the sampling rate is the same at the input and output of the filter. The bases for the comparison are noticeably modified if this constraint disappears, as will be shown in chapter 10 on multirate filtering.

APPENDIX

Program to determine the coefficients of an elliptic filter (CERII)

```
C     PROGRAM CERII (CALCULATION OF THE COEFFICIENTS OF AN
C     ELLIPTIC FILTER)
C
C     PASS-BAND ATTENUATION IN DB:BP
C     STOP-BAND ATTENUATION IN DB:AT
C     SAMPLING FREQUENCY IN HZ:FE
C     END FREQUENCY OF PASS-BAND:FB
C     START FREQUENCY OF STOP BAND:FA
      IMP=7
      PI=3.1415926535
      FE=8000.
      AT=45.
      BP=0.2
      FB=1700.
      FA=2000.
      EPS=SQRT(10.**(BP/10.)-1.)
      A=10.**(AT/20.)
      T=1./FE
      WC=FB*2.*PI
      WR=FA*1.*PI
      WRITE(IMP, 103)
103   FORMAT(1H1,//,79(1H*),//)
      WRITE(IMP, 102)BP,. AT, FB, FA, FE
102   FORMAT(2X,33H PASS BAND OSCILLATIONS=,F10.4,4H DB
     12X, ATTENUATION IN STOP BAND=',F10.2,4H DB/
     22X,25H CRITICAL FREQUENCIES:,3(F15.2,3X)/)
      CALL ELLI(EPS,A,WR,WC,FE)
83    CONTINUE
84    STOP
      END
```

```
      SUBROUTINE ELLI(EPS.A.WR.WC.FE)
C     THE SUB-PROGRAM CALCULATES SUCCESSIVELY THE ORDER
C     N, THE POLES AND ZEROS IN THE PLANE OF LAMBDA,
C     OF S, THEN OF Z, FINALLY GIVES THE COEFFICIENTS
C     A(I),B(I),C(I) AND D(I).
C     THE COMPLETE AND INCOMPLETE INTEGRAL FUNCTIONS
C     ARE CALCULATED IN THE SUB-PROGRAM, AS WELL AS THE
C     TRANSFER FUNCTION OF THE SECTION
      COMPLEX ZERO (16),POLE(16)
      DIMENSION C(16), D(24), CT (16), DT (24)
      IMP=7
      WRITE(IMP, 101)
101   FORMAT(///,79(1H*),//,20X,'ELLIPTIC FILTER'///)
112   FORMAT(9X, 'Z PLANE',//9X,
     1'FUNCTION=(1+BZ-1+AZ-2)(1+DZ-1+CZ-2)'///
     212X,'..X..',14X,'..Y..',15X,'A=X**2+Y**2'/
     351X,'D=-2X', 8X,'C=X**2+Y**2'//)
108   FORMAT(2X,5HZERO=,2(E15.8,3X),'B=',E15.8,2X,'A=',E15.8)
104   FORMAT(5X,'THE MAXIMUM NUMBER OF POLES IS 16'///)
      T=1/FE
      PI=3.1415926535
      DK1=EPS/SQRT(A**2-1.)
      DK=TAN(WC*T/2.)/TAN(WR*T/2.)
      DKP=SQRT(1-DK**2)
      AK1=AK(DK)
      AK2=AK(DK1)
      AK3=AK(DKP)
      AK4=AK(SQRT(1.-DK1**2))
      N=(AK4*AK1)/(AK2*AK3)
      N=(N/2)+1
      NDEU=2*N
      IF(N.LE.16)GO TO 20
      WRITE(IMP,104)
      RETURN
20    U0=(-AK3/AK4)*ALOG((1.+SQRT(1.+EPS**2))/EPS)
```

```fortran
      DO 1 I=1,N
      XMAG=(2*I-1)*AK1/NDEU
      ZERO(I)=CMPLX(-AK3,XMAG)
      POLE(I)=CMPLX(U0,XMAG)
1     CONTINUE
      DO 4 I=1,NDEU
      IF (I.LE.N) GO TO 5
      Q=REAL(POLE(I=N))
      R=AIMAG(POLE(I-N))
      GO TO 6
5     Q=REAL(ZERO(I))
      R=AIMAG(ZERO(I))
      SIGMA=0.
6     A1=SN(Q,DKP,AK3,AK1)
C     A1=SN(Q,KPRIM)
C     B1=SN(R,K)
      B1=SN(R,DK,AK1,AK3)
      DN=SQRT(1-DK*B1)**2)
      DE=1-(A1*DN)**2
      IF(I.LE.N) GO TO 8
      SIGMA=A1*SQRT((1-A1**2)*(1-B1**2))*DN/DE
8     OMEGA=B1*SQRT(1-(DKP*A1)**2)/DE
      C(I)=-2*SIGMA*WC
      D(I)=(SIGMA**2+OMEGA**2)*WC*WC
      SIGMA=SIGMA*TAN(WC*T/2.)
      OMEGA=OMEGA*TAN(WC*T/2.)
      IF(I.LE.N) GO TO 7
      POLE(I-N)=CMPLX(SIGMA,OMEGA)
      GO TO 4
7     ZERO(I)=CMPLX(SIGMA,OMEGA)
4     CONTINUE
      WRITE(IMP,112)
      DO 10 I=NDEU,1,-1
      J=I
      IF(J.LE.NDEU/2) GOTO 22
      J=J-NDEU/2
```

```
22        WRITE(IMP,106)N-J+1
          IF(I.LE.N) GO TO 11
          X=REAL(POLE(I-N))
          Y=AIMAG(POLE(I-N))
          GO TO 12
11        X=REAL(ZERO(I))
          X=0
          Y=AIMAG(ZERO(I))
12        RE=(1-X*X-Y*Y)/((1.-X)**2+Y*Y)
          V=2*Y/((1.-X)**2+Y**2)
          C1=-2*RE
          D1=RE*RE+V*V
          IF(I.LE.N) GO TO 15
          POLE(I-N)=CMPLX(RE,V)
          WRITE(IMP,109) POLE(I-N),C1,D1
          CT(I)=C1
          DT(I)=D1
          GO TO 10

15        ZERO(I)=CMPLX(RE,V)
          WRITE(IMP,108) ZERO(I),C1,D1
          DT(16+I)=C1
10        CONTINUE
          WRITE(IMP,107)
107       FORMAT(//,79(1H*),//)
          DO 13 I=N,1,-1
          CT(3)=CT(I+N)
          DT(3)=DT(I+N)
          DT(24)=DT(16+I)
          CT(1)=C(I)
          DT(1)=D(I)
          CT(2)=C(I+N)
```

```
          DT(2)=D(I+N)
          JJ=1
          WRITE(IMP,106) N-I+1
106       FORMAT(/9X,' SECTION NUMBER',I2)
13        CALL TRANSF(CT,DT,FE,JJ,WC)
23        CONTINUE
          WRITE(IMP,107)
          RETURN
          END
          FUNCTION AK(Y)
C         AK(Y)=SUM OF 0 A PI/2 DE (1-Y*SIN(T)**2)**-1/2
          X=1-Y**2
          AK=1.38629436112+0.09666344259*X+0.03590092383*X**X+
         10.03742563713*X**3+0.01451196212*X**4-ALOG(X)*(0.5+
         20.12498959397*X+0.06880248576*X*X+0.03328355346*
         3X**3+0.00441787012*X**4)
          RETURN
          END
          FUNCTION SN(Y,A,AK1,AK3)
C         FUNCTION SN(Y,A)  AK1=K(K)  AK3=KPRIM(K)
          PI=3.1415926535
          NS=SQRT(50.*AK1/(PI*AK3))+2
          X=Y*0.5/AK1
          Q=EXP(-PI*AK3/AK1)
          SUP=2.*(Q**(1./4.))*SIN(PI*X)
          DEN=1.
          I=-2
          N=1
1         SUP=SUP+ I*(Q**((N+0.5)**2))*SIN((2*N+1)*PI*X)
          DEN=DEN+I*(Q**(N**2))*COS(2.*N*PI*X)
          I=-I
          N=N+1
          IF(N.LE.NS) GO TO 1
          SN=SUP/(DEN*SQRT(A))
          RETURN
          END

          W1=W1/2.
          W=WC*TAN(W1*0.5/FE)/TAN(WC*0.5/FE
          FONCT=1
          DO 2 I=1, N
          FONCT=FONCT*( (D(I)-W*W)**2+
          (C(I)*W)**2)/((D(I+N)-W*W)**2
         1+ (C(I+N)*W)**2)
          FONCT=FONCT*(D(I+N)**2)/(D(I)**2)
2         CONTINUE
          X=1+C(3)+D(3)
          Y=C(3)-D(3)-1
          Z=(1-D(3))**2
          S=TAN(W1*0.5/FE)
          FONP=1.
          IF(N.NE.1) GO TO 4
          FONP=FONP*((1+S*S)**2 )/
          ((X+S*S*Y)**2+4.*S*S*Z)
          FONP=FONP*X*X/(2.+D(24))**2
4         F=FE/PI*ATAN(W/WC*TAN(WC/(2.*FE)))
          FONCT=10*ALOG10(FONCT)
          FONP=10.*ALOG10(FONP)
          IF(N.EQ.1) WRITE(IMP,3)F,FONCT
          IF(N.NE.1) WRITE(IMP,3)F,FONCT
3         FORMAT(24X,F8.2,5X,F8.2)
1         CONTINUE
          RETURN
          END
```

```
      SUBROUTINE TRANSF (C,D,FE,N,WC)
C     THE SUBROUTINE CALCULATES THE TRANSFER FUNCTION FOR A FILTER OF N SECTIONS
      DIMENSION C(16),D(24)
      IMP=7
      WRITE(IMP,5)
5     FORMAT(/,15X,'TRANSFER FUNCTION'/,17X,'FREQUENCY IN HZ',
     15X,'GAIN IN DB')
      DO1NF=1,15
      PI=3.1415926535
      W1=3.14159265*NF*FE/8
```

PASS BAND OSCILLATIONS=0.2000 DB
ATTENUATION IN STOP BAND=45.00 DB
CRITICAL FREQUENCIES: 1700.00 2000.00 8000.00

ELLIPTIC FILTER
Z PLANE
FUNCTION=(1+BZ-1+AZ-2)/(1+DZ-1+CZ-2)
..X..   ..Y..

SECTION NUMBER 1			
POLE=0.1958857E+00	0.92604369E+00	D=-0.39177147E+00	C=0.89592814E+00
SECTION NUMBER 2			
POLE=0.26905876E+00	0.73609573E+00	D=-0.53811753E+00	C=0.61422956E+00
SECTION NUMBER 3			
POLE=0.39402664E+00	0.30706793E+00	D=-0.78805327E+00	C=0.24954771E+00
SECTION NUMBER 4			
ZERO=-0.21079002E-01	0.99977779E+00	B=0.42158004E-01	A=0.10000000E+01
SECTION NUMBER 2			
ZERO=-0.23525374E+00	0.97193396E+00	B=0.47050747E+00	A=0.99999994E+00
SECTION NUMBER 3			
ZERO=-0.81463736E+00	0.57997066E+00	B=0.16292747E+01	A=0.10000000E+01

B=-2X   A=X**2+Y**2
D=-2X   C=X**2+Y**2

TRANSFER FUNCTION

FREQUENCY IN HZ	SECTION NUMBER 1 GAIN IN DB	SECTION NUMBER 2 GAIN IN DB	SECTION NUMBER 3 GAIN IN DB	FILTER GAIN IN DB
250.00	0.05	0.10	-0.09	0.06
500.00	0.20	0.40	-0.45	0.15
750.00	0.50	0.96	-1.31	0.15
1000.00	1.04	1.87	-2.88	0.03
1250.00	2.10	3.12	-5.16	0.06
1500.00	4.67	3.37	-7.94	0.10
1750.00	9.69	-1.38	-11.04	-2.73
2000.00	-22.32	-10.19	-14.41	-46.92
2250.00	-9.54	-28.52	-18.11	-56.17
2500.00	-6.47	-19.48	-22.37	-48.32
2750.00	-5.29	-14.43	-27.75	-47.47
3000.00	-4.69	-12.34	-36.14	-53.17
3250.00	-4.35	-11.23	-52.82	-68.40
3500.00	-4.15	-10.61	-36.99	-51.75
3750.00	-4.04	-10.29	-33.58	-47.91

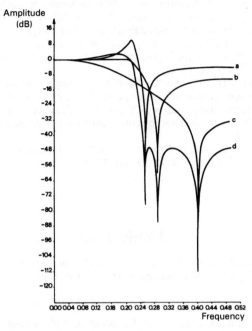

Fig. 7.23   *Frequency response of a sixth-order elliptic filter: (a, b, c) response of sections 1, 2 and 3; (d) response of the whole filter*

REFERENCES

[1] A. OPPENHEIM and R. SCHAFER, *Digital signal processing*, Prentice Hall, Englewood Cliffs NJ, 1974, Chs 5 and 9.

[2] L. RABINER and B. GOLD, *Theory and application of digital signal processing*. Prentice Hall, Englewood Cliffs NJ, 1975, Chs 4 and 5.

[3] R. BOITE et H. LEICH, *Les filtres numériques: analyse et synthèse des filtres unidimensionnels*. Masson, Paris, 1980.

[4] B. GOLD and C. RADER, *Digital processing of signals*, McGraw-Hill, New York, 1969.

[5] A. G. CONSTANTINIDES, Spectral transformation for digital filters. *Proceedings of the IEEE*, **8**, 1970.

[6] A. DECZKY, Synthesis of recursive digital filters using the minimum p-error criterion. *IEEE Transactions on Audio and Electroacoustics*, **20**, Oct 1972.

[7] R. FLETCHER and M. J. D. POWELL, A rapidly convergent descent method for minimization. *Computer Journal*, **6**(2), 1963.

[8] J. P. THIRAN, Equal ripple delay recursive filters. *IEEE Transactions on Circuit Theory*, Nov 1971.

[9] T. DURRANI and R. CHAPMAN, Optimal all-pole filter design based on discrete prolate spheroidal sequences. *IEEE Transactions*, **ASSP32**(4), 716–21, 1984.

[10] P. P. VAIDYANATHAN, S. K. MITRA and Y. NEUVO, A new approach to the realisation of low sensitivity IIR filters. *IEEE Transactions*, **ASSP34**(2), 350–61, 1986.

[11] L. B. JACKSON, Round-off noise analysis for fixed point digital filters in cascade or parallel form. *IEEE Transactions on Audio and Electroacoustics*, June 1970.

[12] A. PELED and B. LIU, *Digital signal processing: theory, design and implementation*, John Wiley, New York, 1976.

[13] Von E. LUEDER, H. HUG and W. WOLF, Minimizing the round-off noise in digital filters by dynamic programming. *Frequenz*, **29**(7), 211–14, 1975.

[14] D. MITRA, Large-amplitude, self-sustained oscillations in difference equations, which describe digital filter sections using saturation arithmetic. *IEEE Transactions*, **ASSP25**(2), 1977.

[15] L. RABINER, J. F. KAISER, O. HERRMANN and M. DOLAN, Some comparison between FIR and IIR digital filters. *BSTJ*, **53**, Feb 1974.

EXERCISES

1 Using the formulae given in Section 7.1, calculate the frequency and phase response and the group delay of the filter section defined by the relation:

$$y(n) = x(n) + 0.7x(n-1) + 0.9y(n-1)$$

Using the same formulae, calculate the frequency and phase response and the group delay of the second-order section with the Z-transfer function:

$$H(Z) = \frac{b_2 + b_1 Z^{-1} + Z^{-2}}{1 + b_1 Z^{-1} + b_2 Z^{-2}}$$

2 It is proposed to use charts of analogue filters in order to calculate a digital band-pass filter. What specifications should be used in order that the digital filter should reject the signal components in the bands (0–0.15) and (0.37–0.5) and should not show attenuation in the band (0.2–0.33), assuming $f_s = 1$?

3 Calculate the coefficients of a Butterworth filter of order 4 whose amplitude has the value $2^{-1/2}$ at the frequency $f_c = 0.25$. Give the decomposition into second-order sections.

4 Use a frequency transformation to transform the band-pass filter in Section 7.2.4 (Figure 7.6) into a high-pass filter with a pass-band limit of $f_H = 0.4$. How are the poles and zeros changed in this operation?

5 Give the decomposition into second-order sections of the filter in Figure 7.10. Calculate the scaling factors of the sections and the internal data wordlength if the round-off noise added by the filter stays below a power of $\frac{1}{10}$ of the noise present at the input and if the numbers at the input have 10 bits.

6 The specification in Figure 7.10 is increased by 0.1 dB in order to permit a rounding of the filter coefficients. How many bits are required to represent the coefficients in the cascade structure? Estimate how many bits are required to represent the coefficients for the parallel structure. Find an optimum for rounding the coefficients. Can the number of bits found earlier be reduced?

7 Does the filter given in Section 7.2.3 exhibit auto-oscillations? What are the frequencies and the amplitudes?

Does the filter given in Figure 7.21 exhibit auto-oscillations? What are the frequencies and the amplitudes?

8 How many multiplications are required by the filter in Figure 7.6? How many memory locations are necessary? How many coefficients are required by an FIR filter for the same specification. Compare the number of multiplications and the memory capacities.

9 It is desired to achieve the channel filtering function in a PCM transmission terminal by digital methods. The telephone signal is sampled at 32 KHz and coded into 12 bits, and the filtering is carried out by a low-pass IIR filter. The pass band is 3300 Hz, and the stop and begins at 4600 Hz.

The ripples in the pass and stop bands have value of:

$$\delta_1 \leqslant 0.015; \quad \delta_2 \leqslant 0.04$$

A computer program for elliptic filters produces the following results:

order of filter: $N = 4$

zeros: $z_1 = 0.09896 \pm j0.995$

$z_2 = 0.5827 \pm j0.8127$

poles: $P_1 = 0.6192 \pm j0.2672$

$P_2 = 0.702 \pm j0.589$

Calculate the transfer function of the filter decomposed into second-order sections.

What is the value of the overall scaling factor, knowing that the amplitude at frequency 0 is 0.99?

The coefficients are quantized into 10 bits. Determine the displacement of the infinite attenuation frequencies and evaluate the additional pass-band ripple.

Calculate the scaling factor to be assigned to each section and estimate the round-off noise produced if the data memories have 16 bits.

Give the complete diagram for the filter.

Evaluate the complexity in terms of:

the number of multiplications and additions per second,
the number of memory bits.

10 The following specifications are given for a low-pass filter:

$$\delta_1 = 0.01; \quad \delta_2 = 0.01; \quad f_1 = 1700\,\text{Hz}; \quad f_2 = 2000\,\text{Hz}; \quad \text{and} \quad f_s = 8000\,\text{Hz}$$

Calculate the filter order. By taking the order $N = 6$, a margin on the in-band ripple is available. Determine the coefficient wordlength.

If the signal-to-noise ratio degradation is limited to 0.1 dB, give the increase of the internal data wordlength with respect to the input data wordlength.

CHAPTER 8

DIGITAL LADDER FILTERS

The filter structures presented in the previous chapters are deduced directly from their Z-transfer functions with the coefficients applied to the multiplying circuits being those of the powers of Z^{-1}. More elaborate structures can be developed.

In analogue filtering structures exist which allow filters with very low ripple and excellent selectivity to be constructed using passive components of limited precision. In digital filtering these properties can be translated into a reduction in the round-off noise and in the number of bits representing the coefficients.

Analogue filter networks are based on cascading two-port circuits whose properties will be considered first [1].

8.1 PROPERTIES OF TWO-PORT CIRCUITS

The general two-port circuit terminated by the resistances R_1 and R_2 is shown in Figure 8.1 together with the currents I and voltages V at the ports 1 and 2. This circuit, assumed to be linear, is defined by its impedance matrix z, which establishes the relations between the variables, generally written in reduced form as

$$R = R_1 = R_2; \quad v = \frac{V}{\sqrt{R}}; \quad i = I\sqrt{R}$$

Then

$$v = zi \qquad (8.1)$$

with

$$z = \begin{bmatrix} z_{11} & z_{12} \\ z_{21} & z_{22} \end{bmatrix}$$

The values z_{12} and z_{21} are the transfer impedances of the two-port circuit. It is reciprocal if $z_{12} = z_{21}$.

If reversing the circuit does not change the external conditions, it is said to be symmetric and we have $z_{11} = z_{22}$.

The transmission and reflection coefficients of the circuit can be demonstrated if another matrix, the distribution matrix, is introduced. If a reference case is defined with unit terminating resistances the incident and reflected waves a and

b can be written

$$a = \tfrac{1}{2}(v + i) \tag{8.2}$$
$$b = \tfrac{1}{2}(v - i) \tag{8.3}$$

The relations between variables a and b can be obtained using equation (8.1):

$$a = \tfrac{1}{2}(z + I_2)i$$
$$b = \tfrac{1}{2}(z - I_2)i; \qquad I_2 = \begin{bmatrix} 1 & 0 \\ 0 & 1 \end{bmatrix}$$

Thus

$$b = Sa \tag{8.4}$$

where

$$S = \begin{bmatrix} S_{11} & S_{12} \\ S_{21} & S_{22} \end{bmatrix}$$

and:

$$S = (z - I_2)(z + I_2)^{-1} \tag{8.5}$$

If the circuit is reciprocal, then

$$S_{12} = S_{21} = \tau \tag{8.6}$$

where τ is the transmission coefficient:

$$\tau = \frac{2V_2}{E} \tag{8.7}$$

If the input and output impedances of the circuit are z_1 and z_2, respectively, one can write

$$S_{11} = \rho_1 = \frac{z_1 - 1}{z_1 + 1}; \quad S_{22} = \rho_2 = \frac{z_2 - 1}{z_2 + 1} \tag{8.8}$$

The values ρ_1 and ρ_2 are the reflection coefficients at the input and output of the two-port circuit.

If the circuit is not dissipative, the power that it absorbs is zero, and it can be shown that the distribution matrix of such a reciprocal two-port network has the form [2]:

$$S = \frac{1}{g}\begin{bmatrix} h & f \\ f & \pm h_* \end{bmatrix} \tag{8.9}$$

where f, g, and h are real polynomials with the following properties:

(1) They are linked by a relation which, on the imaginary axis, corresponds to:

$$|g|^2 = |h|^2 + |f|^2$$

The notation $h_*(p)$ indicates $h(-p)$.

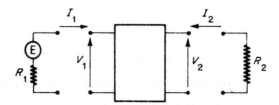

Fig. 8.1   *Two-port element with resistive terminations*

(2) Depending upon whether f is of even or odd degree, the lower or upper sign is taken in equation (8.9).

(3) Each root of g in the complex plane lies in the left-hand half-plane.

The polynomials f, g, and h are the characteristic polynomials of the circuit. The roots of $f(p)$ are generally on the imaginary axis in the stop band and are transmission zeros. The roots of $h(p)$ are attenuation zeros and for a non-dissipative network they are generally on the imaginary axis in the pass band.

For the circuit in Figure 8.1 the transmission coefficient is

$$S_{12} = \frac{2V_2}{E}\sqrt{\frac{R_2}{R_1}} \tag{8.10}$$

The attenuation in decibels is denoted by the function $A_f(\omega)$, where

$$A_f(\omega) = -10\log|S_{12}(\omega)|^2 = 10\log\left|\frac{g(\omega)}{f(\omega)}\right|^2 = 10\log\left(1 + \frac{|h|^2}{|f|^2}\right)$$

The relation

$$\left|\frac{f(\omega)}{g(\omega)}\right|^2 + \left|\frac{h(\omega)}{g(\omega)}\right|^2 = 1 \tag{8.11}$$

expresses the fact that the non-transmitted power is reflected.

For a cascade arrangement it is important to pay equal attention to the transfer matrix t defined by:

$$\begin{bmatrix} b_1 \\ a_1 \end{bmatrix} = t \begin{bmatrix} a_2 \\ b_2 \end{bmatrix} \tag{8.12}$$

The cascade arrangement is represented by the product of the transfer matrices.

The transfer matrix of non-dissipative two-port circuits takes the form:

$$t = \frac{1}{f}\begin{bmatrix} \pm g_* & h \\ \pm h_* & g \end{bmatrix} \tag{8.13}$$

By way of example, Figure 8.2 gives the transfer matrices of several elementary circuits.

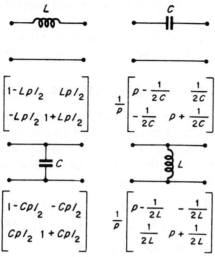

Fig. 8.2 *Transfer matrices for two-port elements*

The fact that the circuit element is non-dissipative has important consequences for the attenuation $A_f(\omega)$. In the pass band, $A_f(\omega)$ cannot take negative values. Consequently, at frequencies where $h(\omega)$ is zero its derivative with respect to any of the parameters must also be zero. In a filter with inductances and capacitances, terminated by resistances, variation of the values of the L and C elements does not affect the attenuation to the first order at frequencies where it is zero.

If the ripple is small it can be assumed that this property applies over the whole pass band. In practice, it can be taken that, in a ladder filter, for example, the interactions between the different branches are such that a perturbation in one element has repercussions for all the other factors of the attenuation function, having an overall compensating effect which minimizes the incidence of the perturbation.

Given this behaviour, it is of interest to find digital filter structures which have similar properties. In effect, in a digital filter where the amplitudes of the ripples in the pass and stop bands are similar the denominator of the transfer function determines the number of bits required to represent the coefficients. Structures derived, for example, from ladder analogue filters can therefore be expected to lead to significant gains in the coefficient wordlengths, in the complexity of the multipliers and also in the power of the round-off noise.

Ladder structures are the most commonly used type in passive analogue filtering. The procedure for obtaining the elements of such a structure using a transfer function is described in detail in Ref. [2]. It consists of factorizing the overall transfer matrix, defined using the calculated transfer function $H(\omega)$ into partial matrices corresponding to the series and parallel arms of the ladder structure.

The most direct approach for obtaining a digital filter structure from an analogue ladder filter is to simulate its voltage–current flow chart.

8.2 SIMULATED LADDER FILTERS

The representation of ladder filters in terms of their flow chart is used to synthesize active filters from integrator or differentiator circuits. The ladder filter shown in Figure 8.3 and terminated with the resistances R_1 and R_2 will be used to demonstrate the voltage–current flow chart. Application of Kirchhoff's laws leads to the following relations

$$I_1 = (E - V_2)R_1^{-1}; \quad I_{K-1} = (V_{K-2} - V_K)Z_{K-1}^{-1}; \quad I_{N+1} = V_N R_2^{-1}$$
$$V_2 = (I_1 - I_3)Y_2^{-1}; \quad V_K = (I_{K-1} - I_{K+1})Y_K^{-1}$$

where the index K takes the values $4, 6, \ldots, N$.

The flow chart is composed of arcs, each of which is associated with a coefficient representing an impedance or an admittance. With each junction there is associated either a voltage at the node or the current in a branch. For each are there is formed the product of the corresponding coefficient and the magnitude associated with its origin. The magnitude associated with each junction is the sum of the products corresponding to its various associated arcs.

The chart for the ladder filter of Figure 8.3 is shown in Figure 8.4. The currents and voltages are associated with alternate junctions. This topology is defined as 'leapfrog'.

Simulated ladder digital filters are the structures obtained by simulating each arc in the group or each branch of the ladder by an element with an equivalent transfer function.

A particularly simple case occurs when the series impedances Z_{K-1} are inductances and the parallel branches Y_K are capacitances ($K = 4, 6, \ldots, N$). Such filters are purely recursive and are without frequencies of infinite attenuation.

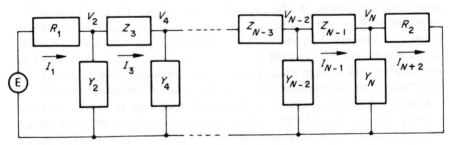

Fig. 8.3 Analogue ladder filter

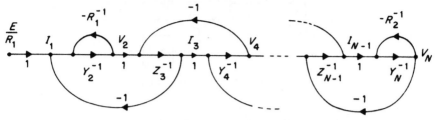

Fig. 8.4   Flow graph of a ladder filter

The transfer functions to account for them take the form:

$$Z_{K-1}^{-1} = \frac{R}{sL_{K-1}}; \quad Y_K^{-1} = \frac{1}{sC_KR}$$

(8.14)

where s is the Laplace variable and R is a normalization constant. Both cases involve the transfer functions of integrators which are easily produced using operational amplifiers and R–C networks. The diagram in Figure 8.5 then appears, being deduced from Figure 8.4. It represents the functions to be implemented and shows the circuit diagram using integrators.

The digital realization consists of replacing each integrator by an equivalent function. In Ref. [3] it is shown that the only digital integrator circuit which is simple to realize and which is equivalent to an analogue integrator is the one represented by Figure 8.6 and whose Z-transfer function, $I(Z)$, is written:

$$I(Z) = \frac{aZ^{-1/2}}{1 - Z^{-1}}$$

(8.15)

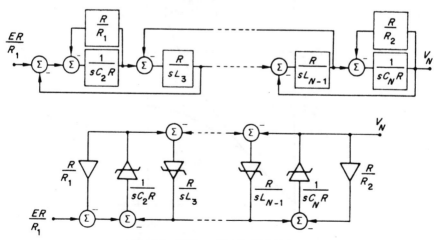

Fig. 8.5   Filter composed of integrators

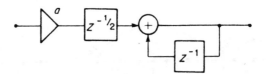

Fig. 8.6   Digital integrator circuit

The equivalence between the analogue and digital integrators is obtained in the same way as for any transfer function by replacing Z by $e^{j\omega T}$. Thus,

$$I(\omega) = \frac{ae^{-j\omega(T/2)}}{1 - e^{-j\omega T}} = \frac{a}{2j} \frac{1}{\sin(\omega(T/2))} \qquad (8.16)$$

where T is the sampling period of the digital circuit. This function is equivalent to the analogue function $a/j\omega T$ with a frequency warping. If f_A denotes the analogue frequency and f_N is the digital frequency, we can show

$$\pi f_A T = \sin(\pi f_N T) \qquad (8.17)$$

The frequency warping thus introduced is different from that obtained with the bilinear transformation introduced in the previous chapter, as shown in Figure 8.7, and it must be taken into account when calculating a filter from a specification.

The circuit in Figure 8.6 shows the disadvantage of introducing the function $Z^{-1/2}$, which corresponds to an additional memory circuit, but the transfer function of a ladder filter is not altered when the impedances of all branches

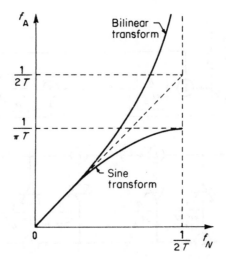

Fig. 8.7   Frequency warping by the sine transform

are multiplied by the same function [3]. This property has already been used to introduce the normalization constant R. If the impedances are multiplied by $Z^{-1/2}$, this term is eliminated from all of the Z-transfer functions of the integrators in the circuit, which then become:

$$I_i(Z) = \frac{TR}{L_i} \frac{1}{1 - Z^{-1}} \quad \text{for } i \text{ odd}$$

and:

$$I_i(Z) = \frac{T}{C_i R} \frac{Z^{-1}}{1 - Z^{-1}} \quad \text{for } i \text{ even}$$

In contrast, the termination resistances are transformed to $R_1 Z^{-1/2}$ and $R_2 Z^{-1/2}$; the terminations are no longer purely resistive, as they have the transfer functions:

$$R_1 e^{-j\pi f T}; \quad R_2 e^{-j\pi f T}$$

This effect can be neglected when the sampling frequency is large compared to the pass band as there is an insignificant change in the transfer function of the filter. Further, the resistances R_1 and R_2 can be chosen as unity, as can the normalization constant R. The circuit of the digital filter obtained under these conditions is given in Figure 8.8.

The coefficients have the following values, for odd order N:

$$a_N = \frac{T}{C_{N+1}}; \quad a_{2i-1} = \frac{T}{C_{2i}}; \quad a_{2i} = \frac{T}{L_{2i+1}}; \quad i = 1, 2, \ldots \frac{N-1}{2} \tag{8.18}$$

The filter thus realized involves N multiplications and N memories for a transfer function of order N. For these parameters the structure is canonic. The number of additions is $2N + 1$.

To summarize, the calculation of a simulated ladder digital filter from an imposed specification involves the following stages:

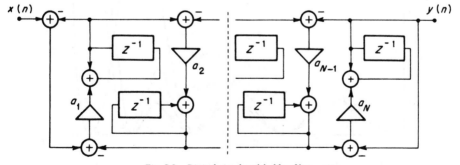

Fig. 8.8 *Digital simulated ladder filter*

(1) Transposition of the specification by modifying the frequency axis using equation (8.17) above;

(2) Calculation of the elements of a filter with passive ladder elements LC, satisfying the transposed specification;

(3) Using the values of the elements so obtained to calculate the coefficients a_i $(i = 1, 2, \ldots, N)$ of the digital filter using equations (8.18).

The chief attraction of the structure obtained in this way is that the coefficients can be represented by a very small number of bits. Also, some multiplications can be replaced by simple additions and, in certain cases, all the multiplications of the filter can be eliminated, resulting in significant savings in the circuitry. To illustrate this property, let us consider a low the pass filter of order $N = 7$ [3], with elements having the following values (Figure 8.5):

$$R = R_1 = R_9 = 1$$
$$C_2 = 1.2597 = C_8$$
$$L_3 = 1.5195 = L_7$$
$$C_4 = 2.2382 = C_6$$
$$L_5 = 1.6796$$

The coefficients a_i $(i = 1, 2, \ldots, N)$ of the corresponding simulated ladder digital filter are calculated with a sampling period $T = 1/f_s = 0.01$ from equations (8.18) above. The ripples of the filter in the pass band are shown in Figure 8.9, where the coefficients are represented by 10, 5, and 3 bits. It is notable that, when represented with 5 bits, the attenuation zeros are conserved. With 3 bits, they are also conserved except for the one closest to the transition band. Thus, the insensitivity to the first order of the attenuation zeros, which was shown in

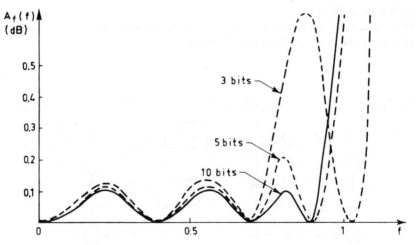

Fig. 8.9   Ripples in the pass band for various coefficient wordlengths

Section 8.1, is demonstrated. In comparison with the cascade structure in the previous chapter, this example shows an estimated gain of 4 or 5 bits for representing the coefficients.

The technique described in this section can be extended to filters other than the purely recursive low-pass filter, but the designs will be more complicated. Also, the need to have a sampling frequency which is large in comparison with the pass band is not very favourable for efficient processing. In practice, the simulated ladder structure is primarily used with a different method of performing the calculations—that used in switched-capacitor devices.

8.3 SWITCHED-CAPACITOR FILTERS

Strictly speaking, because they do not use arithmetic operations filters using switched-capacitor devices are not digital filters. Nevertheless, they do use the same design methods and are complementary to digital filters. They are frequently used in analogue–digital conversion circuits.

The basic principle, which is presented in detail in Ref. [4], is as follows. Switching a capacitance C between two voltages V_1 and V_2 at frequency f_s is equivalent to introducing a resistance R given by

$$R = \frac{1}{C} f_s$$

between the two potentials. In effect, as is shown in Figure 8.10, the capacitance is charged to voltages V_1 and V_2 alternately and a charge transfer $C(V_1 - V_2)$ results. If the operations are performed at frequency f_s, a current i

$$i = C(V_1 - V_2) f_s$$

flows between the voltages V_1 and V_2.

This equivalent resistance is inserted in an integrator circuit as shown in Figure 8.11. The integrator being considered has an adder at its input, like those in Figure 8.5. The equation describing the operation of the analogue integrator is also shown in Figure 8.11. In the switched-capacitor version the capacitance C_1 is alternately connected at frequency f_s across the input of the operational amplifier, and between voltages V_0 and V_1. The equation for the variation ΔV_2

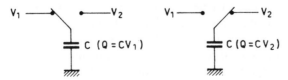

Fig. 8.10   *Switching a capacitance between voltages V_1 and V_2*

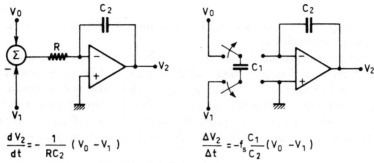

$$\frac{dV_2}{dt} = -\frac{1}{RC_2}(V_0 - V_1)$$

$$\frac{\Delta V_2}{\Delta t} = -f_s\frac{C_1}{C_2}(V_0 - V_1)$$

Fig. 8.11   Switched-capacitor integrator

in the output voltage during the interval Δt, which is assumed to be large in comparison with the period $1/f_s$, is shown in the diagram.

The condition that the two types of integrator are equivalent is

$$C_1 = \frac{1}{f_s R} \tag{8.19}$$

However, in order to completely analyse the switched-capacitor integrator it is necessary to take account of the sampling [5] and to calculate its Z-transfer function. Assume that $v_e(t)$ is the input signal and that $v_2(t)$ is the output one. The sampling period T is assumed to be divided into two equal parts. The capacitance C_1 is connected to the input to the integrator for time $T/2$, and to the voltage $v_e(t)$ itself for time $T/2$. Let us assume that this is between times nT and $(n + \frac{1}{2}T)$. The charge transmitted to the integrator is $Q(nT)$ such that:

$$Q(nT) = C_1 v_e[(n + \tfrac{1}{2})T]$$

Under these conditions at time $(n + 1)T$ the output voltage is

$$v_2[(n + 1)T] = V_2(nT) - \frac{C_1}{C_2}v_e[(n + \tfrac{1}{2})T]$$

By taking the Z-transform of the two components one has

$$\frac{V_2(Z)}{V_e(Z)} = H(Z) = -\frac{C_1}{C_2}\frac{Z^{-1/2}}{1 - Z^{-1}} \tag{8.20}$$

One finds the same type of transfer function as was given by equation (8.15) for digital circuits. The switched-capacitor integrator performs in exactly the same way as the digital circuits described in the previous section and the same warping of the frequency axis is involved. It should be noted that, to ensure that no further delay is introduced and that this function is conserved when cascading two integrators, the capacitances of the two integrators must be switched in antiphase.

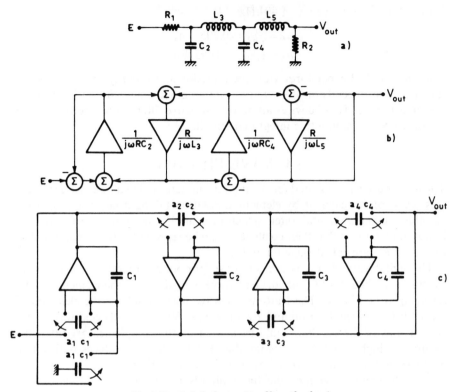

Fig. 8.12 Switched-capacitor filter of order 4

A design involving a switched-capacitor device for a simulated ladder filter like that in Figure 8.5 is obtained by substituting integrator circuits and calculating the value to be given to the switched capacitances in each case.

Example: Assume an implementation using switched capacitors for a Butterworth filter of order 4 with an analogue circuit as shown in Figure 8.12(a).

The procedure described in the previous section results in the design (Figure 8.12(b)) for producing a filter from integrators, assuming unit terminal resistances. The switched-capacitor design is shown in Figure 8.12(c). The coefficients a_i $(i = 1, 2, 3, 4)$ which define the ratios of the capacitances are given by equations (8.18).

If the filter has a 3 dB attenuation frequency f_c equal to 1 kHz, the analogue parameters are as follows:

$$R_1 = R_2 = 1$$
$$C_2 = 121.8 \times 10^{-6}; \quad C_4 = 294.1 \times 10^{-6}$$
$$L_3 = 294.1 \times 10^{-6}; \quad L_5 = 121.8 \times 10^{-6}$$

With a sampling frequency of 40 kHz:

$$a_1 = \tfrac{1}{4.87} = 0.205 = a_4$$
$$a_2 = \tfrac{1}{11.76} = 0.085 = a_3$$

Thus, in switched-capacitor devices the precision and stability of the integrator time constant depends on the external sampling frequency and on the ratio of the capacitances. These devices allow very selective filters to be produced on a silicon chip as a monolithic integrated circuit.

8.4 WAVE FILTERS

The first approach suggested for the digital simulation of analogue ladder filters was based on an element by element simulation taking account of the wave variables. General digital filter structures can be developed this way. An important feature is that their coefficients can be represented by a small number of bits. These are called wave filters [6, 7].

The digital wave filter is assumed to have four ports with four sets of numbers as shown in Figure 8.13. The incident wave is made up of the elements of the set $a_1(n)$ to be filtered. The transmitted wave corresponds to the elements of the set $b_2(n)$ filtered, for example, by a low-pass filter. The reflected wave $b_1(n)$ corresponds to the set that would be filtered by the complementary filter (for example, a high-pass filter). It is notable that the device produces both filters at the same time. The fourth set $a_2(n)$ is assumed to be zero.

As in the previous sections, the digital filter is determined in two stages, using an analogue ladder filter. First, the digital elements corresponding to the analogue elements are defined. At the terminals of an inductance or a capacitance the incident wave A and the reflected wave B are

$$A = V + RI$$
$$B = V - RI \tag{8.21}$$

where V and I are the voltage and the current at the terminals and R is a reference resistance. For an inductance L we obtain:

$$V = LsI$$

A digital equivalent can be obtained in a simple way by using the bilinear

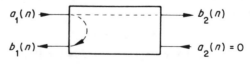

Fig. 8.13   Ports of a wave filter

transform introduced in the previous chapter, except for the scale factor of $2/T$:

$$s = \frac{1 - Z^{-1}}{1 + Z^{-1}}$$

By taking the reference resistance to be $R = L$ the following relation appears between the incident and reflected waves at the terminals of the digital circuit equivalent to the inductance L:

$$B = -Z^{-1}A \tag{8.22}$$

Thus, the digital equivalent involves a unit delay and a multiplication by -1. For a capacitance the digital equivalent is simply a unit delay.

In this procedure it is apparent that each element has a particular reference resistance. The second phase in determining the digital filter consists of suitably interconnecting elements which have different reference resistances. This involves using adaptor devices. To connect two elements in series it is necessary to use a series adaptor whose symbol is shown in Figure 8.14(a).

The relationships between the pairs of wave variables (a_1, b_1), (a_2, b_2), and

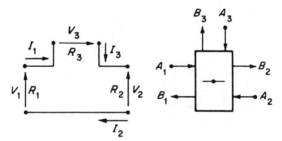

Fig. 8.14(a)   Connection and symbol of a series adaptor

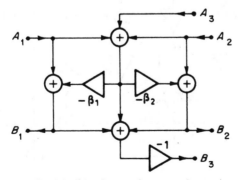

Fig. 8.14(b)   Circuit of a series adaptor

(a_3, b_3) are obtained from the relations which characterize this adaptor:

$$I_1 = I_2 = I_3$$
$$V_1 + V_2 + V_3 = 0 \qquad (8.23)$$

The circuit of the series adaptor is shown in Figure 8.14(b). The reflected waves are expressed by:

$$b_1 = a_1 - \beta_1(a_1 + a_2 + a_3) \quad \text{with} \quad \beta_1 = \frac{2R_1}{R_1 + R_2 + R_3}$$

$$b_2 = a_2 - \beta_2(a_1 + a_2 + a_3) \quad \text{with} \quad \beta_2 = \frac{2R_2}{R_1 + R_2 + R_3} \qquad (8.24)$$

$$b_3 = -(a_1 + a_2 + a_3) - b_1 - b_2$$

For a parallel connection the symbol and circuit are shown in Figure 8.15. The equations which characterize the corresponding adaptor are

$$I_1 + I_2 + I_3 = 0$$
$$V_1 = V_2 = V_3 \qquad (8.25)$$

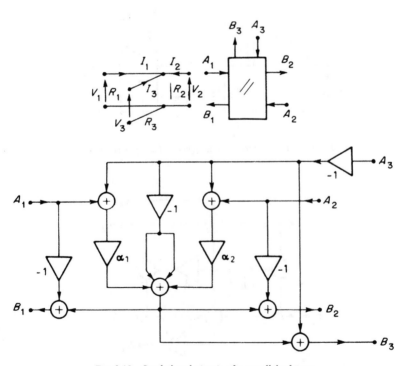

Fig. 8.15   *Symbol and circuit of a parallel adaptor*

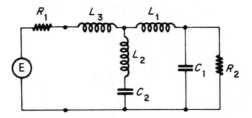

Fig. 8.16   *Example of a reference filter*

As before, these equations result in reflected waves being given by the expressions:

$$b_k = a_0 - a_k \quad \text{with} \quad k = 1, 2, 3$$

and:

$$a_0 = 2a_3 + \alpha_1(a_1 - a_3) + \alpha_2(a_2 - a_3)$$

$$\alpha_k = \frac{2G_k}{G_1 + G_2 + G_3}; \quad G_k = \frac{1}{R_k} \tag{8.26}$$

Zero delay loops may appear in the connection of adaptors. That is, at a port a reflected wave may depend upon the incident wave at the same port, which itself depends on the reflected wave being calculated. This circuit loop can be opened by giving the reference resistance at the port a value such that the reflected wave does not depend on the incident one. This condition is realized at port No. 2 for example, if, in the equation for b_2, the coefficient of a_2 is zero, which implies $R_2 = R_1 + R_3$ for the series adaptor. Then,

$$\beta_1 = \frac{R_1}{R_1 + R_3}; \quad \beta_2 = 1$$

For the parallel adaptor, $G_2 = G_1 + G_3$ with:

$$\alpha_1 = \frac{G_1}{G_1 + G_3} \quad \text{and} \quad \alpha_2 = 1$$

These adaptors are said to be open to the right. All the elements necessary for determining the filter are thus available.

Example: Consider the filter in Figure 8.16 with the following values:

$$R_1 = R_2 = 1\,\Omega$$
$$C_1 = 0.1086\,\text{F}; \quad C_2 = 0.1907\,\text{F};$$
$$L_1 = 0.2091\,\text{H}; \quad L_2 = 0.0202\,\text{H}; \quad L_3 = 0.09021\,\text{H}$$

This filter has five elements and requires five memories and five adaptors.
The circuit for the filter is given in Figure 8.17. The adaptors numbered 1, 3,

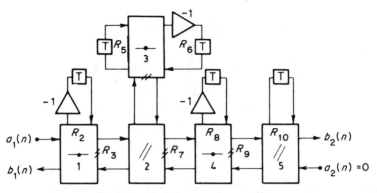

Fig. 8.17   Block diagram of a wave filter

and 4 are of the series type, open to one port, and have the following coefficients:

$$\beta_1 = 0.91725; \quad \beta_3 = 0.99616; \quad \beta_4 = 0.8120$$

Adaptor No. 2 is of the parallel type, open to one port. It has the coefficient $\alpha_2 = 0.8284$. Adaptor No. 5 is of the general parallel type, with two coefficients, $\alpha_5 = 0.8956$ and $\alpha_5' = 0.99618$.

The Z-transfer function corresponding to the transmitted wave is obtained from the elements given in Section 8.1 and takes the form:

$$H(Z) = \frac{(Z + 1)^2 [Z^2 + 2Z(2\beta_3 - 1) + 1]}{D(Z)} \tag{8.27}$$

It has a double zero at the point -1 and two zeros Z_1 and $\overline{Z_1}$ on the unit circle such that:

$$Z_1 = -(2\beta_3 - 1) + j\sqrt{(1 - (2\beta_3 - 1)^2)} \tag{8.28}$$

and which depend only on the coefficient β_3.

Figure 8.18 represents the frequency response of the filter obtained without rounding of the coefficients and with rounding to 6 bits for coefficients other than β_3. The low sensitivity in the pass band is obvious.

This example shows that, as in the simulated ladder filters, the coefficients can be represented by a small number of bits. The effect of rounding the coefficient β_3 is a displacement of the point of infinite attenuation as in a cascade structure.

One interesting property of wave filters is that the calculations and circuits for the wave filter itself simultaneously provide the complementary filter. The filtered output signal then corresponds to the set $b_1(n)$ in Figure 8.13. The variation of the attenuation of this filter with frequency is shown in Figure 8.19.

Wave filters have been obtained by an analogy with transmission lines. An analogy with sound pipes leads to lattice filters.

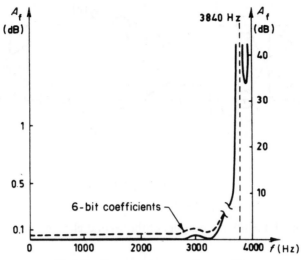

Fig. 8.18   *Frequency response of a wave filter*

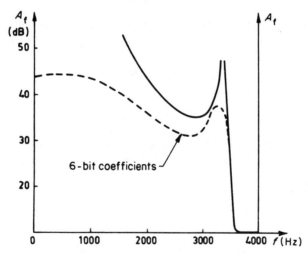

Fig. 8.19   *Attenuation of the reflection filter*

8.5 LATTICE FILTERS

The lattice structure occurs in the analysis and synthesis of speech for simulating the vocal tract and also more generally in systems for linear prediction. It allows the realization of FIR and IIR filters [8].

Consider the structure with M sections as represented in Figure 8.20. The outputs $y_1(n)$ and $u_1(n)$ of the first section are related to the input set $x(n)$ by

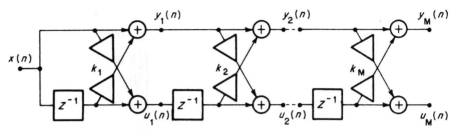

Fig. 8.20   Lattice filter of the FIR type

$$y_1(n) = x(n) + k_1 x(n-1)$$
$$u_1(n) = k_1 x(n) + x(n-1) \tag{8.29}$$

Similarly, $y_2(n)$ and $u_2(n)$, the outputs of the second element, are related to the inputs by:

$$y_2(n) = x(n) + k_1(1 + k_2)x(n-1) + k_2 x(n-2)$$
$$u_2(n) = k_2 x(n) + k_1(1 + k_2)x(n-1) + x(n+2) \tag{8.30}$$

By iteration, $y_M(n)$ and $u_M(n)$, the outputs produced by the filter, are related to the set $x(n)$ by the following equations, which correspond to FIR filtering:

$$y_M(n) = \sum_{i=0}^{M} a_i x(n-i) \tag{8.31}$$

$$u_M(n) = \sum_{i=0}^{M} a_{M-i} x(n-i) \tag{8.32}$$

The two FIR filters thus obtained have the same coefficients but in reverse order. Their Z-transfer functions, $H_M(Z)$ and $U_M(Z)$, are image polynomials. Thus,

$$H_M(Z) = \sum_{i=0}^{M} a_i Z^{-i}$$

$$U_M(Z) = \sum_{i=0}^{M} a_{M-i} Z^{-i} = Z^{-M} H_M(Z^{-1}) \tag{8.33}$$

An iteration process is used to determine the coefficients k_i of the lattice filter from the coefficients a_i $(1 \leqslant i \leqslant M)$. First, the coefficient a_0 is assumed to be equal to unity. Then it is straightforward, using the equations given earlier (and also directly from Figure 8.20) to prove that:

$$k_M = a_M$$

This point is the basis for the calculation. By using $H_m(Z)$ and $U_m(Z)$ $(1 \leqslant m \leqslant M)$ to denote the corresponding transfer functions at the outputs of the mth section, the following matrix equation can be written:

$$\begin{bmatrix} H_m(Z) \\ U_m(Z) \end{bmatrix} = \begin{bmatrix} 1 & k_m Z^{-1} \\ k_m & Z^{-1} \end{bmatrix} \begin{bmatrix} H_{m-1}(Z) \\ U_{m-1}(Z) \end{bmatrix}$$

By assuming $k_m \neq 1$, this equation can also be written as:

$$\begin{bmatrix} H_{m-1}(Z) \\ U_{m-1}(Z) \end{bmatrix} = \frac{1}{1 - k_m^2} \begin{bmatrix} 1 & -k_m \\ -k_m Z & Z \end{bmatrix} \begin{bmatrix} H_m(Z) \\ U_m(Z) \end{bmatrix} \tag{8.34}$$

Thus, the polynomials $H_{m-1}(Z)$ and $U_{m-1}(Z)$ are image polynomials of degree $m-1$ whose coefficients $a_{i(m-1)}$ $(1 \leqslant i \leqslant m-1)$ are calculated using the coefficients a_{im} $(1 \leqslant i \leqslant m)$ of the polynomials $H_m(Z)$ and $U_m(Z)$.

Under these conditions, this becomes:

$$a_{mm} = k_m; \quad a_{(m-1)(m-1)} = k_{m-1}$$

$$a_{(m-1)(m-1)} = \frac{1}{1 - a_{mm}^2} [a_{(m-1)m} - a_{mm} a_{1m}] \tag{8.35}$$

The coefficients k_m $(1 < m < M)$ are thus calculated in M iterations.

Example: Consider the transfer function $H_3(Z)$ such that:

$$H_3(Z) = 1 - 1.990 Z^{-1} + 1.572 Z^{-2} - 0.4583 Z^{-3}$$

Then

$$k_3 = -0.4583$$

Using equation (8.34) one can write:

$$H_2(Z) = 1 - 1.607 Z^{-1} + 0.8355 Z^{-2}$$

Thus:

$$k_2 = 0.8355$$

By application of equation (8.35) we have

$$k_1 = -0.88756 \quad \text{and} \quad H_1(Z) = 1 - 0.8756 Z^{-1}$$

The realization of purely recursive IIR-type filters results in a dual structure, as represented in Figure 8.21. The sets $x_1(n)$, $u(n)$, and $y(n)$ are related by the

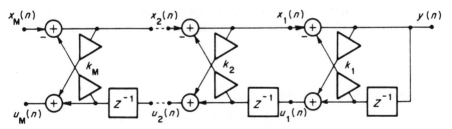

Fig. 8.21   Lattice filter of the purely recursive IIR type

equations:

$$y(n) = x_1(n) - k_1 y(n-1)$$
$$u_1(n) = k_1 y(n) + y(n-1)$$

Similarly, the sets $x_2(n)$, $x_1(n)$, $u_1(n)$, and $u_2(n)$ are related by

$$x_1(n) = x_2(n) - k_2 u_1(n-1)$$
$$u_2(n) = k_2 x_1(n) + u_1(n-1)$$

This results in the transfer function $H_2(Z)$ between the input $x_2(n)$ and the output $y(n)$ given by:

$$H_2(Z) = \frac{1}{1 + k_1(1 + k_2)Z^{-1} + k_2 Z^{-2}}$$

Similarly, the transfer function $U_2(Z)$ relates $u_2(n)$ and $y(n)$, where

$$U_2(Z) = k_2 + k_1(1 + k_2)Z^{-1} + Z^{-2}$$

By iteration it can be seen that the sets $x_M(n)$ and $y(n)$, on the one hand, and $u_M(n)$ and $y(n)$, on the other, are related by the equations:

$$y(n) = x_M(n) - \sum_{i=1}^{M} b_i y(n-i) \tag{8.36}$$

$$u_M(n) = \sum_{i=0}^{M-1} b_{M-i} y(n-i) + y(n-M) \tag{8.37}$$

Consequently, the transfer functions $H_M(Z)$ and $U_M(Z)$ are:

$$H_M(Z) = \frac{1}{1 + \sum_{i=1}^{M} b_i Z^{-i}} = \frac{1}{D_M(Z)}$$

$$U_M(Z) = \sum_{i=0}^{M-1} b_{M-i} Z^{-i} + Z^{-M} = Z^{-M} D_M(Z^{-1}) \tag{8.38}$$

The coefficients k_i $(1 \leqslant i \leqslant M)$ of the lattice filter are calculated by iteration from the coefficients b_i of the IIR filter, after noting that:

$$k_M = b_M$$

By using $H_m(Z)$ and $U_m(Z)$ to denote the relating functions for the set of m sections $(1 \leqslant m \leqslant M)$, it is possible, using the equations of definition:

$$x_{m-1}(n) = x_m(n) - k_m u_{m-1}(n-1)$$
$$u_m(n) = k_m x_{m-1}(n) + u_{m-1}(n-1)$$

to produce the following matrix equation:

$$\begin{bmatrix} D_m(Z) \\ U_m(Z) \end{bmatrix} = \begin{bmatrix} 1 & k_m Z^{-1} \\ k_m & Z^{-1} \end{bmatrix} \begin{bmatrix} D_{m-1}(Z) \\ U_{m-1}(Z) \end{bmatrix}$$

As in the case of the FIR type of filter, this matrix equation is also written, for $k_m \neq 1$:

$$\begin{bmatrix} D_{m-1}(Z) \\ U_{m-1}(Z) \end{bmatrix} = \frac{1}{1-k_m^2} \begin{bmatrix} 1 & -k_m \\ -k_m Z & Z \end{bmatrix} \begin{bmatrix} D_m(Z) \\ U_m(Z) \end{bmatrix} \qquad (8.39)$$

As with FIR filters, this expression allows the calculation of the coefficients k_i $(1 \leqslant i \leqslant M)$ of the IIR lattice filter in M iterations, using the polynomial $D_M(Z)$, where

$$D_M(Z) = 1 + \sum_{i=1}^{M} b_i Z^{-i}$$

The lattice structures given in Figures 8.20 and 8.21 are canonic for the data memories but not for the multiplications. They can be made canonic, for an IIR type filter, for example, by using the single-multiplication section represented in Figure 8.22. However, a further addition is then necessary. The equations of this first-order section are as follows:

$$(1 + k)x_1(n) = y(n) + ky(n - 1)$$
$$(1 + k)u_1(n) = ky(n) + y(n - 1)$$

To within a factor of $(1 + k)$, they are equivalent to the equations for a two-multiplier lattice.

In contrast with the structures described in the previous sections, lattice filters do not have any particular advantages for the number of bits needed to represent the coefficients. Nevertheless, in practice, they have one interesting property in that a necessary and sufficient condition for an IIR filter to be stable and have its poles inside the unit circle is that the coefficients have a modulus of less than unity:

$$|k_i| < 1; \quad 1 \leqslant i \leqslant M$$

This property is obvious for k_1 in Figure 8.21 if the appropriate section is isolated. It can be extended to the other coefficients by considering the subcircuits and using recurrence.

This results in a control of the stability which can be realized quite simply and is of particular use in systems such as adaptive filters, where the values of the coefficients are constantly changing.

The lattice structures considered above are either non-recursive or purely recursive. Note that the purely recursive structure can be completed to make a general filter; it is sufficient to form a weighted sum of the variables $u_m(n)$. That is the expression:

$$v_M = \gamma_0 y(n) + \sum_{m=1}^{M} \gamma_m u_m(n)$$

defines FIR type filtering of the signal $y(n)$ by virtue of equations (8.37). As the coefficients $b_i (1 \leq i \leq M)$ are fixed, the coefficients γ_i can be determined to obtain any numerator for the general filter.

It is useful to observe also that the purely recursive structure consists of the pure all-pass function. In fact, equations (8.37) and (8.38) yield:

$$H_D(Z) = \frac{U_M(Z)}{X(Z)} = \frac{b_M + b_{M-1}Z^{-1} + \ldots + Z^{-M}}{1 + b_1 Z^{-1} + \ldots + b_M Z^{-M}}$$

This expression shows that, as indicated in Section 6.3, the signal $u_M(n)$ is the output of an all-pass network with $x(n)$ as the input. The transfer function $H_D(Z)$ can be expressed directly as a function of the lattice coefficients by a continued fraction:

$$H_D(Z) = k_M + \cfrac{(1 - k_M^2)Z^{-1}}{k_M Z^{-1} + \cfrac{1}{k_{m-1}Z^{-1} + \cfrac{(1 - k_{M-1}^2)Z^{-1}}{k_{M-1}Z^{-1} + \cfrac{\ddots}{\quad + \cfrac{1}{k_1 + \cfrac{(1 - k_1^2)Z^{-1}}{k_1 Z^{-1} + 1}}}}}}$$

This observation can be used to calculate the poles of the lattice filter directly [9].

An interesting application of the above results is the implementation of the notch filter introduced in Section 6.3. The notch filter output $y_N(n)$ is obtained simply by incorporating one more adder into Figure 8.21 to carry out the following operation:

$$y_N(n) = x_M(n) + u_M(n) \tag{8.40}$$

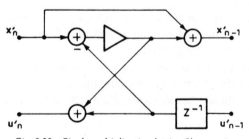

Fig. 8.22 Single-multiplication lattice filter section

For order 2 the transfer function of the all-pass circuit is:

$$H_D(Z) = \frac{k_2 + k_1(1 + k_2)z^{-1} + z^{-2}}{1 + k_1(1 + k_2)z^{-1} + k_2 z^{-2}}$$

A very peculiar property of the approach is that the frequency ω_0 and the 3-dB attenuation band width B_{3N} can be tuned independently [10]. This decoupling effect is due to the relationships:

$$k_1 \simeq -\cos \omega_0$$

$$k_2 \simeq \frac{1 - \tan \pi B_{3N}}{1 + \tan \pi B_{3N}} = (1 - \varepsilon)^2 \tag{8.41}$$

If a subtraction is performed instead of the addition in equation (8.40) the complementary filter is obtained.

8.6 COMPARISON ELEMENTS

Having discussed the various structures for digital filters this is an appropriate point at which to summarize their properties. Reference [11] gives a detailed comparative analysis.

The cascade structure is the most simple to obtain, as the coefficients correspond to a simple factorization of the Z-transfer function. It produces a minimum number of multiplications, additions and memories. However, the representation of the coefficients may require a large number of bits.

The choice between cascade and lattice structures is not normally realistic for filters with fixed coefficients as the lattice structure only corresponds to particular applications.

The structures derived from analogue ladder networks, simulated ladder filters and wave filters offer the possibility of representing the coefficients with few bits, even for very selective filters. Consequently, elimination of the multipliers can be envisaged. As these are generally the most complex circuits, the savings in materials is appreciable. Nevertheless, some complications do occur. The number of additions is increased, as is the complexity of the sequence of operations. For example, in wave filters it is necessary to make calculations for the incident and reflected waves inside the filter and, therefore, to carry out the operations corresponding to the forwards and backwards interchange between input and output. Extra memories are required for storing the intermediate results. Also, multiplexing of the operations between several filters, an important advantage of digital filters, becomes more difficult. In view of these factors a detailed evaluation is necessary before this type of structure can be used.

REFERENCES

[1] V. BELEVITCH, *Classical Network Theory*, Holden-Day, San Francisco, 1968.

[2] J. NEIRYNK et P. van BASTELAER, La synthèse des filtres par factorisation de la matrice de transfert. *Revue MBLE*, **10**(1), 1967.

[3] L. T. BRUTON, Low sensitivity digital ladder filters. *IEEE Transactions*, **CAS22**(3), 1975.

[4] B. HOSTICKA, R. BRODERSEN and P. GRAY, MOS sampled data recursive filters using switched capacitor integrators. *IEEE Journal of Solid-State Circuits*, **12**, Dec 1977.

[5] R. BRODERSEN, P. GRAY and D. HODGES, MOS switched-capacitor filters. *Proceedings of the IEEE*, **67**(1), 1979.

[6] A. FETTWEIS, Digital filter structures related to classical filter networks. *Archiv. Elek. Ubertragung*, **25**, Feb 1971.

[7] J. R. BOITE and H. LEICH, Synthèse des filtres numériques simulant les filtres à inductances et capacités. *Annales des Télécom.*, **31**(3/4), 1976.

[8] S. K. MITRA, P. S. KAMAT and D. C. HUEY, Cascaded lattice realization of digital filters. *Circuit Theory and Applications*, **5**, 1977.

[9] W. B. JONES and A. O. STEINHARDT, Finding the poles of the lattice filter. *IEEE Transactions*, **ASSP33**(4), 1328–31, 1985.

[10] T. SARAMAKI, T. H. YU and S. K. MITRA, Very low sensitivity realization of IIR digital filters using a cascade of complex all-pass structures. *IEEE Transactions*, **CAS34**, 876–86, 1987.

[11] R. E. CROCHIÈRE and A. V. OPPENHEIM, Analysis of linear digital networks. *Proceedings of the IEEE*, **63**(4), 1975.

EXERCISES

1 Give the impedance and distribution matrices for the elementary two-port networks of Figure 8.2. Give the impedance, distribution, and transfer matrices when the elements are resonant *LC* circuits.

2 Consider the Butterworth filter of order 4 given in Figure 8.12(a). Draw the corresponding flow chart. Show the circuit of the digital simulated ladder filter and calculate the coefficients from the given values for the analogue elements for a sampling frequency of 40 kHz. Study the modification of the transfer function in the pass band introduced by a reduction in the sampling frequency from 40 kHz to 10 kHz.

3 Draw the flow chart for the filter in Figure 8.16, which contains a resonant circuit in one branch. Show the circuit for its realization using switched capacitor devices and calculate the capacitance ratios for each integrator from the analogue elements, assuming a sampling frequency of 20 kHz. Compare the frequency response obtained with that of the wave filter.

4 Determine the elements of the wave filter corresponding to a low-pass Chebyshev filter of order 7 whose analogue elements are given in Section 8.2. The sampling frequency is taken as 10 kHz. Show the design for the corresponding simulated ladder filter.

Compare the number of operations to be carried out in each realization. Which is the most advantageous?

Compare the frequency responses when the coefficients are represented by 5 bits.

5 Calculate the frequency response of the lattice filter given as the example in Section 8.5. How does this response evolve when the parameters are represented by 5 bits? Give the circuit of the filter with single multiplication elements. How should the circuit be modified to produce the filter using an inverse Z-transfer function?

CHAPTER 9

COMPLEX SIGNALS AND MINIMUM PHASE FILTERS

Complex signals in the form of sets of complex numbers are currently used in the digital analysis of signals. Some of these sets have been given as examples in the chapters on discrete Fourier transforms. In this chapter, analytic signals, a particular category of complex signal, will be studied. These display some interesting properties and occur primarily in modulation and multiplexing. The properties of the Fourier transforms of real causal sets will be examined first [1–3].

9.1 THE FOURIER TRANSFORM OF A REAL AND CAUSAL SET

Consider a set of elements $x(n)$ whose Z-transform is written:

$$X(Z) = \sum_{n=-\infty}^{\infty} x(n)Z^{-n}$$

The Fourier transform of this set is obtained by replacing Z with $e^{j2\pi f}$ in $X(Z)$:

$$X(f) = \sum_{n=-\infty}^{\infty} x(n)e^{-j2\pi nf}$$

If the elements $x(n)$ are real numbers, we obtain:

$$X(-f) = \overline{X(f)} \tag{9.1}$$

The values of $X(f)$ at negative frequencies are complex conjugates of the values at positive frequencies. The supplementary condition of causality can be imposed on the set $x(n)$ and the consequences for $X(f)$ will now be examined.

The function $X(f)$ can be separated into real and imaginary parts:

$$X(f) = X_R(f) + jX_1(f) \tag{9.2}$$

If the set $x(n)$ is real, then using equation (9.1), the function $X_R(f)$ is even. Thus, it is the Fourier transform of an even set $x_p(n)$. The function $X_1(f)$ is the Fourier transform of an odd set $x_i(n)$ such that:

$$x_p(n) = x_p(-n)$$
$$x_i(n) = -x_i(-n)$$
$$x(n) = x_p(n) + x_i(n) \tag{9.3}$$

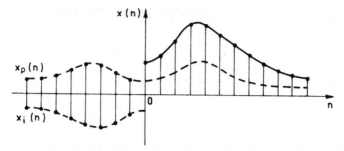

Fig. 9.1 Decomposition of a causal set into even and odd parts

Under these conditions, we derive:

$$X_R(f) = x_p(0) + 2 \sum_{n=1}^{\infty} x_p(n) \cos(2\pi n f) \tag{9.4}$$

$$X_I(f) = -2 \sum_{n=1}^{\infty} x_i(n) \sin(2\pi n f) \tag{9.5}$$

If the set $x(n)$ is causal, that is, if:

$$x(n) = 0 \quad \text{for} \quad n < 0$$

the following equations (Figure 9.1) are derived:

$$x_i(n) = x_p(n) = \tfrac{1}{2}x(n) \quad \text{for} \quad n \geqslant 1$$
$$x_p(0) = x(0)$$

and hence

$$X_R(f) - x(0) = \sum_{n=1}^{\infty} x(n) \cos(2\pi n f) \tag{9.6}$$

$$X_I(f) = -\sum_{n=1}^{\infty} x(n) \sin(2\pi n f) \tag{9.7}$$

It can be seen that these two functions are related. To go from one to the other, it is sufficient to change $\cos(2\pi n f)$ to $-\sin(2\pi n f)$ or conversely. Such an operation is called quadrature and it will now be expressed analytically.

By definition, a causal set is one which satisfies the equality:

$$x(n) = x(n) Y(n)$$

where the set $Y(n)$ is such that:

$$Y(n) = 0 \quad \text{for} \quad n < 0$$
$$1 \quad \text{for} \quad n \geqslant 0$$

This set is a sampling of the unit step function $Y(t)$ which, in terms of distributions, has a Fourier transform FY given in Ref. [2]:

$$FY = \frac{1}{j2\pi} vp\left(\frac{1}{f}\right) + \tfrac{1}{2}\delta(f) \tag{9.8}$$

where $vp(1/f)$ is the distribution defined by

$$\left\langle vp\left(\frac{1}{f}\right), \phi \right\rangle = VP \int_{-\infty}^{\infty} \frac{\phi(f)}{f} df \tag{9.9}$$

The principal value of the Cauchy integral is itself defined by:

$$VP \int_{-\infty}^{\infty} \frac{\phi(f)}{f} df = \int_{-\infty}^{\infty} \frac{\phi(f)}{f} df = \lim_{\varepsilon \to 0} \left[\int_{-\infty}^{-\varepsilon} \frac{\phi(f)}{f} df + \int_{\varepsilon}^{\infty} \frac{\phi(f)}{f} df \right]$$

As the sampling introduces a periodicity to the spectrum it can be shown that the Fourier transform of the set $Y(n)$ such that:

$$Y(n) = 0 \quad \text{for} \quad n < 0$$
$$Y(0) = \tfrac{1}{2}$$
$$Y(n) = 1 \quad \text{for} \quad n > 0$$

is the distribution FY_n:

$$FY_n = \frac{1}{2j} vp[\cot \pi f] + \tfrac{1}{2}\delta(f) \quad \text{for} \quad -\tfrac{1}{2} \leqslant f \leqslant \tfrac{1}{2} \tag{9.10}$$

The convolution product of the Fourier transforms corresponds to the product of the two sets. Hence,

$$X(f) = \left[\frac{1}{2j} vp[\cot \pi f] + \tfrac{1}{2}\delta(f) \right] * X(f) + \tfrac{1}{2}x(0)$$

By separating the equation into real and imaginary parts one obtains

$$X_R(f) + jX_1(f) = vp[\cot \pi f] * [X_1(f) - jX_R(f)] + x(0)$$

The equations which relate the real and imaginary parts of $X(f)$ are

$$X_R(f) = x(0) + \int_{-1/2}^{1/2} X_1(f') \cot [\pi(f - f')] df' \tag{9.11}$$

$$X_1(f) = - \int_{-1/2}^{1/2} X_R(f') \cot [\pi(f - f')] df' \tag{9.12}$$

or further, in a different form, without introducing Cauchy's principal values,

$$X_R(f) = x(0) - \int_{-1/2}^{1/2} [X_1(f) - X_1(f')] \cot [\pi(f - f')] df' \tag{9.13}$$

$$X_I(f) = \int_{-1/2}^{1/2} [X_R(f) - X_R(f')] \cot [\pi(f - f')] df' \qquad (9.14)$$

The real and imaginary parts of the Fourier transform of a causal set are related by equations (9.11) and (9.12) which correspond to the Hilbert transformation for continuous signals.

Example: Consider the set $U_k(n)$ such that:

$$U_k(n) = 0 \quad \text{for} \quad n \neq k$$
$$U_k(k) = 1$$
$$X_R(f) = \cos(2\pi kf); \quad X_I(f) = -\sin(2\pi kf)$$

It can be shown directly that:

$$\int_{-1/2}^{1/2} \cos(2\pi kf') \cot [\pi(f - f')] df'$$

$$= \int_{-1/2}^{1/2} \cos[2\pi k(f - f')] \cot(\pi f') df' = \sin(2\pi kf)$$

$$\int_{-1/2}^{1/2} \sin(2\pi kf') \cot [\pi(f - f')] df'$$

$$= \int_{-1/2}^{1/2} \sin[2\pi k(f - f')] \cot(\pi f') df' = -\cos(2\pi kf)$$

On the other hand, using Parseval's equation, we can write:

$$\int_0^{1/2} X_R^2(f) df = \int_0^{1/2} X_I^2(f) df \qquad (9.15)$$

The real and imaginary parts of $X(f)$ have the same power.

9.2 ANALYTIC SIGNALS

Analytic signals correspond to causal signals in which time and frequency are exchanged. Their spectrum has no negative frequency component and their name derives from the fact that they represent the restriction to the real axis of an analytic function of a complex variable, i.e. they can be expanded in series in a region which contains this axis. The properties of analytic signals are deduced from the properties of causal signals by exchanging time and frequency.

Consider the signal $x(t) = x_R(t) + jx_I(t)$ such that:

$$X(f) = 0 \quad \text{for} \quad f < 0$$

The functions $x_R(t)$ and $x_1(t)$ are Hilbert transforms of each other:

$$x_R(t) = \frac{1}{\pi} \int_{-\infty}^{\infty} \frac{x_1(t')}{t-t'} dt' \tag{9.16}$$

$$x_1(t) = -\frac{1}{\pi} \int_{-\infty}^{\infty} \frac{x_R(t')}{t-t'} dt' \tag{9.17}$$

The Fourier transform of the real function:

$$x_R(t) = \tfrac{1}{2}[x(t) + \bar{x}(t)] \tag{9.18}$$

is the function $X_R(f)$ such that:

$$X_R(f) = \tfrac{1}{2}[X(f) + \bar{X}(-f)] \tag{9.19}$$

that is, $X_R(f) = X(f)/2$ for positive frequencies and $X_R(f) = \bar{X}(-f)/2$ for negative ones.

Similarly:

$$X_1(f) = -j\tfrac{1}{2}[X(f) - \bar{X}(-f)] \tag{9.20}$$

Figure 9.2 shows the decomposition of the spectrum of a signal into real and imaginary parts.

Example

$$x(t) = e^{j\omega t}; \quad x_R(t) = \cos \omega t = \tfrac{1}{2}[e^{j\omega t} + e^{-j\omega t}]$$
$$x_1(t) = \sin \omega t = -j\tfrac{1}{2}[e^{j\omega t} - e^{-j\omega t}]$$

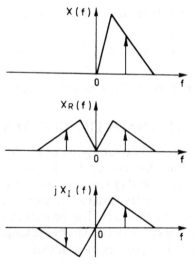

Fig. 9.2   Spectrum of an analytic signal

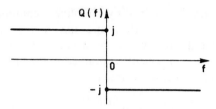

Fig. 9.3 *Frequency response of a quadrature filter*

Finally, it can be seen that the following equations hold between $X_R(f)$ and $X_I(f)$:

$$X_I(f) = -jX_R(f) \quad \text{for} \quad f > 0$$
$$jX_R(f) \quad \text{for} \quad f < 0$$

That is, $X_I(f)$ is obtained from $X_R(f)$ by a rotation of the components through $\pi/2$. The Hilbert transform consists of orthogonalizing the signal components. This is a filtering operation and the frequency response $Q(f)$ is represented in Figure 9.3.

Example

$$x_R(t) = \int_0^\infty [A(f)\cos(2\pi ft) - B(f)\sin(2\pi ft)] \, df \tag{9.21}$$

$$x_I(t) = \int_0^\infty [A(f)\sin(2\pi ft) + B(f)\cos(2\pi ft)] \, df \tag{9.22}$$

The properties of continuous analytic signals can be transferred to discrete signals after certain modifications.

A discrete signal has a periodic Fourier transform. A discrete analytic signal $x(n)$ deduced from a real signal is a discrete signal whose Fourier transform $X_n(f)$, which has the period $f_s = 1$, is zero for $-\frac{1}{2} < f < 0$ (Figure 9.4).

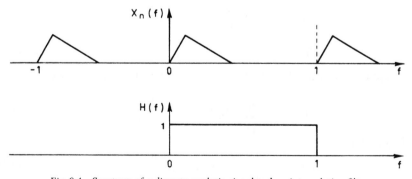

Fig. 9.4 *Spectrum of a discrete analytic signal and an interpolation filter*

If a discrete signal $x(n)$ is obtained by sampling a continuous analytic signal $x(t)$ at a frequency $f_s = 1$ it is worth noting that the continuous signal can be reconstructed from the discrete values by using a reconstruction filter which preserves only the signal components contained in the band $(0, f_s)$, as is shown in Figure 9.4. The reconstruction formula is:

$$x(t) = \sum_{n=-\infty}^{\infty} x(n) \frac{\sin [\pi(t - n)]}{\pi(t - n)} e^{j\pi(t-n)} \tag{9.23}$$

Sampling does not introduce any degradation to an analytic signal $x(t)$ if its spectrum does not contain any components with frequencies larger than or equal to f_s. Thus, the sampling theorem for an analytic signal is:

An analytic signal which does not contain any components with frequencies greater than or equal to f_m is wholly determined by the set of its values sampled at time intervals of $T = 1/f_m$.

The set $x(n)$ is decomposed into a real set $x_R(n)$ and an imaginary set $x_I(n)$, such that:

$$x(n) = x_R(n) + jx_I(n)$$

The corresponding Fourier transforms $X_{nR}(f)$ and $X_{nI}(f)$ are obtained from the Fourier transform $X_n(f)$ by equations (9.19) and (9.20) given above.

$$X_{nI}(f) = -jX_{nR}(f) \quad \text{for} \quad 0 < f < \tfrac{1}{2}$$
$$\qquad\quad jX_{nR}(f) \quad \text{for} \quad -\tfrac{1}{2} < f < 0$$

The relations between the sets $X_R(n)$ and $X_I(n)$ are obtained by considering the quadrature filter whose frequency response is given in Figure 9.5. The impulse response of this filter is the set $h(n)$, such that:

$$h(n) = \int_{-1/2}^{0} j e^{j2\pi nf} \, df + \int_{0}^{1/2} (-j) e^{j2\pi nf} \, df$$

$$h(n) = \frac{2}{\pi n} \sin^2 \left(\frac{n\pi}{2} \right) \quad \text{for} \quad n \neq 0$$

$$h(0) = 0 \tag{9.24}$$

By applying to this filter the set $x_R(n)$ we obtain the set $x_I(n)$, thus:

$$x_I(n) = \frac{2}{\pi} \sum_{\substack{m=-\infty \\ m \neq n}}^{\infty} x_R(n - m) \frac{\sin^2 [\pi(m/2)]}{m} \tag{9.25}$$

and similarly:

$$x_R(n) = -\frac{2}{\pi} \sum_{\substack{m=-\infty \\ m \neq n}}^{\infty} x_I(n - m) \frac{\sin^2 [\pi(m/2)]}{m} \tag{9.26}$$

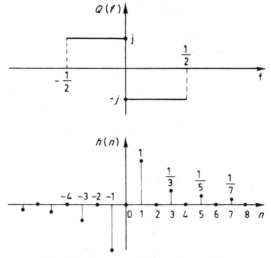

Fig. 9.5 Responses of the quadrature filter

The sets $x_R(n)$ and $x_I(n)$ are related by the discrete Hilbert transform [4].

Examination of the elements of the set $h(n)$ leads to several comments. First, the fact that every other element is zero implies that, if the set $x_R(n)$ also has every other element zero, then so must the set $x_I(n)$, and further, the sets $x_R(n)$ and $x_I(n)$ must be interleaved. An example will be given later.

The impulse response of the quadrature filter corresponds to a linear phase FIR filter as described in Section 5.2. Its frequency response is

$$Q(f) = -j2 \sum_{n=1}^{\infty} h(n) \sin(2\pi n f) \qquad (9.27)$$

In other to realize this filter the number of coefficients must be limited.

9.3 CALCULATING THE COEFFICIENTS OF AN FIR QUADRATURE FILTER

A realizable quadrature filter is easily obtained by limiting the number of terms over which summation (9.27) is performed. The frequency response then deviates from the ideal response. In practice the filter is specified by giving a limit δ for the ripple in a frequency band (f_1, f_2) as shown in Figure 9.6. A satisfactory FIR filter can be obtained using a low-pass filter and the results are obtained in Chapter 5.

One interesting example is that of the filter whose response is given in Figure 9.7. This filter is called a half-band filter because the pass band represents half of the useful band.

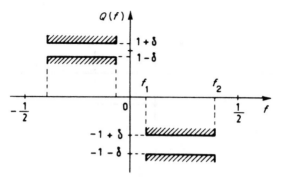

Fig. 9.6   Mask for a quadrature filter

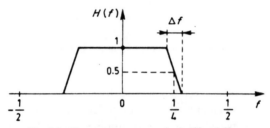

Fig. 9.7   Frequency response of a half-band filter

Further, $H(0.25) = 0.5$ and the response is assumed to be asymmetric about the point $f_c = 0.25$, that is, $H(0.25 + f) = 1 - H(0.25 - f)$. This filter is specified by the transition band Δf and the ripple in the pass and stop bands, which are equal to δ_0. Its coefficients can be calculated using conventional FIR filter design programs. This type of filter is of interest because the symmetry of its frequency response implies that the coefficients h_n are zero for even n except for h_0. Thus, if $n = 2p$, decomposition of $H(f)$ into the sum of two functions, leads to

$$h_{2p} = \int_0^{0.25} \cos(2\pi 2pf)\,df + (-1)^p \left[\int_0^{\Delta f/2} H(0.25 + f)\cos 4\pi pf\,df \right.$$

$$\left. + \int_0^{\Delta f/2} [H(0.25 - f) - 1]\cos 4\pi pf\,df \right]$$

and hence

$$h_{2p} = 0$$

For a number of coefficients $N = 4M + 1$ the frequency response is

$$H(f) = e^{-j2\pi 2Mf} \frac{1}{2}\left[1 + 2\sum_{i=1}^{M} h_{2i-1} \cos[2\pi(2i-1)f] \right] \qquad (9.28)$$

Translation of this response through 0.25 on the frequency axis leads to the function $H'(f)$ such that:

$$H'(f) = H(f - 0.25) = e^{-j2\pi 2Mf}\frac{1}{2}\left[1 - 2\sum_{i=1}^{M}(-1)^{i}h_{2i-1}\sin[2\pi(2i-1)f]\right]$$

The coefficients h'_n of the corresponding filter are

$$h'_{2i-1} = j(-1)^{i}h_{2i-1}; \quad h'_{-(2i-1)} = -j(-1)^{i}h_{2i-1}; \quad 1 \leqslant i \leqslant M$$

They have imaginary values. By comparing the expression for $H'(f)$ with equation (9.27) for $Q(f)$ it is evident that the set of coefficients a_n such that:

$$a_{2i-1} = -2(-1)^{i}h_{2i-1}; \quad a_{-(2i-1)} = 2(-1)^{i}h_{2i-1}; \quad 1 \leqslant i \leqslant M$$

represents the coefficients of a quadrature filter whose ripple equals $2\delta_0$ in the band $[\Delta f/2, \frac{1}{2} - \Delta f/2]$.

Example. For the specification $\delta_0 = 0.01$ and $\Delta f = 0.111$ it is found that $M = 5$.

$$a_1 = 0.6283 \quad a_3 = 0.1880$$
$$a_5 = 0.0904 \quad a_7 = 0.0443$$
$$a_9 = 0.0231$$

The expression $H'(f)$ corresponds to a complex filter which contains two parts, a circuit producing a delay of $2M$ elementary periods and a quadrature filter, as shown in Figure 9.8. The outputs of these two circuits from the real and imaginary parts of the complex signal. The system can be said to comprise two branches, one real and the other imaginary. It allows a real signal to be converted to an analytic signal and is thus an analytic FIR filter.

It should be noted that this property even if the fundamental low-pass filter is not of the half-band type. In this case, the coefficients with even index do not cancel out. In fact, the translation through a frequency of 0.25 corresponds to multiplication of the coefficients by a complex factor, so that the h'_n takes the values:

$$h'_n = e^{-j(\pi/2)n}h_n \tag{9.29}$$

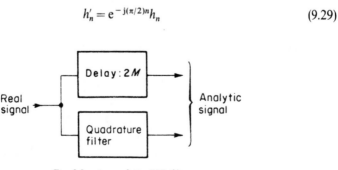

Fig. 9.8   An analytic FIR filter

Under these conditions, the real branch of the system is not a simple delay. A filtering function is achieved at the same time as the generation of the analytic signal.

Circuits with finite impulse response thus allow an ideal quadrature filter to be approximated without error in the phase shift but with approximation of the amplitude in the pass band. Circuits with infinite impulse response, or recursive circuits, provide an alternative approach. By using pure phase-shifters they allow the approximation of the quadrature filter with no error in the amplitude but with an approximation in the phase.

9.4 RECURSIVE 90° PHASE SHIFTERS

A recursive phase-shifter is characterized by the fact that the numerator and denominator of its Z-transfer function are image polynomials. That is, they have the same coefficients but in reverse order. The properties of phase-shifters were introduced in Section 6.3.

It is possible to design a pair of phase-shifters such that the output signals have a phase difference which approximates 90° with an error of less than ε, in a given frequency band (f_1, f_2). The calculation techniques are the same as for IIR filters. The procedure for producing a phase difference with an elliptic behaviour is as follows [5]:

(1) Determination of the order of the circuit:

$$N = \frac{K(k_1)K(\sqrt{(1-k^2)})}{K(k)K(\sqrt{(1-k_1^2)})}$$

with the following values for the parameters:

$$k = \frac{\tan(\pi f_1)}{\tan(\pi f_2)}; \quad k_1 = \left[\frac{1 - \tan(\pi \varepsilon)}{1 + \tan(\pi \varepsilon)}\right]^2$$

(2) Determination of the zeros z_i of the Z-transfer function:

$$A = Sn\left[\frac{(4i+1)K(\sqrt{(1-k^2)})}{2N}, \sqrt{(1-k^2)}\right]$$

(where Sn is the elliptic function).

$$p_i = -\tan(\pi f_1)\frac{A}{\sqrt{(1-A^2)}}$$

$$z_i = \frac{1+p_i}{1-p_i} \quad \text{for } 0 \leqslant i \leqslant N-1$$

Example: Assuming the specification $f_1 = 0.028$; $f_2 = 0.33$ and $\varepsilon = 1°$, it is found that:

$$\tan(\pi f_1) = 0.0875; \quad k = 0.0505; \quad k_1 = 0.9657; \quad N = 4.8$$

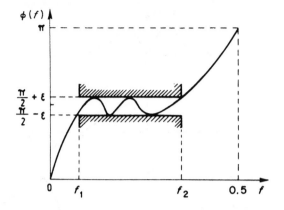

Fig. 9.9 *Characteristic of 90° phase shifter*

By taking $N = 5$ we obtain:

$$p_0 = -0.0395 \quad z_0 = 0.9240$$
$$p_1 = -0.3893 \quad z_1 = 0.4396$$
$$p_2 = -3.8360 \quad z_2 = -0.5864$$
$$p_3 = -1.0039 \quad z_3 = -0.00197$$
$$p_4 = -0.1509 \quad z_4 = 0.7377$$

In forming the circuit the first three zeros z_1 are attributed to one branch and the last two to the other. The variation in the phase difference with frequency is given by the function $\phi(f)$ represented in Figure 9.9.

Recursive phase shifters allow two orthogonal signals to be obtained. It should be noted that in performing this operation they also introduce a phase distortion which is the same for both signals. These circuits can be used in modulation and multiplexing equipment.

9.5 SINGLE SIDE-BAND MODULATION

Modulation of a signal results in a displacement of the spectrum on the frequency axis. It is single side-band (SSB) modulation if, for a real signal, the part of the spectrum which corresponds to positive frequencies is displaced towards positive frequencies and the part which corresponds to negative ones is displaced towards negative frequencies. Thus, the signal $s(t) = \cos \omega t$ has the modulated signal:

$$s_m(t) = \cos(\omega + \omega_0)t$$

This operation can be performed by the following procedure:

(1) Form the analytic signal $s_a(n) = s_R(n) + js_I(n)$ which corresponds to the real signal represented by the set $s(n)$.

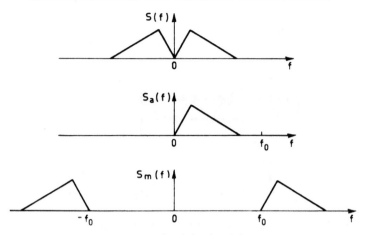

Fig. 9.10   Single side-band modulation

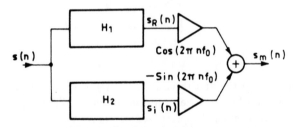

Fig. 9.11   Single side-band modulation circuit

(2) Multiply the set $s_a(n)$ by the set of complex numbers:

$$\cos(2\pi n f_0) - j \sin(2\pi n f_0)$$

and retain the real part $s_m(n)$ of the set thus obtained. Then,

$$s_m(n) = s_R(n) \cos(2\pi n f_0) + s_i(n) \sin(2\pi n f_0)$$

The signal spectrum evolves as shown in Figure 9.10 and the corresponding circuits are shown in Figure 9.11.

If the analytic filter is of the FIR type the set $s_R(n)$ is simply the delayed $s(n)$, which is different from the case of 90° recursive phase-shifters. The set which corresponds to the modulated signal $s_m(n)$ can be added to other modulated sets to provide, for example, a frequency multiplexed signal in telephony.

9.6 MINIMUM PHASE FILTERS

The properties of causal and analytic signals discussed at the beginning of the chapter allow clarification of a point concerning the phase characteristics of filters [3]. The frequency response of a filter $H(f)$ is written:

$$H(f) = A(f)e^{-j\phi(f)} \quad \text{with} \quad A(f) = |H(f)|$$

and

$$\phi(f) = -\arg[H(f)]$$

The term $A(f)$ is the attenuation and $\phi(f)$ is the phase shift produced by the filter on a sinusoidal signal, at frequency f.

If the filter has real coefficients-its impulse response, the set $h(n)$, is real and consequently:

$$H(f) = \overline{H(-f)}; \quad A(f) = A(-f) \quad \text{and} \quad \phi(f) = -\phi(-f)$$

The term $h(n)$ is obtained by:

$$h(n) = 2 \int_0^{1/2} A(f)\cos[2\pi nf - \phi(f)]\,df \tag{9.30}$$

As the response cannot precede the application of the signal to the filter a realizable filter must be causal, with:

$$h(n) = 0 \quad \text{for } n < 0$$

It follows that if the response is decomposed into a real and imaginary part:

$$H(f) = H_R(f) + jH_I(f)$$

the functions $H_R(f)$ and $H_I(f)$ are related by equations (9.11) and (9.12) given in Section 9.1.

Now, a filter is often only specified by the data for the constraints on the amplitude:

$$A^2(f) = H_R^2(f) + H_I^2(f)$$

and this results in a lack of definition in the phase shift.

To express this generally a certain time corresponding to the propagation time through the system is required to process a signal. This is characterized by the variation of the phase shift with frequency. In order to minimize this propagation time it is necessary to find the minimum phase shift characteristic, which removes the lack of definition in the calculation of the filter. Another approach is to specify a linear phase shift.

A stable and realizable filter has a Z-transfer function, $H(Z)$, whose poles are inside the unit circle and whose zeros may be outside it. Assume A_0 is such a zero and $H_1(Z)$ is the function such that:

$$H_1(Z) = \frac{1 - 2\,\mathrm{Re}(Z_0)Z^{-1} + |Z_0|^2 Z^{-2}}{|Z_0|^2 - 2\,\mathrm{Re}(Z_0)Z^{-1} + Z^{-2}}$$

This is the transfer function of a phase-shifter of the second order, which was introduced in Section 6.3 and whose group delay is given by equation (6.45). One can write:

$$H(Z) = H_1(Z)H_2(Z)$$

where $H_2(Z)$ is a function which is zero at Z_0^{-1} and introduces a smaller phase shift than $H(Z)$. By iteration it results that minimum phase shift is achieved with the function $H_m(Z)$, which is obtained by replacing with their inverse all the zeros in $H(Z)$ outside the unit circle.

The formulation of the minimum phase condition is that the function:

$$\log[H(Z)] = \log[A(Z)] - j\phi(Z)$$

should not have any poles outside the unit circle.

Under these conditions, the functions $\log[A(f)]$ and $\phi(f)$ are related by equations (9.11) and (9.12), which correspond to the Hilbert transform. They are the Bayard–Bode equations for discrete systems:

$$\log[A(f)] = K - \int_{-1/2}^{1/2} \phi(f') \cot \pi(f - f')\,df' \tag{9.31}$$

$$\phi(f) = \int_{-1/2}^{1/2} \log[A(f')] \cot \pi(f - f')\,df' \tag{9.32}$$

The constant K is a scaling factor for the amplitude.

Similarly, a function $H_M(Z)$ whose zeros are outside the unit circle is said to be maximum phase.

REFERENCES

[1] E. A. GUILLEMIN, *Theory of Linear Physical Systems*, John Wiley, New York, 1963.

[2] B. PICINBONO, *Principles of Signals and Systems*, Artech House Inc., London, 1988.

[3] A. OPPENHEIM and R. SCHAFER, *Digital Signal Processing*, Chapter 7, Prentice-Hall, Englewood Cliffs NJ, 1974.

[4] B. GOLD, A. OPPENHEIM and C. RADER, Theory and implementation of the discrete Hilbert transform. *Proc. of Symp. Computer Processing in Communications*, Vol. 19. Polytechnic Press, New York, 1970.

[5] B. GOLD and C. RADER, *Digital Processing of Signals*, Chapter 3, McGraw-Hill, New York, 1969.

EXERCISES

1 Calculate the Fourier transform $X(f)$ of the real causal set $x(n)$ such that:

$$x(n) = 0 \quad \text{for } n < 0$$
$$a^n \quad \text{for } n \geq 0$$

with

$$|a| < 1$$

Decompose $X(f)$ into real and imaginary parts.

2 Show that the function $X(Z)$ such that:

$$X(Z) = \frac{1}{1 - aZ^{-1}}$$

can be obtained from its real part $X_R(\omega)$ on the unit circle, given by:

$$X_R(\omega) = \frac{1 - a\cos\omega}{1 - 2a\cos\omega + a^2} \quad \text{with } |a| < 1$$

3 Starting from a real signal represented by the set $x(n)$, a complex signal is formed whose real and imaginary parts are given by:

$$x_R(n) = x(n)\cos\left[2\pi(n/4)\right]$$
$$x_I(n) = x(n)\sin\left[2\pi(n/4)\right]$$

What comments can be made about the sets $x_R(n)$ and $x_I(n)$?

Is the signal obtained an analytic signal?

A half-band filter is applied to each of the sets $x_R(n)$ and $x_I(n)$ and the complex signal filtered in this way is multiplied by the complex set $e^{-j2\pi n/4}$. What operation has been carried out on the real signal $x(n)$? Perform this set of operations on the signal $x(n) = \cos(\pi n/5)$.

4 Study the effect of coefficient wordlength limitation on the quadrature FIR filter. Following the procedure described in Sections 5.7 and 5.10 for linear phase filters, find an expression for the estimate of the coefficient wordlength as a function of the quadrature filter parameters, i.e. ripple and transition band.

5 Give a simplified expression for the order of the 90° IIR phase shifter. Study the effect of the coefficient wordlength limitation and find a formula for estimating the coefficient wordlength as a function of the parameters. Check the results on the example given in Section 9.4.

6 Consider a function $H(Z)$ defined by:

$$H(Z) = [(1 - Z_0 Z^{-1})(1 - \bar{Z}_0 Z^{-1})]^2$$

with

$$Z_0 = 0.5(1 + j)$$

This is a minimum phase function. Give the expression for the linear phase function and the maximum phase function which have the same amplitude characteristic. Compare the impulse responses.

CHAPTER 10

MULTIRATE FILTERING

Multirate filtering is a technique for reducing the calculation rate needed in digital filters and, in particular, the number of multiplications to be performed per second. As will be shown later, this parameter is generally regarded as being a reflection of the complexity of the system.

In a filter the number of multiplications M_R to be performed per second is given by:

$$M_R = Kf_s$$

where f_s is the frequency at which the calculations are made. The parameter f_s generally corresponds to the sampling frequency of the signal represented by the numbers to be processed. The factor K depends on the type of filter and on its performance.

In reducing the value of M_R the factor K can be influenced by choosing the most appropriate type and structure of filter and by optimizing the order of this filter to suit the constraints and required characteristics. Also, f_s can be influenced by changes in the sampling frequency during processing. In many practical cases the advantages thus obtained are considerable.

The sampling frequency for a real signal must be more than twice its bandwidth, which can vary during processing. For example, a filtering operation eliminates undesirable components, so that the useful bandwidth is reduced. Once the useful bandwidth has been decreased, the sampling frequency of the signal can itself be reduced. As a result, the sampling frequency can be adapted to the bandwidth of the signal at each stage of the processing, so as to maximize the computation speed in the filter. Before studying the development and implementation of this fundamental principle it is appropriate first to analyse the effect of a change in the sampling frequency on the signal and its spectrum.

10.1 COMPLEMENTS IN SAMPLING THEORY

Given the signal $s(t)$ whose spectrum $S(f)$ has no component with frequency higher than f_m and assuming that the signal is sampled with period T such that:

$$1/MT > 2f_m$$

where M is a whole number, let us examine the relationship between the Fourier transforms $S_i(f)$ of the sets:

$$s\left[\left(n+\frac{i}{M}\right)MT\right]; \quad i=0,1,2,\ldots,M-1$$

From the results in Section 1.2 the Fourier transform of the distribution $u_0(t)$,

$$u_0(t)=\sum_{n=-\infty}^{\infty}\delta(t-nMT)$$

is the distribution $U_0(f)$ given by

$$U_0(f)=\sum_{n=-\infty}^{\infty}e^{-j2\pi fnMT}=\frac{1}{MT}\sum_{n=-\infty}^{\infty}\delta\left(f-\frac{n}{MT}\right)$$

The Fourier transform of the distribution $u_i(t)$,

$$u_i(t)=\sum_{n=-\infty}^{\infty}\delta\left[t-\left(n+\frac{i}{M}\right)MT\right]; \quad i=0,1,\ldots,M-1 \tag{10.1}$$

is the distribution $U_i(f)$ given by

$$U_i(f)=\sum_{n=-\infty}^{\infty}e^{-j2\pi f(n+i/M)MT}=\frac{1}{MT}e^{-j2\pi fiT}\sum_{n=-\infty}^{\infty}\delta\left(f-\frac{n}{MT}\right)$$

or,

$$U_i(f)=\frac{1}{MT}\sum_{n=-\infty}^{\infty}e^{-j2\pi(in/M)}\delta\left(f-\frac{n}{MT}\right) \tag{10.2}$$

As $S_i(f)$ $(i=0,1,\ldots,M-1)$ is the convolution product of $S(f)$ with the distribution $U_i(f)$, we have

$$S_i(f)=\frac{1}{MT}\sum_{n=-\infty}^{\infty}e^{-j2\pi(in/M)}S\left(f-\frac{n}{MT}\right) \tag{10.3}$$

Let us calculate the spectrum $S^M(f)$, where

$$S^M(f)=\sum_{i=0}^{M-1}S_i(f)=\frac{1}{MT}\sum_{n=-\infty}^{\infty}S\left(f-\frac{n}{MT}\right)\sum_{i=0}^{M-1}e^{-j2\pi(in/M)}$$

As the second summation cancels out except for values of n which are multiples of M, this becomes:

$$S^M(f)=\frac{1}{T}\sum_{n=-\infty}^{\infty}S\left(f-\frac{n}{T}\right) \tag{10.4}$$

which is the spectrum corresponding to a signal sampled at frequency $1/T$.

The $S_i(f)$ terms can also be expressed as a function of $S^M(f)$, and, in equation (10.3) the summation can be decomposed as follows:

$$S_i(f)=(1/MT)\sum_{n=-\infty}^{\infty}\sum_{m=0}^{M-1}S\left[f-\left(n+\frac{m}{M}\right)\frac{1}{T}\right]e^{-j2\pi(im/M)}$$

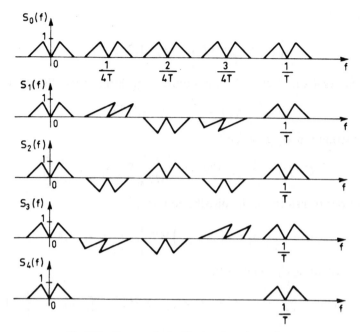

Fig. 10.1   Spectra obtained by interleaved samplings

or:

$$S_i(f) = (1/M) \sum_{m=0}^{M-1} e^{-j2\pi(im/M)} \frac{1}{T} \sum_{n=-\infty}^{\infty} S\left[\left(f - \frac{m}{MT}\right) - \frac{n}{T}\right]$$

and finally:

$$S_i(f) = (1/M) \sum_{m=0}^{M-1} e^{-j2\pi(im/M)} S^M\left(f - \frac{m}{MT}\right) \qquad (10.5)$$

Figures 10.1 illustrates the case for $M = 4$. The spectra $S_i(f)$ correspond to interleaved samplings at frequency $1/MT$ and the spectrum $S^M(f)$ corresponds to sampling at frequency $1/T$. A change in the sampling frequency will exchange these spectra.

It is interesting to note in Figure 10.1 that retarding the set of sampling pulses causes phase rotations through multiples of $2\pi/M$ for image bands around multiples of the sampling frequency $1/MT$.

Addition of all the sets of retarded pulses causes a cancellation of the image bands except around frequencies which are multiples of $1/T$, which becomes the new sampling frequency. This is an application of the linearity properties of the Fourier transform.

It is also useful to establish relations between the Z-transfer functions of the sequences involved. Let $S(Z)$ be the Z-transform of the set $s(nT)$. By definition:

$$S(Z) = \sum_{n=-\infty}^{\infty} s(nT)Z^{-n} \tag{10.6}$$

The spectrum $S^M(f)$ of the signal sampled with period T is obtained by replacing Z by $(e^{j2\pi fT})$. Then:

$$S^M(f) = S(e^{j2\pi fT})$$

Decomposing the summation in $S(Z)$ leads to:

$$S(Z) = \sum_{n=-\infty}^{\infty} \sum_{i=0}^{M-1} S(nMT + iT)Z^{-(nM+i)}$$

or:

$$S(Z) = \sum_{i=0}^{M-1} Z^{-i} \sum_{n=-\infty}^{\infty} S(nMT + iT)Z^{-nM}$$

Defining $S_i(Z^M)$ by:

$$S_i(Z^M) = \sum_{n=-\infty}^{\infty} S(nMT + iT)Z^{-nM}$$

we get:

$$S(Z) = \sum_{i=0}^{M-1} S_i(Z^M)Z^{-i} \tag{10.7}$$

The terms $s_i(Z^M)$ are Z-transforms of the sets $s[(n+i/M)MT]$ for $i = 0, 1, \ldots, M-1$. The factor Z^{-i} reflects the interleaving of these sets.

By analogy with equation (10.5):

$$Z^{-i}S_i(Z^M) = (1/M) \sum_{m=0}^{M-1} e^{-j2\pi(im/M)}S(Ze^{-j2\pi(m/M)}) \tag{10.8a}$$

Setting, as in chapter 2, $W = e^{-j2\pi/M}$, we get:

$$S_i(Z^M) = (1/M) \sum_{m=0}^{M-1} W^{im}Z^i S(ZW^m) \tag{10.8b}$$

Equations (10.7) and (10.8) are fundamental for multirate filtering.

The results obtained, and particularly equations (10.3) and (10.4), are valid for signals $s(t)$ whose spectrum is not limited to the frequency $1/2MT$, but spectrum aliasing then appears [1].

10.2 DECOMPOSITION OF A LOW-PASS FIR FILTER

Multirate filtering will be introduced first for FIR filters, where it appears naturally. Consider a low-pass FIR filter which eliminates components with a frequency greater than or equal to the frequency f_c in a signal sampled at frequency f_s. The filtered signal only requires a sampling frequency equal to

$2f_c$ and in fact it is sufficient to provide the output numbers at this frequency.

In an FIR filter of order N, the relation which determines the numbers of the output set $y(n)$ from the set of input numbers $x(n)$ is written:

$$y(n) = \sum_{i=0}^{N-1} a_i x(n-i) \qquad (10.9)$$

Each output number $y(n)$ is calculated from a set of N input numbers by weighted summation with the coefficients a_i ($i = 0, 1, \ldots, N-1$). Under these conditions the input and output rates are independent and a decrease in the output sampling rate by a factor $k = f_s/2f_c$, assumed to be an integer, results in a reduction in the computation rate by the same factor.

The same reasoning applies to raising the sampling rate, or interpolation. In this case, the output rate is greater than the input one. To show the savings in computation it is sufficient simply to regard the rates as being equal by incorporating a suitable amount of null data into the input set.

The independence of the output from the input in FIR filters can be exploited in narrow band-pass filters, even if the input and output rates are identical, by dividing the filtering operation into two phases [2]:

(1) Reducing the sampling frequency from the value f_s to an intermediate value f_0 such that:

$$f_0 \geqslant 2f_c$$

(2) Raising the sampling frequency or interpolating from f_0 to f_s.

Figure 10.2 illustrates this decomposition. If the two operations are carried out with similar filters of order N, the number of multiplications M_D to be performed per second is given by:

$$M_D = Nf_0 \times 2 \qquad (10.10)$$

This value is to be compared with the direct realization by a single filter, which results in the value M_R such that:

$$M_R = Nf_s \qquad (10.11)$$

Consequently, decomposition is advantageous as soon as $k = f_s/2f_c$ is greater

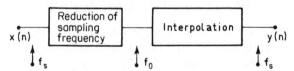

Fig. 10.2   Filter with sample rate reduction and increase

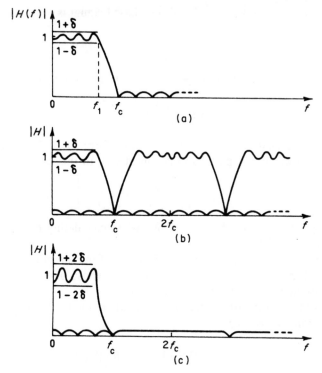

Fig. 10.3 *Frequency response of multirate filter. (a) Direct filter response; (b) sample rate reduction filter response; (c) multirate filter response*

than 2, that is, as soon as:

$$f_c < \frac{f_s}{4}$$

This approach, therefore, appears to be well suited to narrow band-pass filters.

It should be noted, however, that the filtering function obtained in the two cases is not exactly the same and that distortions are introduced by the decomposition, as shown in Figure 10.3. In effect, the intermediate subsampling at frequency $f_0 \geqslant 2f_c$ has three consequences:

(1) Aliasing of residual components with frequencies greater than $f_0/2$ into the band below $f_0/2$. Harmonic distortion results. Its power B_R depends on the attenuation of the sampling rate reduction filter and is calculated from its transfer function $H(f)$ by using the results given in the previous chapters.

For example, if the input signal has a uniform spectral distribution and unit power, the total power B_T of the aliased signal is

$$B_T = \frac{1}{f_s} \int_{f_0/2}^{f_s - f_0/2} |H(f)|^2 \, df \qquad (10.12)$$

If f_1 is the filter pass-band edge, an upper bound on B_T is provided by

$$B_T < \sum_{i=0}^{N-1} |a_i|^2 - \frac{2f_1}{f_s}$$

The distortion can be assumed to have a uniform spectral distribution and only the power in the pass band is considered. In this case we obtain:

$$B_R < \frac{2f_1}{f_0} \left[\sum_{i=0}^{N-1} |a_i|^2 - \frac{2f_1}{f_2} \right] \qquad (10.13)$$

This degradation must be allowed for when calculating the sampling rate reduction filter [3].

(2) The periodicity in frequency of the response of the sampling rate reduction filter, with period f_0, introduces a distortion whose power B_i is a function of the attenuation of the interpolation filter.

   If this filter is the same as the sampling rate reduction filter, with the same assumptions, we obtain:

$$B_i = (1/f_s) \int_{f_0/2}^{f_s - f_0/2} |H(f)|^2 \, df$$

This distortion outside the pass band can be troublesome when other signals are to be added to the filtered signal.

(3) Cascading two filters increases the ripple in the pass band. For example, the ripple is doubled if identical filters are used in both operations.

Finally, the sub-units of the multirate filter should be designed so that the complete system satisfies the overall specifications imposed on the filter.

The circuit in Figure 10.2 is simplified if the sampling frequencies of the signal before and after filtering can be different. The principle can also be applied to high-pass and band-pass filters, provided, for example, modulation and demodulation stages are introduced.

The principle of decomposition can be extended to the sampling rate reduction subunit and to the interpolation subunit, which introduces a further advantage. The FIR half-band filter is a particularly efficient element for implementing these subunits.

10.3 HALF-BAND FIR FILTERS

The half-band FIR filter was introduced in Section 9.3. This is a linear phase filter and the frequency response $H(f)$ takes the value $\frac{1}{2}$ at the frequency $f_s/4$.

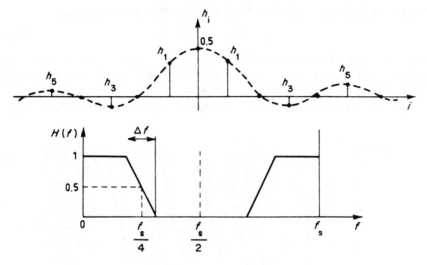

Fig. 10.4   Half-band FIR filter

It is antisymmetric about this point, i.e. the function $H(f)$ satisfies the equations:

$$H(f_s/4) = 0.5; \quad H\left(\frac{f_s}{4}+f\right) = 1 - H\left(\frac{f_s}{4}-f\right) \qquad (10.14)$$

For a number of coefficients $N = 4M + 1$,

$$H(f) = e^{-j2\pi 2Mf} \times \frac{1}{2}\left[1 + 2\sum_{i=1}^{M} h_{2i-1}\cos\left[2\pi(2i-1)f\right]\right] \qquad (10.15)$$

The coefficients h_n are zero for even n except for h_0. Figure 10.4 illustrates the properties of this filter. The specification is defined by the ripple (δ) in the pass and stop bands and by the width Δf of the transition band. Given these parameters, the formulae of Section 5.7 can be used to estimate the filter parameters. The filter order is

$$N \simeq \tfrac{2}{3}\log(1/10\delta^2)f_s/\Delta f$$

By taking the attenuation A_f as

$$A_f = 10\log(1/\delta^2)$$

and bearing in mind the particular significance of the frequency $f_s/4$ in this type of filter, one can write

$$M \simeq (2/3)(A_f/10 - 1)f_s/4\Delta f$$

Hence, the relationship between the attenuation (in decibels) and the transition band for a given number of coefficients can be expressed simply as

$$A_f = 10 + 15M(4\Delta f/f_s) \tag{10.16}$$

This approximation is valid when M exceeds a few units, as shown in Figure 10.5, which gives the actual correspondence between the attenuation and the transition band for the first few values of M.

Equation (10.16) and the curves of Figure 10.5 are very useful for determining

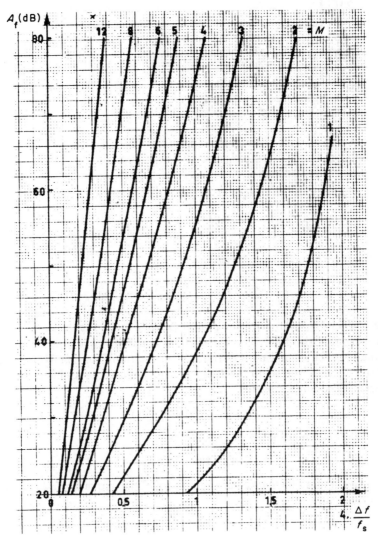

Fig. 10.5   *Variation of attenuation with transition bandwidth for half-band FIR filters*

the various elements of a cascade of half-band filters. The coefficients are calculated using the general program for FIR filters, with appropriate data. The relationship between the filter input and output sets is

$$y(n) = \frac{1}{2}\left[x(n - 2M) + \sum_{i=1}^{M} h_{2i-1}[x(n - 2M + 2i - 1) + x(n - 2M - 2i + 1)] \right]$$

$$(10.17)$$

and M multiplications have to be performed for each element of the output set $y(n)$. It should be noted that these operations are performed only on elements of the input set with odd index. It follows that if such a filter is used to reduce the sampling frequency from f_s to $f_0 = f_s/2$, the number of multiplications to be performed per second is Mf_0. The same is true if the sampling frequency f_s is increased from f_0 to $2f_0$, which is achieved simply by calculating a sample between two samples of the input set.

Finally, the number of multiplications to be performed per second in a half-band filter with a change of sampling frequency is

$$M_R = \left[\tfrac{2}{3}\log\left(\frac{1}{10\delta^2} \right) \frac{f_s}{\Delta f} \frac{1}{4} \right] f_s/2 \qquad (10.18)$$

The number of data memories needed with a reduction of the sampling frequency is the number required to store the data over which the weighted summation is carried out, i.e. $MM_D = 2M$. In an interpolation it is not only necessary to calculate the weighted sum but also to interleave the results with the input set delayed by M periods. Then

$$MM_D = 3M$$

In both cases the number of memories needed for the coefficients is

$$MM_C = M$$

Example: A group of half-band filters with useful application characteristics [4] is shown in Table 10.1, which gives the quantized coefficients, the quantization step being taken as unity.

The frequency response can be calculated simply from equation (10.15). Filters F4, F6, F8, and F9 correspond to a unit value of the parameter $4\Delta f/f_s$, with ripples of 37, 50, 67, and 79 dB, respectively. Filters F2, F3, and F5 have monotonic responses. The advantages of the particular filter structure described in this section can be applied to general multirate filtering.

10.4 DECOMPOSITION WITH HALF-BAND FILTERS

The properties of the elementary half-band filter can be used to produce the multirate filter design shown in Figure 10.6. The intermediate frequency f_0 is

Table 10.1. HALF-BAND FIR FILTER

Filter	h_0	h_1	h_3	h_5	h_7	h_9
F 1	1	1				
F 2	2	1				
F 3	16	9	-1			
F 4	32	19	-3			
F 5	256	180	-25	3		
F 6	346	208	-44	9		
F 7	512	302	-53	7		
F 8	802	490	-116	33	-6	
F 9	8 192	5 042	$-1 277$	429	-116	18

related to the sampling frequency f_s by a power of:

$$f_s = 2^P f_0 \tag{10.19}$$

Reducing and increasing the sampling frequency are carried out by a cascade of P half-band filters. The complete unit comprises a basic filter which operates at frequency f_0 and has a cascade of half-band filters on either side.

The overall low-pass filter is specified by the following parameters:

(1) Ripple in the pass band: δ_1
(2) Ripple in the stop band: δ_2
(3) Width of the transition band: Δf
(4) Edge of the pass band: f_1
(5) Edge of the stop band: f_2

To calculate the half-band filters their specifications have to be defined. The ripple in the pass band is assumed to be divided between the half-band filters and the basic filter. Also, each filter has to have a ripple in the stop band smaller than δ_2. As a result, for each half-band filter the ripple δ_0 is given by:

$$\delta_0 = \min\left\{\frac{\delta_1}{4P}, \delta_2\right\} \tag{10.20}$$

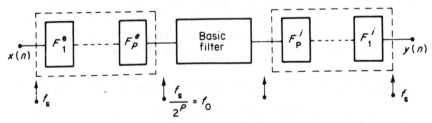

Fig. 10.6 Decomposition with half-band filters

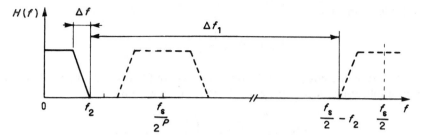

Fig. 10.7 *Transition band of first half-band filter*

The first half-band filter F_1^e of the cascade can be determined if its transition band Δf_1 is fixed (Figure 10.7). To fix Δf_1 it must be taken that the role of the first filter is to eliminate those components of the signal which can be folded in the useful band after having the sampling frequency. Thus:

$$\Delta f_1 = \frac{f_s}{2} - 2f_2$$

From equation (10.18) the number of multiplications M_C to be performed in the first filter is

$$M_{C_1} = \tfrac{2}{3}\log\left(\frac{1}{10\delta_0^2}\right)\left(\frac{f_s}{\Delta f_1}\right)\frac{1}{4}$$

By introducing the parameter k, which represents the difference between f_2 and $\frac{1}{2}f_s/2^p$

$$k = \frac{f_2}{f_s} \times 2^{P+1}$$

and taking:

$$D(\delta_0) = \tfrac{2}{3}\log\left(\frac{1}{10\delta_0^2}\right)$$

we have

$$M_{C_1} = D(\delta_0)\left[\frac{1}{2} - \frac{k}{2^P}\right]^{-1} \times \frac{1}{4}$$

The transition band of the ith filter of the cascade is written as

$$\Delta f_i = \frac{f_s}{2^i} - 2f_2 = f_s\left[\frac{1}{2^i} - \frac{k}{2^P}\right]$$

As the sampling frequency at the input of this filter is $f_s/2^{i-1}$ the number of multiplications M_C is

$$M_{C_i} = D(\delta_0)\frac{1}{2^{i-1}}\left(\frac{1}{2^i} - \frac{k}{2^P}\right)^{-1}\frac{1}{4}$$

As the sampling frequency at the output of the ith filter is $f_s/2^i$, the total number of multiplications M_C in the cascade of P filters becomes

$$M_C = \sum_{i=1}^{P} D(\delta_0)\frac{1}{2^{i-1}}\left(\frac{1}{2^i} - \frac{k}{2^P}\right)^{-1}\frac{1}{4}\frac{f_s}{2^i}$$

Hence,

$$M_C = \tfrac{1}{2}D(\delta_0)f_s\sum_{i=1}^{P}\frac{1}{2^i}\left[1 - \frac{k}{2^{P-i}}\right]^{-1}$$

The function $S(k)$ given by

$$S(k) = \sum_{i=1}^{P}\frac{1}{2^i}\left[1 - \frac{k}{2^{P-i}}\right]^{-1} \tag{10.21}$$

has values close to unity except for low values of P and when k approaches unity. This should be avoided, for example, by reducing the number of filters in the cascade. Consequently, a simple approximate expression can be given as:

$$M_C \simeq \tfrac{1}{3}\log\left(\frac{1}{10\delta^2}\right)f_s \tag{10.22}$$

The number of memories for the coefficients in each filter is equal to the number of multiplications M_{C_i}, thus

$$MM_{CC} = \sum_{i=1}^{P} M_{C_i} = \tfrac{1}{2}D(\delta_0)\sum_{i=1}^{P}\left[1 - \frac{k}{2^{P-i}}\right]^{-1}$$

This value can be approximated by:

$$MM_{CC} \simeq \tfrac{1}{2}D(\delta_0)\left[P + k\left(\frac{1}{1-k} + \frac{1}{1-2k}\right)\right] \tag{10.23}$$

The number of data memories is twice that for the coefficients with suitable choices of structure, as will be seen later.

To estimate the volume of calculations to be performed per second in the complete filter the order N of the basic filter must be determined:

$$N \simeq D\left(\frac{\delta_1}{2}, \delta_2\right)\frac{1}{\Delta f}\frac{f_s}{2^P}$$

with

$$D\left(\frac{\delta_1}{2}, \delta_2\right) = \tfrac{2}{3}\log\left(\frac{1}{5\delta_1\delta_2}\right)$$

The values of the parameters showing the complexity of the complete filter in Figure 10.6 are finally:

(1) The number of multiplications per second:

$$M_R = f_s \left[D(\delta_0) + \frac{1}{2^{2P+1}} \frac{f_s}{\Delta f} D\left(\frac{\delta_1}{2}, \delta_2\right) \right] \tag{10.24}$$

(2) The number of data memories:

$$MM_D = D\left(\frac{\delta_1}{2}, \delta_2\right) \frac{f_s}{\Delta f} \frac{1}{2^P} + 2D(\delta_0) \left[P + k\left(\frac{1}{1-k} + \frac{1}{2-k}\right) \right] \tag{10.25}$$

(3) The number of memories for the coefficients:

$$MM_C = D\left(\frac{\delta_1}{2}, \delta_2\right) \frac{f_s}{\Delta f} \frac{1}{2^{P+1}} + \tfrac{1}{2}D(\delta_0) \left[P + k\left(\frac{1}{1-k} + \frac{1}{2-k}\right) \right] \tag{10.26}$$

This estimate is based on the assumption that the two cascades of half-band filters are identical. Equations (10.24)–(10.26) are of practical use for evaluating the advantages of multirate filtering in a given case. The two examples below illustrate their application.

Example 1: Assume a narrow low-pass filter defined by the following:

$$f_s = 1; \quad f_2 = 0.05; \quad \Delta f = 0.025; \quad \delta_1 = 0.01; \quad \delta_2 = 0.001$$

The parameters have the values:

$$P = 3; \quad \delta_0 = \min\{0.01/12; 0.001\} = 0.00083; \quad D(\delta_0) = 3.3$$
$$D(\delta_1/2, \delta_2) = 2.76; \quad k = 0.8; \quad S(k) = 1.6$$

Thus:

$$M_R = 6.2; \quad MM_D = 65; \quad MM_C = 20$$

A direct realization results in a filter of order $N = 110$, which corresponds to the values:

$$M_R = 55; \quad MM_D = 110; \quad MM_C = 55$$

Example 2: Assume a very narrow filter such that:

$$f_s = 1; \quad f_2 = 0.005; \quad f = 0.00025; \quad \delta_1 = 0.001; \quad \delta_2 = 0.0001$$

The parameters have the values:

$$P = 6; \quad \delta_0 = \min\{0.001/24; 0.0001\} = 0.0000416;$$
$$D(\delta_0) = 5.38; \quad D(\delta_1/2, \delta_2) = 4.72; \quad k = 0.64; \quad S(k) = 1.07$$

Thus:

$$M_R = 7.76; \quad MM_D = 350; \quad MM_C = 152$$

A direct realization would result in a filter with an order N of several thousands, which cannot be envisaged in practice.

These examples clearly show the importance of multirate filtering techniques. It is also interesting to compare them with IIR filters. This comparison will be made on a filter satisfying the following specification:

$$\delta_1 = \delta_2 = 10^{-2}$$

$$\Delta f = \frac{f_1 + f_2}{2} \times 10^{-1}$$

and for which the position of the transition band, i.e. the relation:

$$\frac{2f_s}{f_1 + f_2}$$

can be varied.

The IIR filter is assumed to be of the elliptic type and to be implemented using second-order sections, requiring four multiplications each. Assuming that $f_s = 1$, the number of multiplications to be performed per second and the number of data memories are given in Table 10.2.

The results show that multirate filtering is clearly more advantageous than the FIR filter both for data memories and multiplications. In contrast, the IIR filter, which has a minimum phase shift, requires the smallest number of memories.

It can be seen that if the basic filter has a higher order, and if phase linearity is not indispensable, then it is advantageous to replace the basic filter, which operates at the frequency f_0 at both input and output, with an IIR filter. A substantial reduction in the volume of calculation can thus be obtained.

Multirate filtering using cascades of half-band filters is thus shown to reduce the number of multiplications and memories, but the implementation of a filter as a cascade of sub-units each operating at different rates can complicate the

Table 10.2. COMPARISON OF COMPLEXITIES FOR FIR, MULTIRATE AND IIR FILTERS

$\dfrac{2f_s}{f_1 + f_2}$	Multiplications			Memories		
	FIR	Multirate	IIR	FIR	Multirate	IIR
2	45	23	15	90	90	7
3	65	19	15	130	80	7
5	110	12	15	220	95	7
10	220	8	15	440	105	7

sequencing of the operations, and thus increase the complexity of the system control unit or the size of the program memory in the processing computers. To assess the importance of the latter point, the techniques for implementation must be studied, with particular attention being paid to half-band filters.

10.5 IMPLEMENTATION STRUCTURES

As the multirate filters being discussed are basically FIR elements their structures are deduced from those given in Section 5.9. Nevertheless, modifications are required and suitable structures must be chosen so as to avoid, for example, having to introduce extra buffer memories between the stages. Some such structures are given in Figure 10.8 for the half-band filter for both reducing and increasing the sampling frequency.

The technique employed is that of the transposed structure, making use of the symmetry of the coefficients and of the fact that the central coefficient h_0 has the value $\frac{1}{2}$.

The filter illustrated in Figure 10.8(a) has three different coefficients and requires six data memories. The input set $x(n)$ is decomposed into two interleaved sets $x(2p)$ and $x(2p-1)$ which correspond to even and odd indices. The elements of the set $x(2p)$ are added at a suitable point during the course of the calculation.

Table 10.3 gives the contents of the memory registers at a given time. They contain partial sums, and the last register contains the final result. This result is available throughout a complete sampling period and can be used by the following stage. The filter in Figure 10.8(b) performs the same calculations, always at the lowest frequency, but it has to have an extra data memory to interleave the input numbers and the calculated numbers.

The method described for performing the calculations is characterized by the fact that multiplications by the filter coefficients are performed before all other operations. It can be particularly advantageous, instead of using separate registers as in Figure 10.8, to use a single very long register, for example, in serial arithmetic. The memory is then divided into sections which contain the various partial sums.

This approach is illustrated in Figure 10.9, which gives the design corresponding to the summation:

$$y(n) = \sum_{i=0}^{N-1} a_i x(n-i-1) \tag{10.27}$$

A single register contains all the partial sums, which are shifted at the frequency of the data. This structure lends itself particularly well to a completely serial realization with a single arithmetic unit. The control unit simply provides a signal for the input and output of the data.

The design can be adapted to suit the symmetry of the coefficients of the linear phase filter and to an increase or decrease in the sampling frequency. In

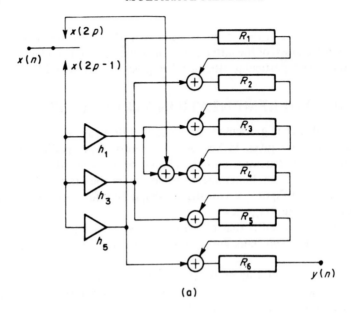

(a)

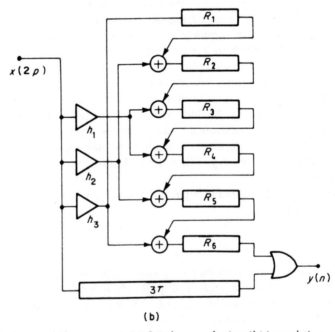

(b)

Fig. 10.8   Half-band filter structures. (a) Sample rate reduction; (b) interpolation

Table 10.3. CONTENTS OF THE MEMORIES FOR A HALF-BAND FILTER

R_1	R_2	R_3	R_4	R_5	R_6
$h_5 x_{2p-1}$	$h_5 x_{2p-3}$ $+$ $h_3 x_{2p-1}$	$h_5 x_{2p-5}$ $+$ $h_3 x_{2p-3}$ $+$ $h_1 x_{2p-1}$	$h_5 x_{2p-7}$ $+$ $h_3 x_{2p-5}$ $+$ $h_1 x_{2p-3}$ $+$ $h_1 x_{2p-1}$ $+$ x_{2p-2}	$h_5 x_{2p-9}$ $+$ $h_3 x_{2p-7}$ $+$ $h_1 x_{2p-5}$ $+$ $h_1 x_{2p-3}$ $+$ $h_3 x_{2p-1}$ $+$ x_{2p-4}	$h_5 x_{2p-11}$ $+$ $h_3 x_{2p-9}$ $+$ $h_1 x_{2p-7}$ $+$ $h_1 x_{2p-5}$ $+$ $h_3 x_{2p-3}$ $+$ $h_5 x_{2p-1}$ $+$ x_{2p-6}

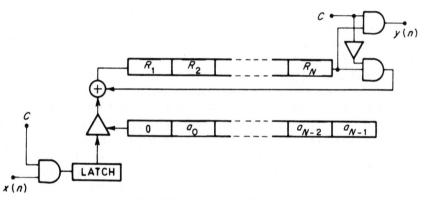

Fig. 10.9   *Filter structure with long register*

the latter case an accumulator has to be incorporated. Further, this approach is suitable for multiplexed filters where it is sufficient to interleave the sets of numbers to be filtered and to present the coefficients in the correct order.

Structures other than half-band filters can be envisaged for multirate filtering. First, one can seek a means of reducing the number of stages by varying the sampling frequencies by factors greater than two. Techniques for choosing the best factors are given in Ref. [5]. The number of calculations required is higher than with half-band filters but the interconnection is simpler. This can be appreciable if the use of a general purpose computer is envisaged. Reference [6] gives a program for calculating a multirate filter using these techniques.

Minimization of the circuitry can lead to rounding the intermediate results.

This introduces round-off noise, which appears at each stage and is accumulated in the filtering chain. Structures such as multirate filtering which reduce the number of calculations are in that respect very attractive. It is straightforward to evaluate the round-off noise by using the results given in Chapter 5 and adapting them to the properties of the particular elements used.

Multirate filtering, as described above, is based upon a decomposition of the filters with the introduction of non-recursive subunits. The discussion can be generalized by raising the question as to whether it might be possible to make computational savings through the use of recursive subunits. In fact, it is important, in order to optimize multirate processing in more general cases than that of a single filter, to better understand the physical processes which are involved.

10.6 ADVANTAGES OF THE PHASE SHIFT FUNCTION

Gains in computation speed correspond to a reduction in redundancy. As the computation rate is proportional to the sampling frequency redundancy exists when there are points in the processing at which the signal is sampled at a frequency greater than is strictly necessary. In order to eliminate this redundancy the sampling frequency must be minimized at each point in the system, which can constitute an optimization criterion. Such a condition, which is necessary but not sufficient, requires that for each elementary subunit the sampling frequency at input and output should not exceed twice the highest frequency component in the useful signal. This condition can be satisfied easily in non-recursive structures, as the sampling frequencies at the input and output are independent. In recursive circuits there is a feedback loop between the output and the input; the output numbers are added to the input numbers and the sampling frequencies at the input and output must be equal. Under these conditions the only way of satisfying the optimization criterion given above is to use only subunits which do not remove any components of the signal. Then, if the sampling frequency is minimal at the input it will also be minimal at the output. Subunits which do not affect the amplitude of the signal can operate only on the phase; thus, they are phase-shifters.

The phase relations between different samplings of the same signal were developed in Section 10.1. The results will be used to analyse multirate filtering from this point of view.

Assume that the sampling frequency f_s is reduced by a factor N. Let $X(Z)$ be the Z-transform of the input set $x(n)$; the Fourier transform is obtained by replacing Z with $e^{j2\pi f/f_s}$, forming $X(e^{j2\pi f/f_s})$. The output set $y(Nn)$ sampled at frequency f_s/N has $Y(Z^N)$, a function of Z^N, as its Z-transform. Consequently, if phase shifting circuits are involved in this operation their transfer function is also a function of the variable Z^N and can be calculated from the overall filter function.

The structure of a phase-shifter will be discussed initially for the single element making up the half-band FIR filter, which is defined by equation (10.17). This relation can be rewritten as

$$y(n) = \frac{1}{2}\left[x(n - 2M) + \sum_{i=1}^{2M} a_i x(n - 2i + 1) \right] \tag{10.28}$$

with:

$$a_i = h_{(2M - 2i + 1)} = a_{2M - i + 1} \quad \text{for } 1 \leqslant i \leqslant M$$

The corresponding Z-transfer function is

$$H(Z) = \frac{1}{2}\left[Z^{-2M} + Z^{-1} \sum_{i=0}^{2M-1} a_{i+1} Z^{-2i} \right] \tag{10.29}$$

or, again:

$$H(Z) = \frac{1}{2}[H_0(Z^2) + Z^{-1} H_1(Z^2)] \tag{10.30}$$

The corresponding frequency response is

$$H(f) = \frac{1}{2}[e^{-j2\pi(f/f_s)2M} + e^{-j2\pi(f/f_s)} H_1(f)] \tag{10.31}$$

The function $H_0(f)$ corresponds to a delay and so is the characteristic of a purely linear phase-shifter.

Because of the symmetry of the coefficients, $H_1(f)$ is also a linear phase function. Since this part of the filter is operating at the frequency $f_s/2$, $H_1(f)$ displays the periodicity $f_s/2$. As the number of the coefficients is even, we can use the results of Section 5.2, to write:

$$H_1(f) = \exp\left(-j2\pi f(M - \tfrac{1}{2})\frac{2}{f_s} \right)|H_1(f)|$$

or, again:

$$H_1(f) = e^{-j2\pi(f/f_s)2M} e^{-j\phi(f)} |H_1(f)| \tag{10.32}$$

The function $\phi(f)$ is linear and has the periodicity $f_s/2$. Consequently, it is expressed by:

$$\phi(f) = \pi\left(\left[\frac{2f}{f_s} + \frac{1}{2} \right] - \frac{2f}{f_s} \right) \tag{10.33}$$

where $[x]$ represents the largest whole number contained in x.

Figure 10.10 shows the functions $|H_1(f)|$ and $\phi(f)$ which characterize the filter. For the part of the filter with transfer function $H_1(Z^2)$ the amplitude is constant beyond the transition band except for the ripple. It cancels at $f_s/4$, the frequency which corresponds to a change in phase of π. The phase $\Phi(f)$ such that:

$$\Phi(f) = \phi(f) + 2\pi(f/f_s) \tag{10.34}$$

is constant and has the value 0 or π.

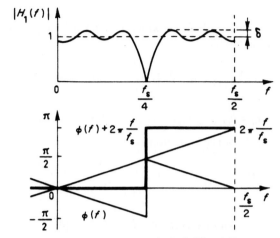

Fig. 10.10   Function $H_1(f)$ for half-band filter

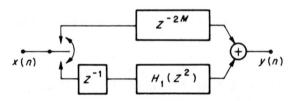

Fig. 10.11(a)   Half-band filter with phase shifters

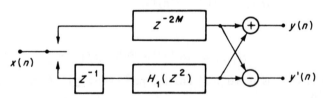

Fig. 10.11(b)   Filter bank with two filters

Finally, the circuit whose frequency response is $e^{-j2\pi f/f_s}H_1(f)$ approximates a pure linear phase-shifter in the pass and stop bands. Its number of coefficients and its complexity depend on the degree of approximation, i.e. on the transition band Δf and on the amplitude δ of the ripple. These results are similar to those of Section 9.3.

The half-band filter appears as a network with two branches, as in Figure 10.11(a). The overall response corresponds to the sum of the responses of the two branches as shown in Figure 10.12.

If the sign of function $H_1(f)$ is changed, then a high-pass filter is obtained,

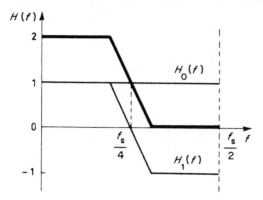

Fig. 10.12   Amplitude of half-band filter frequency response

as shown in Figure 10.12. Thus, a pair of filters is determined by the calculations for one of them. Let $B_0(z)$ and $B_1(z)$ be the transfer functions of these filters. The corresponding system is shown in Figure 10.11(b). It is characterized by the matrix equation:

$$\begin{bmatrix} B_0(Z) \\ B_1(Z) \end{bmatrix} = \begin{bmatrix} 1 & 1 \\ 1 & -1 \end{bmatrix} \begin{bmatrix} Z^{-2M} \\ Z^{-1}H_1(Z^2) \end{bmatrix} \tag{10.35}$$

It is worth noting that the matrix of the discrete Fourier transform of order 2 appears in the expression. The generalization to a bank of N filters leads to a DFT of order N. This is presented in Section 10.8.

It can also be noted that, by changing the sign of every other coefficient in $H_1(f)$, the frequency response is displaced by $f_s/4$ and the quadrature filter of Section 9.3 is obtained.

The results obtained for the half-band filter can be generalized quite simply by using an FIR filter to reduce or increase the sampling frequency by a factor of N. Assume $H(Z)$ is the Z-transfer function of such a filter. By assuming that it has KN coefficients, we can write:

$$H(Z) = \sum_{i=1}^{KN} a_i Z^{-i} = \sum_{n=0}^{N-1} Z^{-n} H_n(Z^N) \tag{10.36}$$

with:

$$H_n(Z^N) = \sum_{k=0}^{K-1} a_{kN+n}(Z^{-N})^k$$

This filter can be implemented by a network with N paths, as in Figure 10.13, which is called a polyphase network because each path has a frequency response which approximates that of a pure phase-shifter. The phase shifts are constant in frequency and are whole multiples of $2\pi/N$. When there is a change in sampling

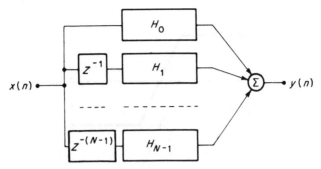

Fig. 10.13   *Filtering by polyphase network*

frequency by a factor of N, the circuits in the different paths of the network operate at the frequency f_s/N.

As infinite impulse response filters have greater selectivities than finite impulse response filters it is important to study multirate filters with recursive elements.

10.7 MULTIRATE FILTERING WITH IIR ELEMENTS

The basic technique for calculating a multirate filter with IIR elements is to perform the same type of decomposition on the transfer function $H(Z)$ of the overall IIR filter as was used to produce equation (10.36). The function $H(Z)$ is assumed to be a rational fraction in which the denominator and the numerator have the same degree.

Such a decomposition is obtained by finding the poles of $H(Z)$:

$$H(Z) = a_0 \frac{\prod_{k=1}^{K}(Z - Z_k)}{\prod_{k=1}^{K}(Z - P_k)} \qquad (10.37)$$

From the identity:

$$Z^N - P_k^N = (Z - P_k)(Z^{N-1} + Z^{N-2}P_k + \cdots + P_k^{N-1}) \qquad (10.38)$$

one can write

$$H(Z) = a_0 \frac{\prod_{k=1}^{K}(Z - Z_k)(Z^{N-1} + P_k Z^{N-2} + \cdots + P_k^{N-1})}{\prod_{k=1}^{K}(Z^N - P_k^N)}$$

which, in another form, is

$$H(Z) = \frac{\sum_{i=0}^{KN} a_i Z^{-i}}{1 + \sum_{k=1}^{K} b_k Z^{-Nk}}$$

Thus

$$H_n(Z^N) = \frac{\sum_{k=0}^{K} a_{kN+n} Z^{-Nk}}{1 + \sum_{k=1}^{K} b_k Z^{-Nk}} \qquad (10.39)$$

or

$$H_n(Z^N) = \frac{N_n(Z^N)}{D(Z^N)}$$

(10.40)

Each path of the polyphase network is thus determined. They all have the same recursive part and are distinguished by the non-recursive one as shown in equation (10.40). In principle, when compared with the previous section the difference is that the individual IIR phase shifters obtained do not have linear phase.

The circuit of a multirate filter in this form is given in Figure 10.14 for the case of an increase in the sampling rate. All the computations are made at frequency $1/NT$, and it can be seen that a total of $KN + K + 1$ multiplications is required. A direct realization of the IIR filter requires $(2K + 1)N$ multiplications under the same conditions. Consequently, the decomposition contributes an improvement which is only about a factor of 2. The real value of this decomposition lies in its use with banks of filters.

In the procedure above it should always be noted that the recursive part of the circuit in Figure 10.14 corresponds to poles raised to the power N, P_k^N. This

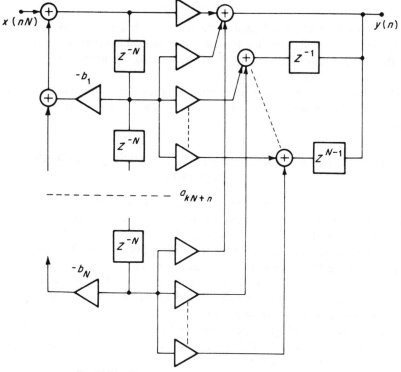

Fig. 10.14 *Recursive structure for sample rate increase*

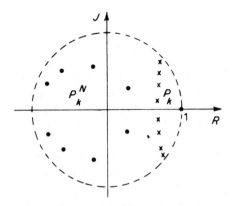

Fig. 10.15   Poles of polyphase network

transformation is shown in Figure 10.15 for $k = 8$. The poles are seen to be distributed inside the unit circle and at a distance from it [7]. Indeed if we have $|P_k| = 1 - \varepsilon$ with ε very small and positive then, after raising to the power N,

$$|P_k^N| \simeq 1 - N\varepsilon$$

As shown in Section 7.5, and particularly in equation (7.56), this is very favourable for the effects on the transfer function of the denominator coefficient wordlength limitation. In fact, the structure is realizable in the direct form even though the initial filter is not.

The effects of coefficient wordlength limitations and the evaluation of the round-off noise power can be derived in a simple way from the results in Chapter 7.

10.8 FILTER BANKS USING POLYPHASE NETWORKS AND DFT

A discrete Fourier transform computer is a bank of filters (Section 2.4) which are suitable for multirate filtering. However, it should be noted that filtering functions achieved in this way have a large degree of overlap. To improve the discrimination between the components of the signal the numbers are weighted before applying the discrete Fourier transform. The weighting coefficients are samples of functions called spectral analysis windows. The production of filter banks by means of a polyphase network and a DFT represents the generalization of such spectral analysis windows [8].

Assume that a bank is to be created with N filters which cover the band $[0, f_s]$ and are obtained by shifting a basic filter function along the frequency axis by mf_s/N with $1 < m < N - 1$.

If $H(Z)$ is the basic filter Z-transfer function, a change in frequency by mf_s/N

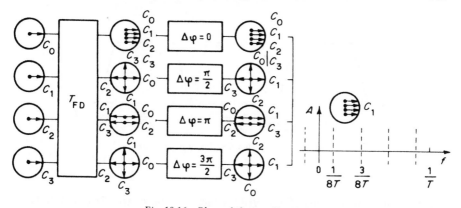

Fig. 10.16 *Phase shifts in a filter bank*

appears as a change in the variable from Z to $Z\,\mathrm{e}^{\mathrm{j}2\pi m/N}$. That is, the filter with index m has a transfer function $B_m(Z)$ given by

$$B_m(Z) = H(Z\,\mathrm{e}^{\mathrm{j}2\pi m/N})$$

By applying the decomposition of $H(Z)$ introduced in the earlier sections this becomes:

$$B_m(Z) = \sum_{n=0}^{N-1} Z^{-n}\,\mathrm{e}^{-\mathrm{j}2\pi(mn/N)}H_n(Z^N)$$

By allowing for the fact that the functions $H_n(Z^N)$ are the same for all the filters a factorization can be introduced which results in the following matrix equation:

$$
\begin{bmatrix} B_0 \\ B_1 \\ \vdots \\ B_{N-1} \end{bmatrix}
=
\begin{bmatrix}
1 & 1 & 1 & \cdots & 1 \\
1 & W & W^2 & \cdots & W^{N-1} \\
\vdots & \vdots & \vdots & & \vdots \\
1 & W^{N-1} & W^{2(N-1)} & \cdots & W^{(N-1)^2}
\end{bmatrix}
\begin{bmatrix} H_0(Z^N) \\ Z^{-1}H_1(Z^N) \\ \vdots \\ Z^{-(N-1)}H_N(Z^N) \end{bmatrix}
\tag{10.41}
$$

where $W = \mathrm{e}^{-\mathrm{j}2\pi/N}$.

The square matrix is the matrix of the discrete Fourier transform. The bank of filters is realized by forming a cascade of the polyphase network described in Figure 10.13 and a discrete Fourier transform.

The operation of this device is illustrated in Figure 10.16, which shows, for the case where $N = 4$, the phase shifts introduced at the various points in the system so as to preserve only the signal in the band $[\frac{1}{2}NT, \frac{3}{2}NT]$.

The polyphase network has the effect of correcting, in the useful part of the elementary band $[\frac{1}{2}NT, \frac{3}{2}NT]$, for the interleaving of the numbers at the output of the discrete Fourier transform computer. This allows overlap between the filters to be avoided and leads to the filter function in Figure 10.17. This function

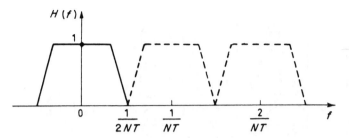

Fig. 10.17   *Filtering function of a filter bank*

depends only on the basic filter $H(Z)$, which can be of the FIR or IIR type, and which can be specified so that the filters in the bank have no overlap or have a crossover point, for example, at 3 dB or 6 dB.

The impulse response of the basic filter is the spectral analysis window of the system. If the filters are specified to have no frequency overlap the sampling frequency at the output of the filters, or at the input according to the method of use, can have the value $1/NT$ and the overall calculation can be performed at this frequency.

If N is assumed to be a power of 2, the fast Fourier transform algorithm can be used to calculate the discrete Fourier transform and the number of real multiplications M_R to be performed during a sampling period in the complete system is

$$M_R = N \times 2K + 2N \log_2 \frac{N}{2} = 2N \left[K + \log_2 \frac{N}{2} \right] \qquad (10.42)$$

This value can be compared with the value $2KN^2$ required by N IIR filters of the same order operating at frequency $1/T$. An application in telecommunications will be discussed in a later chapter.

10.9 CONCLUSION

FIR filters have the property that the sampling frequencies at the input and output are independent. This can be used to adapt the computation rate to the bandwidth of the signal during processing. This property can be extended to recursive structures through the use of a suitable transformation. Phase shifting is the basic function involved.

Multirate filtering can be applied to narrow band filters. It can provide considerable savings in computation when there is a factor of more than an order of magnitude between the sampling frequency and the pass band of the filter, as often occurs in practice.

The use of these techniques requires more detailed analysis of the processing. The limitations on their use result primarily from complications in the computation sequencing produced by the cascading of stages operating at different frequencies. This point should be examined carefully for each potential application of multirate filtering so that an excessive increase in the control unit or in the instruction program does not outweigh the computational gains.

Theoretically, the most important result is that given by equations (10.24)–(10.26), which illustrate the origins of the savings. The calculation rate M_R can be written in simplified form as:

$$M_R = f_s D(\delta_0) + \tfrac{1}{2} f_0 N_0 \qquad (10.43)$$

with the following parameters:

(1) f_s is the sampling frequency at the input or output of the filter;
(2) $D(\delta_0)$ is the parameter specifying the amount of aliasing distortion;
(3) f_0 is the operating frequency of the basic filter, approximately twice the pass band of the filter;
(4) N_0 is the number of coefficients of the basic filter, which represents the filter specifications.

In comparison with earlier filtering techniques, multirate filtering introduces new parameters, the passband width and the degree of harmonic distortion. Thus, determination of the computation rate involves parameters which are used to calculate the information rate in a transmission channel as indicated in Section 1.13.

The results obtained provide an example of how to use signal characteristics to determine the amount of processing needed to perform a given filtering function.

REFERENCES

[1] R. W. Schafer and L. B. Rabiner, A digital signal processing approach to interpolation. *Proceedings of the IEEE*, **61**(6), 1973.
[2] M. Bellanger, J. Daguet and G. Lepagnol, Interpolation, extrapolation and reduction of computation speed in digital filters. *IEEE Transactions*, **ASSP22**(4), 1974.
[3] F. Mintzer and B. Liu, Aliasing error in the design of multirate filters. *IEEE Transactions*, **ASSP26**, Feb 1978.
[4] D. J. Goodman and M. J. Carey, Nine digital filters for decimation and interpolation. *IEEE Transactions*, **ASSP25**, Apr 1977.
[5] R. E. Crochière and L. R. Rabiner, *Multirate digital signal processing*, Prentice Hall, Englewood Cliffs NJ, 1983, Ch. 5.
[6] R. E. Crochière, *Programs for digital signal processing*, IEEE and ASSP Society, 1979, Ch. 8.

[7] M. BELLANGER, G. BONNEROT and M. COUDREUSE, Digital filtering by polyphase network: application to sample rate alteration and filter banks. *IEEE Transactions, ASSP24*(2), 1976.

[8] P. P. VAIDYANATHAN, *Multirate systems and filter banks*, Prentice Hall, Englewood Cliffs NH, 1993.

[9] N. J. FLIEGE, *Multirate digital signal processing*, John Wiley, Chichester, 1994.

EXERCISES

1 Give the number of bits assigned to the coefficients of the half-band filters in Table 10.1.

Place the filters F_6, F_8 and F_9 on the curves of Figure 10.5. Use the results in Section 5.8 to evaluate the extra ripple introduced in the half-band filters by limiting the number of bits representing the coefficients. Test this evaluation on the filters F_6, F_8, and F_9.

2 Use the curves in Figure 10.5 to calculate the number of coefficients of the three filters in the cascade in Example 1, Section 10.4.

As each multiplication result is rounded, analyse the round-off noise produced by the reduction in sampling rate and give an expression for its power. Perform the same analysis for an increase in sampling rate and the basic filter. Compare the results with the direct realization.

Give an estimate of the aliasing distortion.

3 A filter for a telephone channel has a pass band which extends from 0 to 3400 Hz with a ripple of less than 0.25 dB. Above 4000 Hz, the attenuation is greater than 35 dB. For a sampling frequency $f_s = 32$ kHz, design a multirate filter as a cascade of half-band filters. Compare the number of multiplications and additions to be performed per second and the size of the memory with the values obtained for direct FIR filtering.

The signal applied to the filter is composed of numbers with 13 bits, and the computations are made using 16 bit registers. Evaluate the power of the round-off noise with a reduction and an increase in the sampling rate.

4 A discrete Fourier transform computer is a bank of uniform filters with characteristics as shown in Section 2.4. Study the phase shifts introudced in the odd Fourier transforms and in the doubly odd Fourier transforms in Section 3.3. What are the characteristics of the banks of filters so obtained?

5 Consider a bank of two filters to be produced using the IIR filter of order 4 given as an example in Section 7.2. The zeros and the poles of the upper half-plane have co-ordinates (Figure 7.6):

$$Z_1 = -0.816 + j0.578; \quad Z_2 = -0.2987 + j0.954$$
$$P_1 = 0.407 + j0.313; \quad P_2 = 0.335 + j0.776$$

Using the equations in Section 10.7, calculate the Z-transfer functions of the polyphase network paths. Give the co-ordinates of the poles and the zeros in the complex plane.

Use the results of Chapter 7 to determine the effect on the frequency response of limiting the number of bits in the coefficients of the denominator of the transfer function. Compare with the direct realization.

Draw the circuit diagram of this bank of two filters and determine the number of multiplications required, assuming that the sampling frequency at the output of each filter is half the value at the input.

CHAPTER 11

FILTER BANKS

Filter banks are used to analyse a signal or to construct a signal from discrete components. It is useful, in particular, to be able to decompose a signal into elementary components called sub-bands, which are subsampled and then used to reconstruct the initial signal without distortion. This operation arises, for example, in certain techniques of coding and compression of speech, sound and images. The principle of decomposition and reconstruction of a signal by banks of filters will be examined before considering lossless techniques.

11.1 DECOMPOSITION AND RECONSTRUCTION

In the realization of filter banks using polyphase networks and DFT as explained in Section 10.8, the operations involved are reversible and this leads to the arrangement of Figure 11.1 for the decomposition and reconstruction of a signal [1].

The difficulty, in practice, consists of applying the operations associated with the functions $H_i^{-1}(Z^N)$.

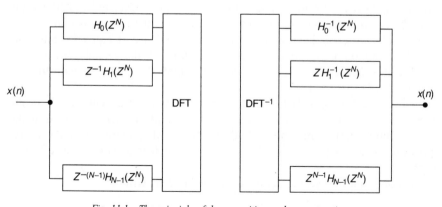

Fig. 11.1 *The principle of decomposition and reconstruction*

The filter $H(Z)$ which serves as a basis for the process, sometimes called the prototype filter, has a polyphase decomposition whose elements satisfy relation (10.8):

$$Z^{-i}H_i(Z^N) = \frac{1}{N}\sum_{m=0}^{N-1} e^{-j(2\pi/N)im} H(Ze^{-j(2\pi/N)m}); \quad 0 \le i \le N-1 \qquad (11.1)$$

If the prototype filter $H(Z)$ has a cut-off frequency less than $f_s/2N$ and infinite attenuation for frequencies greater than or equal to $f_s/2N$, that is if aliasing of the spectrum due to sampling at the rate $1/NT$ is negligible, one can write:

$$H[\exp j2\pi(f/f_s - m/N)]H[\exp j2\pi(f/f_s - k/N)] = 0; \qquad m \ne k \qquad (11.2)$$

Under these conditions the following equation is satisfied on the unit circle, except for the factor Z^{-N}:

$$H_i(Z^N)H_{N-i}(Z^N) = \frac{1}{N^2}\sum_{m=0}^{N-1} H^2(Ze^{-j(2\pi/N)m}) = H_0^2(Z^N)$$

Hence:

$$\frac{H_i(Z^N)}{H_0(Z^N)}\frac{(H)_{N-i}(Z^N)}{H_0(Z^N)} = 1; \quad 0 \le i \le N-1 \qquad (11.3)$$

These equations simply convey the phase relations illustrated in Figure 10.16.

One can then take $H_{N-i}(Z^N)$ to realise $H_i^{-1}(Z^N)$, the same filter bank is thus used for decomposition and reconstruction of the signal; the overall operation corresponds to multiplication by $H_0^2(Z^N)$.

In certain applications it is not possible to neglect aliasing; this is the case, for example, when it is required to decompose a signal, sample it at the rate f_s/N and then reconstruct it with the greatest possible accuracy over the band $(0, f_s)$. Let $G(Z)$ be the transfer function of the basic filter for the reconstruction. As the product of a discrete Fourier transform and its inverse is equal to unity, the overall operation corresponds to decomposition of the signal $x(n)$ into N interleaved sequences $x(pN + i)$ to which are applied N operators with transfer functions $G_i(Z^N)H_i(Z^N)$.

Figure 11.2 corresponds to a reduction of sampling frequency by N in the decomposition part, sometimes called analysis, and an increase by N in the reconstruction part, sometimes called synthesis. All processing in the corresponding device is performed at a rate $1/N$, which is a particularly effective approach.

The condition for reconstruction with a delay D is written:

$$G_i(Z^N)H_i(Z^N) = Z^{-D}; \quad 0 \le i \le N-1 \qquad (11.4)$$

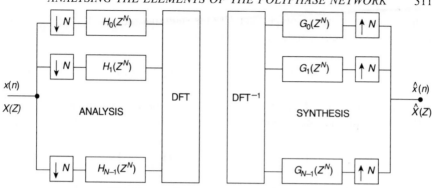

Fig. 11.2   *Polyphase filter banks for signal analysis and synthesis*

The delay D must be the same in all branches of the polyphase network for which the interleaving corresponds to an increase of sampling frequency by N and this can delay the initial signal, $x(n - D)$.

To determine the inverse functions $G_i(Z^N)$, it is necessary to perform a detailed analysis of the frequency response for the elements of the polyphase network.

11.2 ANALYSING THE ELEMENTS OF THE POLYPHASE NETWORK

The frequency response of the elements $H_i(Z^N)$ of the polyphase network follow directly from equation (11.1). However, simplifications can be made. Indeed, the filters of the bank generally provide a limited coverage, as in Figure 11.3. Under these conditions, a given filter is superimposed only on its immediate neighbours

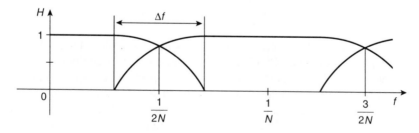

Fig. 11.3   *Overlap of adjacent filters*

if the response of the prototype filter $H(Z)$ is such that $H(f) = 0$ for $|f| > 1/N$. Then for the branch of index i one can write:

$$H_i(f) = H(f) + e^{-j(2\pi/N)i}H\left(f - \frac{1}{N}\right) \tag{11.5}$$

Since the periodicity of this response is $1/N$, if the coefficients are real, it is sufficient to consider the response in the interval $0 \le f \le 1/2N$ and this gives:

$$H_i(f) = H(f) + e^{-j(2\pi/N)i}\bar{H}\left(\frac{1}{N} - f\right) \tag{11.6}$$

Assuming that the response of the prototype filter is a monotonically decreasing curve in the transition band Δf, $H_i(f)$ cannot be zero for $f \ne 1/2N$. For the value $1/2N$, this gives:

$$H_i\left(\frac{1}{2N}\right) = H\left(\frac{1}{2N}\right)(1 + e^{-j(2\pi/N)i}) \tag{11.7}$$

This response is zero for the branch $i = N/2$.

Hence, with the decomposition (11.1), the same result as equation (10.36), the branch of index $N/2$ cannot be inverted since its Z-transfer function has a zero in the Z-plane at the point -1. To obtain a set of invertible branches, it is necessary to use another polyphase decomposition.

In Chapter 5 it was shown that a linear phase FIR filter with an even number of coefficients is an interpolator at mid sample period. It can be considered to result from an FIR filter with an odd number of coefficients, having the same frequency response by a subsampling of a factor 2. It is therefore necessary to start with a polyphase decomposition with $2N$ branches and to retain only one branch in two. One then obtains this equation for $H(Z)$:

$$H(Z) = \sum_{i=0}^{N-1} Z^{-(i+1/2)}H_{i+1/2}(Z^N) \tag{11.8}$$

With this decomposition the minimum amplitude $H_{\min}$ in a branch is given by:

$$H_{\min} = \left|H_{(N+1)/2}\left(\frac{1}{2N}\right)\right| = |(1 - e^{-j\pi/N})| = 2\sin\frac{\pi}{2N} \tag{11.9}$$

As a result, to have an invertible polyphase network, it is sufficient to require that the linear phase prototype FIR filter has an even number of coefficients.

The zeros of the functions $H_{i+1/2}(Z^N)$ in the Z^N-plane with respect to the unit circle, notice that they are divided equally between the interior and the exterior of the unit circle. The justification is found in the fact that the elements of $H_{i+1/2}(Z^N)$ are almost of linear phase. Furthermore, the amplitude of the frequency response remains close to unity, the zeros are far from the unit

Table 11.1. COEFFICIENTS OF A PROTOTYPE FILTER

	h_i	coefficients
0	$h_0 = h_{15}$	0.002898
1	$h_1 = h_{14}$	-0.009972
2	$h_2 = h_{13}$	-0.001921
3	$h_3 = h_{12}$	0.035969
4	$h_4 = h_{11}$	-0.016119
5	$h_5 = h_{10}$	-0.095302
6	$h_6 = h_9$	0.106799
7	$h_7 = h_8$	0.477347

circle, except for the branches which present a high attenuation at the frequency $1/2N$ when N is large.

As an example, consider a bank of $N = 2$ branches with a prototype filter of 16 coefficients according to Table 11.1. The transfer function defined by:

$$H_0(Z^2) = \sum_{i=0}^{7} h_{2i}(Z^{-2})^i \qquad (11.10)$$

has 7 zeros distributed as shown in Figure 11.4.

It is evident that they are remote from the unit circle. The second function $H_1(Z^2)$ has the same coefficients, but in the reverse order. The zeros are thus the inverse of the previous ones.

11.3 DETERMINING THE INVERSE FUNCTIONS

By placing the transfer functions in the Z-plane, determination of the inverse function for the polyphase elements starts with a factorization where the L_1 zeros within the unit circle are separated from the L_2 zeros which are outside:

$$H_i(Z) = h_{i0} \prod_{h=1}^{L_1}(1 - Z_h Z^{-1}) \prod_{l=0}^{L_2}(1 - Z_l Z^{-1}) \qquad (11.11)$$

The term h_{i0} is a scaling factor.

For all zeros Z_l outside the unit circle one can write:

$$\frac{1}{1 - Z_l Z^{-1}} = \frac{-Z}{Z_l} \sum_{l=0}^{\infty} (Z_l^{-1})^i Z^i \qquad (11.12)$$

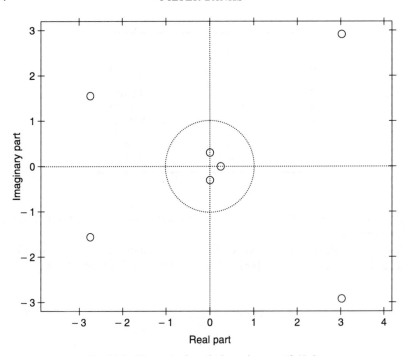

Fig. 11.4   The zeros of a polyphase element with N=2

and, consequently, the inverse of the second factor of (11.12) can be approached with an arbitrary accuracy within a finite number of terms. Consider the function $G_i(Z)$ defined by:

$$G_i(Z) = \frac{\displaystyle\sum_{l=0}^{L_3} a_l Z^{-l}}{\displaystyle\prod_{h=1}^{L_1}(1 - Z_h Z^{-1})} = \frac{\displaystyle\sum_{l=0}^{L_3} a_l Z^{-l}}{1 + \displaystyle\sum_{l=0}^{L_1} b_l Z^{-l}} \tag{11.13}$$

where L_3 is an integer.

The condition for inversion is satisfied if:

$$\left(\sum_{l=0}^{L_2} c_l Z^{-l}\right)\left(\sum_{l=0}^{L_3} a_l Z^{-l}\right) = Z^{-(L_2 + L_3)} \tag{11.14}$$

The choice of delay $L_2 + L_3$ is justified by the fact that the coefficients of the expansion of (11.12) are decreasing and that the second factor in (11.11) is of maximum phase. The inversion relation is written in matrix form:

$$MA = \begin{bmatrix} C_0 & 0 & 0 & \ldots & 0 \\ C_1 & C_0 & 0 & \ldots & 0 \\ \vdots & \vdots & \vdots & & \vdots \\ 0 & 0 & 0 & \ldots & C_{L_2} \end{bmatrix} \begin{bmatrix} a_0 \\ a_1 \\ \vdots \\ a_{L_3} \end{bmatrix} = \begin{bmatrix} 0 \\ 0 \\ \vdots \\ 1 \end{bmatrix} \quad (11.15)$$

where the vector A has the unknown coefficients a_l as elements. The system is overdetermined and permits a solution in the least squares sense, given by the relation:

$$A = (M^t M)^{-1} M^t \begin{bmatrix} 0 \\ 0 \\ \vdots \\ C_{L_2} \end{bmatrix} \quad (11.16)$$

Taking the previous example again, the coefficients a_1 and b_1 of $G_0(Z)$ which arise in equation (11.13), are given in Table 11.2 for $L_3 = 5$. The deviation from the ideal value of the inverse, i.e. the reconstruction error, is calculated as the magnitude of the vector $(MA)^t - [0, 0, \ldots, 1]$. As a quadratic value one obtains:

$$J = \|(MA)^t - [0, 0, \ldots, 1]\|^2 \quad (11.17)$$

In the above example, it is found that $J = 1.02 \times 10^{-7}$ with $L_3 = 5$ and $J = 8.0 \times 10^{-7}$ for $L_3 = 4$.

The polyphase synthesis elements $G_i(Z)$, have the structure of a general IIR filter. As the poles Z_h are far from the unit circle, a realization with a direct structure is possible. It is represented in Figure 11.5. With $L_3 = 5$ it needs 9 multiplications and 5 memories to realize one element, and the delay required by the combined analysis and synthesis is equal to $L_2 + L_3 = 9$.

Table 11.2. COEFFICIENTS OF A POLYPHASE SYNTHESIS ELEMENT

	a_i	b_i
0	− 0.000017	1
1	− 0.001214	− 0.207030
2	0.000053	− 0.033224
3	0.000662	0.002642
4	− 0.004129	0.005796
5	− 0.019944	

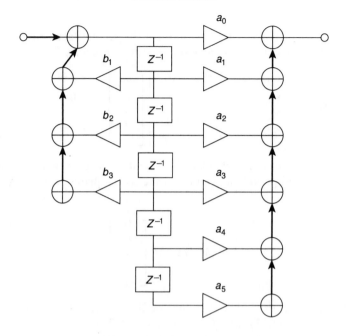

Fig. 11.5   The structure of a polyphase synthesis element

The method of calculating the polyphase synthesis elements described above is general and applies to all analysis filters, with the sole condition that there are no zeros on the unit circle. It is necessary to perform one calculation per branch, since different coefficients are obtained. This enables the analysis filter to be specified relatively independently of the synthesis filter. However, it can be useful to sacrifice a little flexibility in order to obtain a simpler and more systematic calculation. This is the case with QMF filters.

11.4   QMF FILTERS

Quadrature mirror filters (QMF) correspond to a bank of $N = 2$ filters. In order to address this particular case, it is useful to clarify the relationship between the output $X(Z)$ and the input $X(Z)$ for the general case by means of Figure 11.2.

By using the results of Section 10.1, in particular equations (10.7) and (10.8), one can write the transform of the series $x(n)$ for $X(Z)$:

$$X(Z) = \sum_{i=0}^{N-1} X_i(Z) = \frac{1}{N} \sum_{i=0}^{N-1} \sum_{m=0}^{N-1} e^{-j(2\pi/N)im} X(Ze^{-j(2\pi/N)m})$$

For the transform $\hat{X}(Z)$ of the reconstructed signal this gives:

$$\hat{X}(Z) = \sum_{i=0}^{N-1} G_i(Z^N) H_i(Z^N) X_i(Z) \tag{11.18}$$

Hence:

$$\hat{X}(Z) = \sum_{m=0}^{N-1} X(Ze^{-j(2\pi/N)m}) \frac{1}{N} \sum_{i=0}^{N-1} G_i(Z^N) H_i(Z^N) e^{-j(2\pi/N)im} \tag{11.19}$$

In this sum it is useful to isolate the term corresponding to $m = 0$, since then:

$$\hat{X}(Z) = X(Z) \frac{1}{N} \sum_{i=0}^{N-1} G_i(Z^N) H_i(Z^N)$$

$$+ \sum_{m=1}^{N-1} (Ze^{-j(2\pi/N)m}) \frac{1}{N} \sum_{i=0}^{N-1} G_i(Z^N) H_i(Z^N) e^{-j(2\pi/N)im}$$

Reconstruction of the signal is obtained with a delay of K sampling periods, if the following equations are satisfied:

$$\frac{1}{N} \sum_{i=0}^{N-1} G_i(Z^N) H_i(Z^N) = Z^{-K)} \tag{11.20}$$

$$\sum_{i=0}^{N-1} G_i(Z^N) H_i(Z^N) e^{-j(2\pi/N)im} = 0; \quad 1 \leq m \leq N-1 \tag{11.21}$$

Equation 11.20 is the condition for absence of amplitude distortion and the set of equations (11.21) is the condition for the absence of aliasing due to intermediate sampling at the rate $1/NT$. A simple solution for this system exists for the case where $N = 2$. One then has:

$$\frac{1}{2}[G_0(Z^2) H_0(Z^2) + G_1(Z^2) H_1(Z^2)] = Z^{-K} \tag{11.22}$$

$$G_0(Z^2) H_0(Z^2) - G_1(Z^2) H_1(Z^2) = 0 \tag{11.23}$$

To satisfy condition (11.23) it is sufficient to take:

$$G_0(Z^2) = H_1(Z^2) \qquad G_1(Z^2) = H_0(Z^2)$$

The condition for the absence of amplitude distortion becomes:

$$H_1(Z^2) H_0(Z^2) = Z^{-K} \tag{11.24}$$

The equation for decomposition of $H(Z)$ gives:

$$H_0(Z^2) = \frac{1}{2}[H(Z) + H(-Z)]$$

$$Z^{-1}H_1(Z^2) = \frac{1}{2}[H(Z) - H(-Z)]$$

Hence:

$$Z^{-1}H_1(Z^2)H_0(Z^2) = \frac{1}{4}[H^2(Z) - H^2(-Z)] \tag{11.25}$$

If the basic filter is a linear phase FIR type with $2P$ real coefficients, using equation (5.14), its frequency response can be written as:

$$H(f) = e^{-j2\pi f(P-1/2)}2H_R(f) \tag{11.26}$$

where $H_R(f)$ is a real even function. Also:

$$H\left(f - \frac{1}{2}\right) = e^{-j2\pi f(P-1/2)}e^{j\pi(P-1/2)}2H_R\left(f - \frac{1}{2}\right)$$

Under these conditions, on the unit circle one obtains:

$$\frac{1}{4}[H^2(Z) - H^2(-Z)]_{Z=e^{j2\pi f}} = e^{-j2\pi f(2P-1)}\left[H_R^2(f) + H_R^2\left(\frac{1}{2}-f\right)\right] \tag{11.27}$$

The condition for the absence of amplitude distortion corresponds to the following relation:

$$H_R^2(f) + H_R^2\left(\frac{1}{2}-f\right) = 1 \tag{11.28}$$

The delay due to the combined operations of decomposition and reconstruction is $K = 2P - 2$ sampling periods for the branches of the polyphase networks, and it is appropriate to add on a period for synchronization of the values at the input of the Fourier transform, which results in $2P - 1$ periods for the total delay of the system.

To reconcile this with the previous section, it is necessary to consider the frequency responses of the polyphase elements, denoted here by $H_0(f)$ and $H_1(f)$. Applying equation (11.6), taking account of the decomposition (11.8)

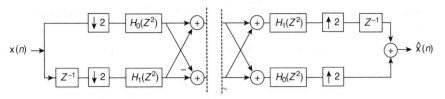

Fig. 11.6   QMF filters

which is applied in the case of a linear phase basic filter and an even number of coefficients, with (11.28) this gives:

$$|H_0(f)| = \left| H(f) + e^{-j\pi/2}H\left(\frac{1}{2}-f\right) \right| = 1$$

$$|H_1(f)| = \left| H(f) - e^{-j\pi/2}H\left(\frac{1}{2}-f\right) \right| = 1$$

(11.29)

That is, the amplitude is constant and the polyphase elements are pure phase shifters if condition (11.28) is satisfied. To invert the phases, it is sufficient in the synthesis section to replace H_0 by H_1 and vice versa. This leads to the arrangement of Figure 11.6.

The coefficients of Table 11.1 have been calculated with the above specifications. The coefficients of the analysis and synthesis filters can be deduced immediately. The quadratic reconstruction error obtained is then equal to 1.8×10^{-6}, consistent with the general method with $L_3 = 4$.

Equation (11.28) is analogous to equation (10.14), which characterizes half-band filters. The basic filter $H(f)$ is thus a low-pass filter whose transfer function squared is a filtering function of the half-band type. This function is sometimes called Nyquist and the filter $H(f)$ semi-Nyquist.

Calculation of the coefficients of $H(f)$ is made from a template specifying the limits of the passband f_1 and the attenuation band f_2, and the ripples δ_1 and δ_2, and by imposing the $\frac{1}{2}\sqrt{2}$ amplitude condition at the frequency $f_s/4$. For $H^2(f)$ to approach the property (11.27) with a ripple δ, the following conditions can be imposed:

$$- f_2 = \frac{f_s}{2} - f_1$$

$$- H(0) = 1$$

$$- \delta_1 = \frac{\delta}{2} \quad \text{and} \quad \delta_2 = \sqrt{\delta}$$

Example: Consider a low-pass filter with $N = 8$ coefficients whose parameters are as follows:

$$\Delta f = 0.24; \quad f_1 = 0.13; \quad f_2 = 0.37; \quad \delta = 0.01$$

Calculation yields the following coefficients:

$$h_1 = h_8 = 0.015235; \quad h_2 = h_7 = -0.085187$$
$$h_3 = h_6 = 0.081638; \quad h_4 = h_5 = 0.486502$$

The filter $H^2(Z) = \sum_{i=1}^{15} h_i' Z^{-i}$ must have its even coefficients equal to zero, except for h_8'. It is found that:

$$\sum_{i=1; i \neq 4}^{7} (h_{2i}')^2 = 1.7 \times 10^{-5}$$

Two iterative methods can, if necessary, complete the calculation and permit a better approach to symmetry [2]. The two branches of the polyphase network obtained, $H_0(Z^2)$ and $H_1(Z^2)$, have the same coefficients but in the reverse order. The above QMF technique can be exploited in a tree structure to decompose a band into N sub-bands, if N is a power of 2. Another generally more efficient approach to decomposing a band into N sub-bands is to use pseudo-QMF filters [3]. This will be described for a uniform bank of N real filters.

11.5 BANKS OF PSEUDO-QMF FILTERS

The principle relies on the assumption that, for a given filter, the attenuation is such that aliases originate only in the adjacent bands. Let $H(Z)$ be the transfer function of a prototype low-pass linear phase filter having the frequency response represented in Figure 11.7. Consideration of a number of coefficients equal to LN gives:

$$H(Z) = \sum_{k=0}^{LN-1} h_k Z^{-k} \tag{11.30}$$

In the bank, the filter of index i, centred on the frequency $(2i + 1)/4N$, has a transfer function $H(Ze^{-j2\pi(2i+1)/4N})$. During the analysis a component of the signal at the frequency $(2i + 1)/4N + \Delta f$, with perhaps $1/4N < \Delta f < 3/4N$, will be attenuated by a factor $H(\Delta f)$. Subsampling at a rate $1/N$ will produce a replica of this component at a frequency:

$$\frac{2i + 1}{4N} + \frac{3}{4N} - \left(\frac{2i + 1}{4N} + \Delta f \right) = \frac{3}{4N} - \Delta f$$

During synthesis this component, aliased in the band of the filter of index i, will be found at the frequency $(2i + 1)/4N + 1/2N - \Delta f$ and it will be subject to

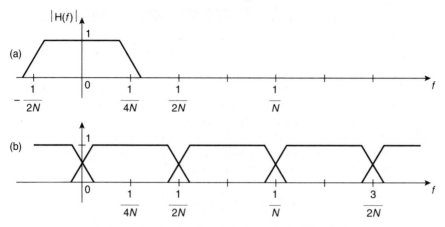

Fig. 11.7   (a) Prototype filter; (b) uniform bank of N real filters

attenuation by the synthesis filter of index i, i.e. $G(1/2N) - \Delta f$, if $G(Z)$ designates the prototype synthesis filter. Finally, the replicated component will have suffered attenuation:

$$H(\Delta f)G\left(\frac{1}{2N} - \Delta f\right)$$

The same component of the signal will now be processed by the filter of index $i + 1$, since it falls in the pass band. Sampling then produces an image component which, during synthesis, will be added to the previous aliased component with attenuation $H(1/2N - \Delta f)G(\Delta f)$. The process is illustrated in Figure 11.8.

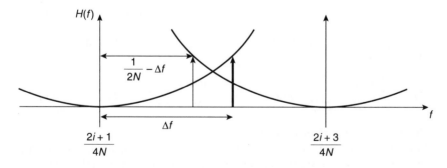

Fig. 11.8   Aliasing of a component in the filter bank

Hence the condition for these components to compensate each other is:

$$\left[H(\Delta f)G\left(\frac{1}{2N}-\Delta f\right)\right]_i + \left[H\left(\frac{1}{2N}-\Delta f\right)G(\Delta f)\right]_{i+1} = 0 \qquad (11.31)$$

This condition for the absence of aliasing can be obtained by taking $G(f) = H(f)$ and applying a phase difference of $\pi/2$ between the filters of index i and $i+1$, during analysis and synthesis.

The necessary phase difference can be obtained by introducing phase shifts into the modulation functions, for example by taking the following values for the coefficients h_{ik} of the analysis filter of index i:

$$h_{ik} = 2h_k \cos\left[\left((2i+1)\frac{\pi}{2N}\right)\left(k-\frac{LN-1}{2}\right)+\theta_i\right] \qquad (11.32)$$

with $0 \le i \le N-1$ and $0 \le k \le NL-1$.

For the synthesis filter:

$$g_{ik} = 2h_k \cos\left[\left((2i+1)\frac{\pi}{2N}\right)\left(k-\frac{LN-1}{2}\right)-\theta_i\right] \qquad (11.33)$$

Putting:

$$a_i = e^{j\theta_i} \qquad C_i = e^{-j(2i+1)\pi/2N}\frac{LN-1}{2} \qquad (11.34)$$

gives, for the corresponding transfer functions:

$$H_i(Z) = a_i c_i H(Ze^{-j(2i+1)\pi/2N}) + \bar{a}_i \bar{c}_i H(Ze^{j(2i+1)\pi/2N}) \qquad (11.35)$$

$$G_i(Z) = \bar{a}_i c_i H(Ze^{-j(2i+1)\pi/2N}) + a_i \bar{c}_i H(Ze^{j(2i+1)\pi/2N}) \qquad (11.36)$$

With the symmetry of the coefficients h_k and alignment of the modulation functions, the following expressions are satisfied:

$$g_{ik} = h_{i(LN-1-k)} \qquad G_i(Z) = Z^{-(LN-1)}H_i(Z^{-1}) \qquad (11.37)$$

Under these conditions the total response of the analysis and synthesis systems can be written:

$$T(Z) = \frac{\hat{X}(Z)}{X(Z)} = \frac{1}{N}\sum_{i=0}^{N-1}H_i(Z)G_i(Z) = \frac{Z^{-(LN-1)}}{N}\sum_{i=0}^{N-1}H_i(Z)H_i(Z^{-1}) \qquad (11.38)$$

So the global system is of linear phase.

With the assumption that the filters have sufficient attenuation for only the adjacent bands to cause significant aliasing, it is now necessary to determine the values for the angles θ_i to obtain the desired cancellation.

At the output of the filter $H_i(Z)$, after subsampling, in accordance with equation (10.8) the signal is:

$$X_i(Z^N) = \frac{1}{N} \sum_{m=0}^{N-1} H_i(ZW^m)X(ZW^m) \tag{11.39}$$

The output signal of the system becomes:

$$\hat{X}(Z) = \sum_{i=0}^{N-1} G_i(Z)X_i(Z_N) = \frac{1}{N} \sum_{m=0}^{N-1} X(ZW^m) \sum_{i=0}^{N-1} G_i(Z)H_i(ZW^m) \tag{11.40}$$

The condition for perfect reconstruction can then be written:

$$\sum_{i=0}^{N-1} G_i(Z)H_i(Z) = Z^{-k} \tag{11.41}$$

$$\sum_{i=0}^{N-1} G_i(Z)H_i(ZW^m) = 0; \qquad 1 \le m \le N-1 \tag{11.42}$$

To make the aliased components appear to cancel, it is necessary to examine the output signal of each of the synthesis filters. At the output of the filter $G_i(Z)$, the signal $X_i(Z)$, using equation (11.39), can be written:

$$\hat{X}_i(Z) = G_i(Z)\frac{1}{N} \sum_{m=0}^{N-1} H_i(ZW^m)X(ZW^m) \tag{11.43}$$

But, following definition (11.36) and the assumptions concerning attenuation, the filter $G_i(Z)$ allows passage of only the band centred on frequency $(2i+1)/4N$ and the two adjacent bands. Taking account of the distribution of the bands on the frequency axis, the indices m associated with these adjacent bands correspond to a frequency translation such that:

$$\frac{2i+1}{4M} \pm \frac{m}{N} = -\frac{2i+1}{4M} \pm \frac{1}{2N} \tag{11.44}$$

In fact, the subsampling leads to frequency translations which are integral multiples of the frequency $1/N$. Under these conditions the values of m can be written:

$$m = \pm i \quad \text{and} \quad m = \pm(i+1)$$

For example, considering the case of Figure 11.8, the aliased component arises from a component at the frequency $-(2i+1)/4N - \Delta f$ shifted by $(i+1)/N$, hence:

$$-\left(\frac{2i+1}{4N} + \Delta f\right) + \frac{i+1}{N} = \frac{2i+3}{4N} - \Delta f$$

Consequently, $\hat{X}_i(Z)$ is limited to the following expansion by using equation (11.35) to define $H_i(Z)$:

$$\hat{X}_i(Z) = G_i(Z)\frac{1}{N}[a_ic_iH(ZW^{-(2i+1)/4})]X(Z) + \bar{a}_i\bar{c}_iH(ZW^{(2i+1)/4})X(Z)$$

$$+ a_ic_iH(ZW^{(2i-1)/4})X(ZW^i) + \bar{a}_i\bar{c}_iH(ZW^{(1-2i)/4})X(ZW^{-i})$$

$$+ a_ic_iH(ZW^{(2i+3)/4})X(ZW^{i+1}) + \bar{a}_i\bar{c}_iH(ZW^{-(2i+3)/4})X(ZW^{-(i+1)})]$$

$$(11.45)$$

Since:

$$\hat{X}(Z) = \sum_{i=0}^{N-1}\hat{X}_i(Z) \qquad (11.46)$$

the aliases in $\hat{X}(Z)$ cancel if the high band of $\hat{X}_i(Z)$ compensates for the low band of $\hat{X}_{i+1}(Z)$. The corresponding condition is obtained by inserting definition 10.62 for $G_i(Z)$ into the expression for $\hat{X}_i(Z)$ and writing the same expression for $\hat{X}_{i+1}(Z)$. The factors of $X[ZW^{i+1}]$ and $X[ZW^{-(i+1)}]$ cancel if the following condition is satisfied:

$$a_i^2 c_i\bar{c}_iH(ZW^{(2i+3)/4})H(ZW^{(2i+1)/4}) + a_{i+1}^2c_{i+1}\bar{c}_{i+1}H(ZW^{(2i+1)/4})H(ZW^{(2i+3}/4) = 0$$

that is if:

$$a_{i+1}^2 = -a_i^2 \qquad (11.47)$$

It is therefore necessary that the phase shifts satisfy:

$$\theta_{i+1} = \theta_i + \frac{\pi}{2}; \quad 0 \le i \le N-1 \qquad (11.48)$$

The first condition for perfect reconstruction (11.41) can thus be written:

$$\frac{1}{N}\sum_{i=0}^{N-1}[c_iH(ZW^{(2i+1)/4})]^2 + [\bar{c}_iH(ZW^{-(2i+1)/4})]$$

$$+ (a_0^2 + \bar{a}_0^2)H(ZW^{1/4})H(ZW^{-1/4})$$

$$+ (a_{N-1}^2 + \bar{a}_{N-1}^2)H(ZW^{(N-1)/4})H(ZW^{-(N-1)/4}) = 1$$

$$(11.49)$$

since the cross products do not cancel at the origin and at the sampling half-frequency as the filters are adjacent.

For the cross products to vanish, it is necessary to take:

$$\theta_i = (-1)^i\frac{\pi}{4} \qquad (11.50)$$

and the imposed condition for calculation of the frequency response of the prototype filter can finally be written:

$$|H(f)|^2 + \left|H\left(\frac{1}{2N} - f\right)\right|^2 = 1 \qquad 0 \leq f < \frac{1}{2N}$$

$$H(f) = 0 \qquad \frac{1}{2N} \leq f \leq \frac{1}{2}$$

(11.51)

Note that the following are also possible for the phase shifts in the analysis and synthesis banks:

(1) $\theta_i = (i+1)\pi/2$—the overall system response cancels at frequencies 0 and 1/2;
(2) $\theta_i = i\pi/2$—the overall response doubles at frequencies 0 and 1/2.

In summary the design procedure for a bank of N real pseudo-QMF filters consists of the following two operations:

(1) Design of the prototype linear phase filter meeting the specification (11.51), with LN coefficients:
(2) Determination of the transfer functions of the analysis and synthesis filters using the expressions:

$$H_i(Z) = 2 \sum_{k=0}^{LN-1} \cos\left[2\pi\left(\frac{2i+1}{4N}\right)\left(k - \frac{LN-1}{2}\right) + (2i+1)\frac{\pi}{4}\right]h_k Z^{-k} \quad (11.52)$$

$$G_i(Z) = 2 \sum_{k=0}^{LN-1} \cos\left[2\pi\left(\frac{2i+1}{4N}\right)\left(k - \frac{LN-1}{2}\right) - (2i+1)\frac{\pi}{4}\right]h_k Z^{-k} \quad (11.53)$$

The specification of the prototype filter reflects the required separation between the sub-bands. Several approaches to the design can be envisaged.

11.6 DETERMINING THE COEFFICIENTS OF THE PROTOTYPE FILTER

A simple method consists of reapplying the approach of Section 11.4 by adapting the specification. As before, this method can be completed using iterative techniques to approach the symmetry condition (11.51) as closely as possible. An example is given in Figure 11.9, which shows the frequency response of an FIR prototype filter with 32 coefficients used to obtain a bank of 16 filters, together with the superposition of the responses of the set of filters obtained. This system can be used to decompose the signal band into 16 sub-bands and then to reconstruct the original signal. A distinctly more selective filter with constant ripples in the attenuation band is given in Figure 11.10. It has 64 coefficients, provides an attenuation of 58 dB and its transition band is 1/16.

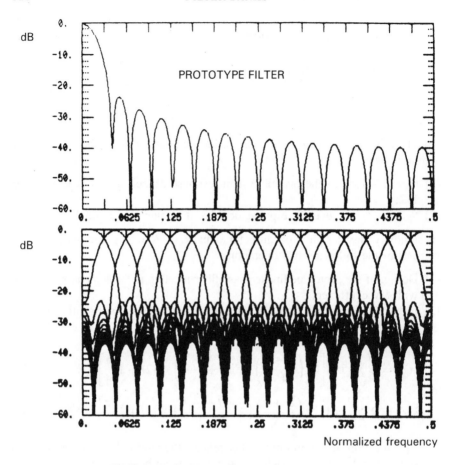

Coefficients of prototype filter

$h_1 = 0.011367 = h_{32}$	$h_9 = 0.032707 = h_{24}$
$h_2 = 0.013889 = h_{31}$	$h_{10} = 0.035004 = h_{23}$
$h_3 = 0.016542 = h_{30}$	$h_{11} = 0.037063 = h_{22}$
$h_4 = 0.019277 = h_{29}$	$h_{12} = 0.038840 = h_{21}$
$h_5 = 0.022065 = h_{28}$	$h_{13} = 0.040312 = h_{20}$
$h_6 = 0.024849 = h_{27}$	$h_{14} = 0.041434 = h_{19}$
$h_7 = 0.027586 = h_{26}$	$h_{15} = 0.042200 = h_{18}$
$h_8 = 0.030220 = h_{25}$	$h_{16} = 0.042583 = h_{17}$

Fig. 11.9   Frequency response of a filter bank for signal analysis and synthesis

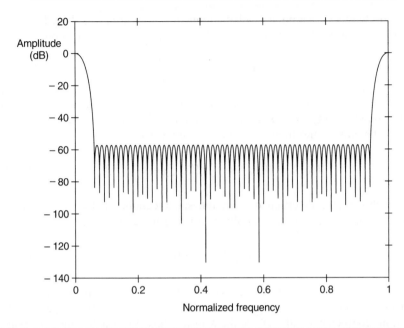

h_i	Coefficients	h_i	Coefficients
$h_1 = h_{64}$	9.2839870e − 004	$h_{17} = h_{48}$	−6.4021587e − 003
$h_2 = h_{63}$	5.9685008e − 004	$h_{18} = h_{47}$	−4.7447370e − 003
$h_3 = h_{62}$	6.8568814e − 004	$h_{19} = h_{46}$	−2.1424791e − 003
$h_4 = h_{61}$	6.8789993e − 004	$h_{10} = h_{45}$	1.4680266e − 003
$h_5 = h_{60}$	5.6482087e − 004	$h_{21} = h_{44}$	6.0982460e − 003
$h_6 = h_{59}$	2.8097967e − 004	$h_{22} = h_{43}$	1.1699623e − 002
$h_7 = h_{58}$	−1.9044096e − 004	$h_{23} = h_{42}$	1.8159361e − 002
$h_8 = h_{57}$	−8.6214757e − 004	$h_{24} = h_{41}$	2.5302338e − 002
$h_9 = h_{56}$	−1.7275867e − 003	$h_{25} = h_{40}$	3.2898876e − 002
$h_{10} = h_{55}$	−2.755776 − e − 003	$h_{26} = h_{39}$	4.0672058e − 002
$h_{11} = h_{54}$	−3.8879464e − 003	$h_{27} = h_{38}$	4.8309834e − 002
$h_{12} = h_{53}$	−5.0372397e − 003	$h_{28} = h_{37}$	5.5491156e − 002
$h_{13} = h_{52}$	−6.0890424e − 003	$h_{29} = h_{36}$	6.1893832e − 002
$h_{14} = h_{51}$	−6.9067117e − 003	$h_{30} = h_{35}$	6.7223291e − 002
$h_{15} = h_{50}$	−7.3368967e − 003	$h_{31} = h_{34}$	7.1226098e − 002
$h_{16} = h_{49}$	−7.2203731e − 003	$h_{32} = h_{33}$	7.3708998e − 002

Fig. 11.10   Prototype filter for 16 band filters

A non-iterative technique, called frequency weighting, is based on the properties of the DFT given in Section 2.4. By reconsidering equation (2.23b):

$$\Phi_k(f) = (-1)^k e^{j\pi k/N} \frac{1}{N} \frac{\sin \pi(f + k/N)N}{\sin \pi(f + k/N)} \tag{11.54}$$

the desired filter can be expressed as a weighted sum of the outputs of a discrete Fourier transform [4]. By taking $2I + 1$ outputs of a DFT and weighting them with coefficients α_i, the filter response can be written:

$$H(f) = \Phi_0(f) + \sum_{i=1}^{I} \alpha_i [\Phi_i(f) + \Phi_{-i}(f)] \tag{11.55}$$

The weights α_i are determined from the specification. Firstly the maximum attenuation at high frequencies is obtained using the expression:

$$1 + 2\sum_{i=1}^{I} \alpha_i = 0 \tag{11.56}$$

Then other relations arise from the fact that all the functions $\Phi_k(f)$ are zero at frequencies which are multiples of $1/N$. Except for one which separates the weighting coefficients and enables them to be chosen to satisfy the constraints imposed by $H^2(f)$ and particularly symmetry.

For a bank of 16 filters, by making a transform with $N = 64$ points, the weighting coefficients are obtained from the following equations with $I = 3$:

$$1 + 2\alpha_1 + 2\alpha_2 + 2\alpha_3 = 0$$
$$\alpha_1^2 + \alpha_3^2 = 1$$
$$\alpha_2^2 = 0.5$$

which leads to the values:

$$\alpha_1 = -0.97196$$
$$\alpha_2 = 0.70711$$
$$\alpha_3 = -0.23515$$

The coefficients are then deduced from the definition of the DFT. This gives:

$$h_{j+1} = 1 + 2\sum_{i=1}^{I} \alpha_i \cos 2\pi ij; \quad 1 \le j \le 63 \tag{11.57}$$
$$h_1 = 0$$

The filter obtained has an odd number of coefficients. The frequency response, whose transition band is equal to $1/16$, is shown with the coefficient values in Figure 11.11. Once the coefficients are determined, the calculations should be set out and performed so that the number of arithmetic operations is minimal [5].

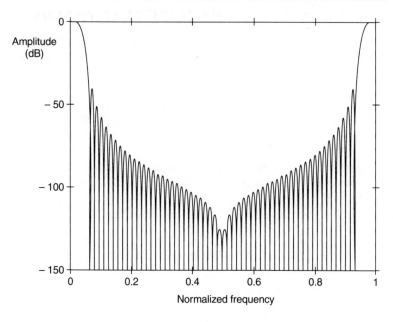

h_i	Coefficients	h_i	Coefficients
h_1	1.9999999e − 09	$h_{17} = h_{49}$	−4.1421356e − 01
$h_2 = h_{64}$	2.4374869e − 03	$h_{18} = h_{48}$	−3.3302151e − 01
$h_3 = h_{63}$	8.9600890e − 03	$h_{19} = h_{47}$	−1.8860434e − 01
$h_4 = h_{62}$	1.7309178e − 02	$h_{20} = h_{46}$	2.4872388e − 02
$h_5 = h_{61}$	2.4078729e − 02	$h_{21} = h_{45}$	3.0941094e − 01
$h_6 = h_{60}$	2.5214011e − 02	$h_{22} = h_{44}$	6.6263312e − 01
$h_7 = h_{59}$	1.6635762e − 02	$h_{23} = h_{43}$	1.0775305e + 00
$h_8 = h_{58}$	−5.0758287e − 03	$h_{24} = h_{42}$	1.5425480e + 00
$h_9 = h_{57}$	−4.2010768e − 02	$h_{25} = h_{41}$	2.0420108e + 00
$h_{10} = h_{56}$	−9.4346766e − 02	$h_{26} = h_{40}$	2.5568746e + 00
$h_{11} = h_{55}$	−1.5992266e − 01	$h_{27} = h_{39}$	3.0657564e + 00
$h_{12} = h_{54}$	−2.3402304e − 01	$h_{28} = h_{38}$	3.5461759 + 00
$h_{13} = h_{53}$	−3.0941094e − 01	$h_{29} = h_{37}$	3.9759213e + 00
$h_{14} = h_{52}$	−3.7662359e − 01	$h_{30} = h_{36}$	4.3344420e + 00
$h_{15} = h_{51}$	−4.2452159e − 01	$h_{31} = h_{35}$	4.6041658e + 00
$h_{16} = h_{50}$	−4.4105818e − 01	$h_{32} = h_{34}$	4.7716422e + 00

Fig. 11.11 Prototopye filter obtained by frequency weighting for a bank of 16 filters and 64 coefficients

11.7 REALIZING A BANK OF REAL FILTERS

Consider a bank of real filters having the frequency responses of Figure 10.19 and an even number of coefficients $2LN$, in which the filter of index i has the coefficients:

$$h_{ik} = h_k \cos\left[\frac{2\pi}{2N}\left(i + \frac{1}{2}\right)\left(k + \frac{1}{2}\right)\right] \tag{11.58}$$

with $0 \le i \le N, 0 \le k \le 2LN - 1$.

A decomposition into a polyphase network and a discrete Fourier transform can be obtained by putting $k = 2Nl + m$ with $0 \le l \le L - 1$ and $0 \le m \le 2N - 1$.

The output $x_i(n)$ for the filter of index i can be written:

$$x_i(n) = \sum_{k=0}^{2LN-1} x(n - k)h_k \cos\left[\frac{2\pi}{2N}\left(i + \frac{1}{2}\right)\left(k + \frac{1}{2}\right)\right] \tag{11.59}$$

or by replacing k With $2Nl + m$ and simplifying:

$$x_i(n) = \sum_{m=0}^{2N-1} \cos\frac{2\pi}{2N}\left(i + \frac{1}{2}\right)\left(m + \frac{1}{2}\right)\sum_{l=0}^{L-1}(-1)^l h_{2Nl+m}x(n - 2Nl - m) \tag{11.60}$$

Applying the general transfer function decomposition (10.36) to the prototype filter gives:

$$H(Z) = \sum_{m=0}^{2N-1} Z^{-m}H_m(Z^{2N}) \tag{11.61}$$

and the filters $H_m(Z^{2N})$ are those which arise in the second summation of expression (11.60). To take account of the factor $(-1)^l$ it is sufficient to introduce the functions $H_m(-Z^{2N})$, and the diagram corresponding to the analysis filters is shown in Figure 11.12. The decomposition produces a polyphase network with $2N$ branches and a cosine transform.

Figure 11.12 can be further simplified since, in the cosine transform considered, two symmetrical inputs are subjected to the same operations, apart from the sign. Factorizing yields an odd-time odd-frequency cosine transform— a special case mentioned in Section 3.3.2. Furthermore, with subsampling, the system operates at a rate $1/N$. As the system consists of $2N$ branches, a given $x(n)$ is processed by the filters H_i and H_{i+N} at two successive instants. Under these conditions, the $2N$ branches of the polyphase network can be regrouped as $N/2$ subgroups having the lattice structure of Figure 11.13. The overall configuration is therefore as in Figure 11.14. In the case of QMF filters (Section 11.6) this arrangement is applicable with the introduction of phase shifts. In fact,

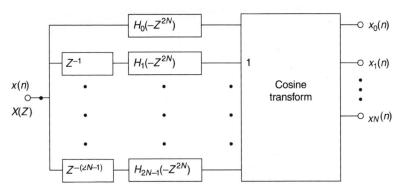

Fig. 11.12 Arrangement of a bank of N real filters

by taking equation (11.59) with the coefficients of the filter of equation (11.52), the output $x_i(n)$ of the filter of index i with $2LN$ coefficients can be written:

$$x_i(n) = 2 \sum_{k=0}^{2LN-1} x(n-k)h_k \cos\left[2\pi \frac{2i+1}{4N}\left(k - \frac{2LN-1}{2}\right) + (2i+1)\frac{\pi}{4}\right]$$

(11.62)

By putting $k = lN + m$ this time, the following double summation is obtained:

$$x_i(n) = 2 \sum_{m=0}^{N-1} \sum_{l=0}^{2L-1} \cos\left[(2i+1)(2m+1+N)\frac{\pi}{4N}\right.$$
$$\left. + (2i+1)(l-L)\pi/2\right]h_{lN+m}x(n-lN-m)$$
(11.63)

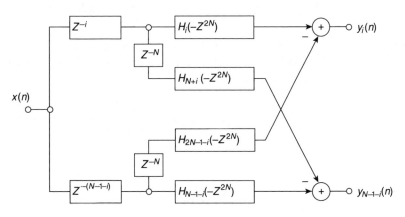

Fig. 11.13 Lattice structure for the polyphase network

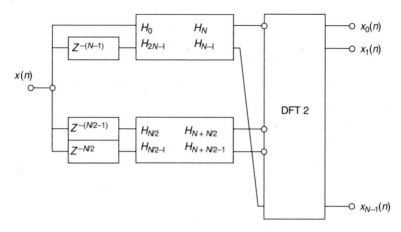

Fig. 11.14   Optimized structure of a bank of N real filters

A cosine expansion produces these terms:

$$\cos(2i + 1)(l - L)\frac{\pi}{2} = \cos(l - L)\frac{\pi}{2}$$

and:

$$\sin(2i + 1)(l - L)\frac{\pi}{2} = (-1)^i \sin(l - L)\frac{\pi}{2}$$

Furthermore, the following expression is satisfied:

$$\cos\left[(2i + 1)[2(N - 1 - m) + 1]\frac{\pi}{4N}\right] = (-1)^i \sin\left[(2i + 1)(2m + 1)\frac{\pi}{4N}\right]$$

Finally, to take account of term N in the cosine argument of (11.63), it is sufficient to scale the data by a factor of $N/2$. By combining all these results, the output $x_i(n)$ is determined as follows:

$$x_i(n) = 2\sum_{m=0}^{N-1} \cos\left[(2i + 1)(2m + 1)\frac{\pi}{4N}\right] y_m(n) \qquad (11.64)$$

with

$$y_m(n) = [-y_{2,N/2-1-m}(n) + y_{2,N/2+m}(n)]; \quad 0 \le m \le \tfrac{1}{2}N - 1$$
$$y_m(n) = [y_{1,m-N/2}(n) - y_{1,3N/2-m-1}(n)]; \quad N/2 \le m \le N - 1$$

and:

$$y_{1,m}(n) = \sum_{m=0}^{2L-1} \cos(l-L)\frac{\pi}{2} h_{lN+m} x(n-lN-m)$$

$$y_{2,m}(n) = \sum_{m=0}^{2L-1} \sin(l-L)\frac{\pi}{2} h_{lN+m} x(n-lN-m)$$

The sequences $y_{1,m}(n)$ and $y_{2,m}(n)$ are interleaved, with sampling frequency $f_s/2N$ and the bank of analysis filters is determined with the help of an odd-time odd-frequency cosine transform. Finally, the phase shifts introduced by the pseudo-QMF technique have been taken into account simply by rearranging the data before the transform.

11.8 FILTER BANKS WITH PERFECT DECOMPOSITION AND RECONSTRUCTION

By increasing the number of calculations, it is possible to realize a bank of filters with perfect decomposition and reconstruction; that is, without aliasing, without amplitude and phase distortion and only a delay of the output with respect to the input [6–9].

With $N=2$ filters, one solution is provided by the so-called conjugate quadrature mirror filters (CQF). The corresponding arrangement is given in Figure 11.15. Subsampling by a factor 2 is involved at the output of the decomposition filters $H_0(Z)$ and $H_1(Z)$. Here the expressions which define perfect reconstruction can be written:

$$G_0(Z)H_0(Z) + G_1(Z)H_1(Z) = AZ^{-K} \tag{11.65}$$

$$G_0(Z)H_0(-Z) + G_1(Z)H_1(-Z) = 0 \tag{11.66}$$

where A is a positive constant.

One solution to this system can be obtained from the half-band filter $H(Z)$ examined in Section 10.6, and by factorizing it as follows:

$$H(Z) = H_0(Z)Z^{-K}H_0(Z^{-1}) \tag{11.67}$$

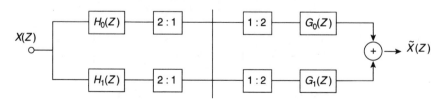

Fig. 11.15   Mirror filters in conjugate quadrature

and assuming that it has a number of coefficients $N = 2K + 1$. With respect to M, defined in Section 10.6 as the number of different coefficients, one then has $K = 2M - 1$, so K is an odd number.

By putting:

$$G_0(Z) = Z^{-K} H_0(Z^{-1}) \tag{11.68}$$

Note that the following expression is satisfied since K is odd:

$$G_0(-Z) = -Z^{-K} H_0(-Z^{-1}) \tag{11.69}$$

Under these conditions, by taking:

$$H_1(Z) = G_0(-Z) \qquad G_1(Z) = -H_0(-Z) \tag{11.70}$$

the conditions for perfect reconstruction are satisfied. To perform the factorization (11.67) it is sufficient to start with the frequency response of the half-band FIR filter whose central coefficient has the value 0.5 as in expression (10.17), and to add to this coefficient the value δ of the ripple. The factorization of Section 5.11 is used again and the constant A is equal to $1 + \delta$.

Compared with the QMF method, it requires approximately twice the amount of calculations. Notice that the phase of the analysis and synthesis filters is no longer linear and the pass band and attenuation band ripples are the same.

Lattice structures are another useful approach. The configuration of Figure 11.16 provides a perfect reconstruction for a bank of two filters. The total transfer function of the system can be written:

$$F(Z) = Z^{-(2k-1)} \prod_{i=0}^{K-1} (1 + \alpha_i^2) \tag{11.71}$$

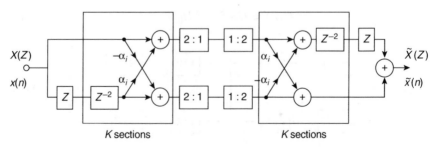

Fig. 11.16   Bank of two filters in a lattice structure

The coefficients α_i are determined from the specification of the analysis filters. For $k = 3$:

$$H_0(Z) = 1 + \alpha_1 Z^{-1} - \alpha_1\alpha_2 Z^{-2} + \alpha_2 Z^{-3}$$
$$H_1(Z) = -\alpha_2 - \alpha_1\alpha_2 Z^{-1} - \alpha_1 Z^{-2} + Z^{-3}$$

(11.72)

which corresponds to $H_1(Z) = Z^{-K}H_0(-Z^{-1})$.

Calculation of the coefficients can be achieved by the method of least squares. Note that rounding of the coefficients does not affect the property of perfect reconstruction [8]. The approach can be extended to a bank of N filters by reconsidering the arrangement of Figure 11.13 and taking the inverse of the lattice for the analysis part to obtain each lattice of the polyphase network in the synthesis part.

For two-dimensional signals, such as images, a simple and effective method consists of designing separable filter banks. The techniques presented are directly applicable. Note that QMF and CQF filters, in a tree architecture, permit non-uniform decomposition and the associated reconstruction. The filters in the bank then have widths which are multiples of an elementary width. This is sometimes called decomposition by wavelets or transformation by wavelets.

REFERENCES

[1] M. BELLANGER and J. DAGUET, TDM-FDM transmultiplexer: digital polyphase and FFT. *IEEE Transactions on Communications*, **22**(9), 1199–1205, 1974.

[2] R. E. CROCHIÈRE and L. R. RABINER, *Multirate digital signal processing*, Prentice Hall, Englewood Cliffs NJ, 1983.

[3] P. L. CHU, Quadrature mirror filter design for an arbitrary number of equal bandwidth channels. *IEEE Transactions,* **ASSP33**(1), 203–18, 1985.

[4] K. W. MARTIN, Small sidelobe channel-banks suitable for multitone data communications. In *Proceedings of DSP'97* Santorini, Greece, July 1997, pp. 1071–73.

[5] J. MASSON, Bancs de filtres numériques pour l'analyse et la synthèse des signaux. In *Proceedings of GRETSI-87*, Nice, June 1987.

[6] M. VETTERLI, Filter banks allowing perfect reconstruction. *Signal Processing*, **10**(3), 219–44, 1986.

[7] M. SMITH and T. BARNWELL, Exact reconstruction techniques for tree-structured sub-band coders. *IEEE Transactions,* **ASSP34**(3), 434–41, 1986.

[8] P. P. VAIDYANATHAN, *Multirate systems and filter banks*, Prentice Hall, Englewood Cliffs NJ, 1993.

[9] N. J. FLIEGE, *Multirate digital signal processing*, John Wiley, Chichester, 1994.

CHAPTER 12

ADAPTIVE FILTERING

Adaptive filtering is involved when it is necessary to realize, simulate or model a system whose characteristics develop with time. It leads to the use of filters with coefficients which change with time. The variations in the coefficients are defined by an optimization criterion and are realized according to an adaptation algorithm, both of which are determined according to the application. Many different criteria and algorithms are possible [1]. This chapter will be concerned with the simple but, in practice, most important case in which the criterion of mean square error minimization is associated with the gradient algorithm.

While fixed coefficient filtering is in general associated with specifications in the frequency domain, adaptive filtering corresponds to specifications in time and it is natural to introduce it by considering the calculation of filter coefficients in these conditions. The FIR filter is examined first.

12.1 CALCULATION OF THE COEFFICIENTS OF AN FIR FILTER FOR SPECIFICATIONS IN TIME

Assume that a set A of N coefficients a_i $(0 \leqslant i \leqslant N-1)$ is to be determined such that for a given set $x(n)$ the deviation $e(n)$ between a given set $y(n)$ and the set $\tilde{y}(n)$ such that:

$$\tilde{y}(n) = \sum_{i=0}^{N-1} a_i x(n-i) \qquad (12.1)$$

is a minimum. This situation occurs, for example, when a system is being modelled and is illustrated in Figure 12.1.

Although the signals $x(n)$ and $y(n)$ can be complex they will be taken as real in this chapter. Extension of the analysis to include complex signals does not present any particular difficulties.

Using the least mean squares criterion on the domain of the indices $(0, 1, \ldots, N_0 - 1)$ with $N_0 > N$, the error function $E(A)$ is minimized, where

$$E(A) = \frac{1}{N_0} \sum_{n=0}^{N_0-1} [y(n) - \tilde{y}(n)]^2 \qquad (12.2)$$

The procedure is similar to that described in Section 5.4, and the required

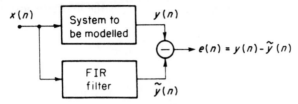

Fig. 12.1 Modelling with FIR filter

coefficients are such that:

$$\frac{\partial E}{\partial a_i} = 0; \quad 0 \leqslant i \leqslant N - 1$$

From the equation of definition for $E(A)$ and the expression for $\tilde{y}(n)$, this becomes:

$$\frac{\partial E}{\partial a_i} = -\frac{2}{N_0} \sum_{n=0}^{N_0-1} x(n-i)[y(n) - \tilde{y}(n)]; \quad 0 \leqslant i \leqslant N - 1 \tag{12.3}$$

In matrix form, the N by N_0 matrix X is introduced, where

$$X = \begin{bmatrix} x(0) & x(1) & x(2) & \cdots & x(N_0-1) \\ x(-1) & x(0) & x(1) & \cdots & x(N_0-2) \\ x(-2) & x(-1) & x(0) & \cdots & x(N_0-3) \\ \vdots & \vdots & \vdots & & \\ x(-N+1) & x(-N+2) & x(-N+3) & \cdots & x(N_0-N) \end{bmatrix} \quad N \text{ rows}$$
$$N_0 \text{ columns}$$

If $[y(n)]$ is the column vector of the $y(n)$ with $o \leqslant n \leqslant N_0 - 1$ and A is the column vector of the a_i with $0 \leqslant i \leqslant N - 1$, the least-squares condition is written:

$$X[y(n)] = [XX^t]A \tag{12.4}$$

This is the Wiener–Hopf equation. Thus, the coefficients are determined by:

$$A = [XX^t]^{-1}X[y(n)]$$

The matrix XX^t can be seen to be a square matrix of order N whose elements γ_{ij} are

$$\gamma_{ij} = \sum_{n=0}^{N_0-1} x(n-i)x(n-j); \quad \begin{matrix} 0 \leqslant i \leqslant N - 1 \\ 0 \leqslant j \leqslant N - 1 \end{matrix}$$

If $x(n)$ is a random stationary discrete signal, or a periodic signal, or can be regarded as such, then from Section 1.8, the $(1/N_0)\gamma_{ij}$ are estimations of the elements of its autocorrelation function.

If σ_x^2 is the power of this signal and $\tilde{R}_N$ is the matrix whose elements r_{ij} are

$$r_{ij} = \frac{\gamma_{ij}}{N_0 \sigma_x^2} \tag{12.5}$$

then

$$A = \frac{1}{\sigma_x^2} \tilde{R}_N^{-1} \frac{1}{N_0} X[y(n)] \tag{12.6}$$

Designating by r_i the autocorrelation function elements, the matrix

$$R_N = \begin{bmatrix} 1 & r_1 & r_2 & \cdots & r_{N-1} \\ r_1 & 1 & r_1 & \cdots & \vdots \\ r_2 & r_1 & 1 & \cdots & \vdots \\ \vdots & \vdots & & \ddots & \vdots \\ r_{N-1} & \vdots & & \cdots & 1 \end{bmatrix} \tag{12.7}$$

is called the normalized autocorrelation matrix of the signal $x(n)$. It is symmetric and all the diagonal elements are identical. It is said to have Toeplitz symmetry [1]. If the signal $x(n)$ is a discrete white noise, the terms are uncorrelated and the matrix R_N is the N-dimensional unit matrix I_N. If the signal $x(n)$ is the unit set $u_0(n)$, i.e.

$$x(n) = 1 \quad \text{for } n = 0$$
$$0 \quad \text{for } n \neq 0$$

we also have $R_N = I_N$ and the coefficients a_i ($0 \leq i \leq N - 1$) are simply the first N terms of the impulse response of the system.

It is important to study the properties of this matrix. More generally, it is necessary to examine autocorrelation and intercorrelation, which play similar roles in the analysis of time systems to those of spectra and transfer functions in the analysis of frequency systems [2].

12.2 AUTOCORRELATION AND INTERCORRELATION

The degree of similarity between two signals can be described using a correlation coefficient, which should logically take the values $+1$ for two identical signals, zero for two signals which have no relationship to each other and -1 for signals in opposition to each other. When time-dependent signals are compared, the correlation coefficient becomes a function of time. It is called the intercorrelation function if the signals are different and the autocorrelation function if they are the same.

Some definitions and properties will now be restated in order to recapitulate and complete Sections 1.8 and 4.4.

As shown in Section 1.8, the autocorrelation function of a random discrete signal $x(n)$ is the set $r_{xx}(n)$ such that:

$$r_{xx}(n) = E[x(i)x(i-n)] \qquad (12.8)$$

where $E[x]$ denotes the expectation value of x.

With the assumption of ergodicity this becomes

$$r_{xx}(n) = \lim_{N \to \infty} \frac{1}{2N+1} \sum_{i=-N}^{N} x(i)x(i-n) \qquad (12.9)$$

The function $r_{xx}(n)$ is even. Its value at the origin is the power of the signal and for any n is:

$$|r_{xx}(n)| \leqslant r_{xx}(0) \qquad (12.10)$$

Consider a set of N coefficients a_i $(1 \leqslant i \leqslant N)$. Calculation of the variance of the variable $y(n)$ such that

$$y(n) = \sum_{i=1}^{N} a_i x(n-i)$$

results in

$$E[y^2(n)] = \sum_{i=1}^{N} \sum_{j=1}^{N} a_i a_j E[x(n-i)x(n-j)]$$

or

$$E[y^2(n)] = \sum_{i=1}^{N} \sum_{j=1}^{N} a_i a_j r_{xx}(i-j) \qquad (12.11)$$

As this variance is positive or zero, we have

$$\sum_{i=1}^{N} \sum_{j=1}^{N} a_i a_j r_{xx}(i-j) \geqslant 0 \qquad (12.12)$$

This property characterizes functions of the positive type. If, in the definition of equation (12.8), $x(i-n)$ is replaced by another signal, a function is obtained which allows two different signals to be compared. The intercorrelation function between two discrete signals $x(n)$ and $y(n)$ is the set $r_{xy}(n)$ such that

$$r_{xy}(n) = E[x(i)y(i-n)] \qquad (12.13)$$

With the assumption of ergodicity,

$$r_{xy}(n) = \lim_{N \to \infty} \frac{1}{2N+1} \sum_{i=-N}^{N} x(i)y(i-n) \qquad (12.14)$$

Similarly,

$$r_{xy}(-n) = E[x(i)y(i+n)] = r_{yx}(n) \qquad (12.15)$$

For example, if the signals are the input and output of a filter:

$$y(n) = \sum_{j=0}^{\infty} h_j x(n-j)$$

then, as shown in Section 4.4, one has

$$r_{yx}(n) = E[y(i)x(i-n)] = \sum_{j=0}^{\infty} h_j r_{xx}(n-j)$$

or

$$r_{yx}(n) = r_{xx}(n) * h(n) \tag{12.16}$$

Similarly,

$$r_{xy}(n) = r_{xx}(n) * h(-n) \tag{12.17}$$

and also

$$r_{yy}(n) = r_{xx}(n) * h(n) * h(-n) \tag{12.18}$$

If two random signals are independent their intercorrelation functions are zero. Further, the following inequality is always valid:

$$|r_{xy}(n)| \leqslant \tfrac{1}{2}[r_{xx}(0) + r_{yy}(0)] \tag{12.19}$$

It is worth mentioning that the autocorrelation and intercorrelation functions can in some cases by computed without multiplication, the signals being replaced by their signs. Thus if $x(n)$ is a Gaussian signal [2]:

$$r_{xx}(n) = \sqrt{\frac{\pi}{2}} \sqrt{r_{xx}(0)} E[x(i) \, \text{sign} \, (x(i-n))] \tag{12.20}$$

$$r_{xx}(n) = r_{xx}(0) \sin\left[\frac{\pi}{2} E[\text{sign} \, [x(i)x(i-n)]]\right] \tag{12.21}$$

These expressions can lead to important simplification in equipment.

The Fourier transform $\Phi_{xy}(f)$ of the intercorrelation function $r_{xy}(n)$ is called the interspectrum

$$\Phi_{xy}(f) = X(f)\overline{Y(f)}$$

where $X(f)$ denotes the spectrum of the set $x(n)$ and $\overline{Y(f)}$ is the conjugate spectrum of the set $y(n)$.

If the set $y(n)$ is the output of a filter with transfer function $H(f)$, then

$$H(f) = \frac{Y(f)}{X(f)} = \frac{Y(f)\overline{X(f)}}{X(f)\overline{X(f)}}$$

Hence,

$$\Phi_{yx}(f) = \Phi_{xx}(f)H(f)$$

which corresponds to equation (12.16). Similarly, for equation (12.17) we have

$$\Phi_{xy}(f) = \Phi_{xx}(f)\overline{H(f)} \tag{12.22}$$

and finally

$$\Phi_{yy}(f) = \Phi_{xx}(f)|H(f)|^2 \tag{12.23}$$

These results apply to the spectrum analysis of random signals in general and are useful for the study of adaptive systems.

12.3 AUTOCORRELATION MATRIX—EIGENVALUES

Throughout the remainder of this chapter the autocorrelation function of the signal $x(n)$ will be denoted by $r(p)$ and the N-dimensional square matrix, which is called the autocorrelation matrix of the signal $x(n)$, will be denoted by R_N:

$$R_N = \begin{bmatrix} r(0) & r(1) & \cdots & r(N-1) \\ r(1) & r(0) & \cdots & r(N-2) \\ \vdots & \vdots & \vdots & \vdots \\ r(N-1) & r(N-2) & \cdots & r(0) \end{bmatrix} \tag{12.24}$$

with

$$r(p) = E[x(n)x(n-p)]$$

As the autocorrelation function is of the positive type (12.12), the autocorrelation matrix is positive and symmetric by definition. In fact, it has a double symmetry as it is also symmetric about the second diagonal. This leads to a set of fundamental properties for adaptive systems.

The eigenvalues λ_i ($0 \leqslant i \leqslant N-1$) of the autocorrelation matrix of order N will be considered first. The characteristic equation

$$\det(\lambda I_N - R_N) = 0$$

leads to the equations

$$\det(R_N) = \prod_{i=0}^{N-1} \lambda_i \tag{12.25}$$

$$Nr(0) = \sum_{i=0}^{N-1} \lambda_i = N\sigma_x^2 \tag{12.26}$$

That is, if the determinant of the matrix is non-zero, each eigenvalue is non-zero and their sum is equal to N times the power of the signal. The positive nature of the matrix R_N further implies that they are all positive:

$$\lambda_i > 0; \quad 0 \leqslant i \leqslant N-1 \tag{12.27}$$

To ensure this, it is necessary and sufficient that the following determinants are all positive:

$$r(0); \quad \det\begin{bmatrix} r(0) & r(1) \\ r(1) & r(0) \end{bmatrix}; \quad \cdots; \quad \det\begin{bmatrix} r(1) & r(1) & \cdots & r(N-1) \\ r(1) & r(0) & \cdots & \\ \vdots & \vdots & \vdots & \vdots \\ r(N-1) & \cdots & \cdots & r(0) \end{bmatrix}$$

The corresponding matrices are the autocorrelation matrices of order less than or equal to N.

Under these conditions the matrix R_N can be diagonalized so that

$$R_N = M^t \operatorname{diag}(\lambda_i) M \tag{12.28}$$

where M is a square matrix of dimension N, such that $M^t = M^{-1}$, and $\operatorname{diag}(\lambda_i)$ is the diagonal matrix of the eigenvalues.

It is useful to examine successive powers of the matrices R_N and R_N^{-1}. By using both the Cayley–Hamilton theorem (according to which a matrix satisfies its own characteristic equation) and the Lagrange interpolation formula (which has already been used in Section 5.5), the power of a matrix can be expressed as a function of the powers of its eigenvalues:

$$R_N^P = \sum_{i=0}^{N-1} \lambda_i^P \prod_{\substack{j=0 \\ j \neq i}}^{N-1} \frac{R_N - \lambda_j I_N}{\lambda_i - \lambda_j} \tag{12.29}$$

For large values of the integer P, with

$$\lambda_{\max} = \max_{0 \leqslant i \leqslant N-1} (\lambda_i)$$

and, if this maximum corresponds to the value for $i = 0$, one can write:

$$R_N^P \simeq \lambda_{\max}^P \prod_{j=1}^{N-1} \frac{R_N - \lambda_j I_N}{\lambda_{\max} - \lambda_j} \tag{12.30}$$

Consequently, for large values of P one can make the approximation

$$R_N^P \simeq \lambda_{\max}^P K_N \quad (K_N: \text{Square matrix of dimension } N) \tag{12.31}$$

where K_N is the square matrix of order N of equation (12.30); as the matrix R_N can be diagonalised and satisfies (12.28), K_N can also be expressed more simply as the product of M^{-1} and a matrix deduced from M and setting all rows to zero except those which correspond to the index of the highest eigenvalue.

Similarly, using equation (12.28), with the same conditions, one can write:

$$(R_N^{-1})^P \simeq (\lambda_{\min}^{-1})^P K_N' \quad (K_N': \text{Square matrix of dimension } N) \tag{12.32}$$

with

$$\lambda_{\min} = \min_{0 \leqslant i \leqslant N-1} (\lambda_i)$$

It will be shown below that these two extreme eigenvalues, $\lambda_{\max}$ and $\lambda_{\min}$, condition the behaviour of adaptive systems.

The physical interpretation of the eigenvalues of the autocorrelation matrix is not readily apparent from their definition, but it can be illustrated by comparing them with the spectrum of the signal $x(n)$.

The case where the signal $x(n)$ is periodic and has period N will be considered first. In this case, the set $r(n)$ is also periodic, and is also symmetrical with:

$$r(N - i) = r(i); \quad 0 \leqslant i \leqslant N - 1$$

Under these conditions, the matrix R_N is a rotating matrix in which each row is derived from the preceding one by shifting. If the set $\Phi_{xx}(n)$ $(0 \leqslant n \leqslant N - 1)$ denotes the Fourier transform of the set $r(n)$, then it can be shown directly that

$$R_N T_N = T_N \operatorname{diag}(\Phi_{xx}(n))$$

where T_N is the matrix of the discrete Fourier transform of order N. Hence

$$R_N = T_N \operatorname{diag}(\Phi_{xx}(n)) T_N^{-1} \qquad (12.33)$$

Comparison with equation (12.28) shows that in this case the eigenvalues of the matrix R_N are the discrete Fourier transform of the autocorrelation function, i.e. the values of the signal power spectral density. M is the cosine transform matrix.

This relation is also valid for discrete white noise as the spectrum is constant and, since the autocorrelation matrix is a unit matrix (to a factor), the eigenvalues are equal.

Real signals generally have a spectral density with non-constant power and their autocorrelation function $r(p)$ decreases as the index p increases. For sufficiently large N, the significant elements of the N-dimensional matrix can be regrouped around the principal diagonal. Under these conditions let R'_N be the autocorrelation matrix of a signal $x(n)$ which is assumed to be periodic with period N. Its eigenvalues $\Phi_{xx}(n)$ form a sample of the power spectral density. The discrepancy between R_N and R'_N is due to the fact that R'_N is a rotating matrix and the difference appears primarily in the upper right-hand and lower left-hand corners. Thus, R_N can be better approximated by a diagonal matrix than R'_N, and consequently its eigenvalues are less dispersed. Indeed, under certain conditions which commonly occur in practice, it can be shown that [3]:

$$\min_{0 \leqslant n \leqslant N-1} \Phi_{xx}(n) \leqslant \lambda_{\min} \leqslant \lambda_{\max} \leqslant \max_{0 \leqslant n \leqslant N-1} \Phi_{xx}(n) \qquad (12.34)$$

and for sufficiently large N:

$$\lambda_{\min} \simeq \min_{0 \leqslant f \leqslant 1} \Phi_{xx}(f); \quad \lambda_{\max} \simeq \max_{0 \leqslant f \leqslant 1} \Phi_{xx}(f) \qquad (12.35)$$

In conclusion, it can be considered in practice that, when the dimension of this matrix is sufficiently large, the extreme eigenvalues of the autocorrelation matrix approximate the extreme values of the power spectral density of the signal.

12.4 RECURSIVE ESTIMATION

Adaptive systems apply estimation techniques to determine the evolution of their parameters. It is the case, for example, to assess the first and second moments of a random signal or to measure the mean power. Given a set of values $x(n)$, the mean $M(n)$ is calculated by

$$M(n) = \frac{1}{N} \sum_{i=0}^{N-1} x(n-i) \qquad (12.36)$$

However, it can also be obtained approximately in a recursive way, through first-order low-pass filtering:

$$y(n) = x(n) + by(n-1)$$

Assume $x(n)$ is the sum of a fixed value m and a discrete white noise $e(n)$, with zero mean and variance σ_e^2. Then, if $y(n) = 0$ for $n \leqslant 0$, applying the results of Section 6.1 leads to:

$$y(n) = m\frac{1 - b^{n+1}}{1-b} + \sum_{i=0}^{n} b^i e(n-i)$$

Now, taking the expectation of $y(n)$:

$$E[y(n)] = m\frac{1-b^{n+1}}{1-b} \qquad (12.37)$$

Since the positive coefficient b is smaller than unity to ensure stability, an estimation of the mean m is provided by the product $(1-b)\,y(n)$, that is, through filtering with the Z-transfer function

$$H(Z) = \frac{1-b}{1-bZ^{-1}} \qquad (12.38)$$

When a zero mean white noise with power σ_e^2 is fed to such a filter the output power is given by equation (6.26) as:

$$\sigma_s^2 = \sigma_e^2\frac{1-b}{1+b}$$

Consequently, the coefficient b has to be close to unity if the estimation of the mean m is to be accurate:

$$b = 1 - \delta \text{ with } 0 < \delta \ll 1$$

In these conditions the following approximation holds:

$$\sigma_s^2 \approx \sigma_e^2\frac{\delta}{2} \qquad (12.39)$$

The recursive estimator is depicted in Figure 12.2. Its equation is

$$M(n) = (1-\delta)M(n-1) + \delta x(n) \qquad (12.40)$$

It must be pointed out that, according to equation (12.37), this estimator is biased. The smaller δ, the larger the duration needed to reach a good estimation.

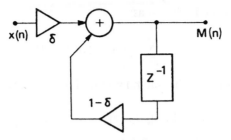

Fig. 12.2   The recursive estimates

In fact, the filter time constant τ is involved; according to equation (6.5) it is:

$$\tau = \frac{1}{\delta}$$

Finally the selection of a value for the parameter δ is a trade-off between estimation speed and residual error. It is a typical situation for the adaptive systems studied below.

12.5 THE GRADIENT ALGORITHM

The principle of adaptive filtering is shown in Figure 12.3. It consists in processing the input signal $x(n)$ to produce an output $\tilde{y}(n)$, whose difference with the reference $y(n)$ is minimized. For every new set of data, reference and input signal, the filter coefficients are updated.

At time n, assuming n data have been received, the cost function for the

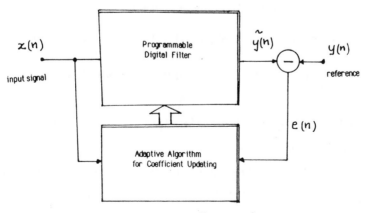

Fig. 12.3   Adaptive filter principle

optimization procedure is chosen as:

$$J(n) = \sum_{p=1}^{n} [y(p) - A^t(n)X(p)]^2 \qquad (12.41)$$

where $X(p)$ now designates the column vector of the N most recent input samples at time p:

$$X^t(p) = [x(p), x(p-1)\cdots x(p+1-N)]$$

Applying the results of Section 12.1 leads to the following expression for the filter coefficients:

$$A(n) = R_N^{-1}(n)R_{yx}(n) \qquad (12.42)$$

The input signal AC matrix can be conveniently expressed as:

$$R_N(n) = \sum_{p=1}^{n} X(p)X^t(p) = \sum_{p=1}^{n} \begin{bmatrix} x(p) \\ x(p-1) \\ \vdots \\ x(p+1-N) \end{bmatrix} [x(p), x(p-1), \ldots, x(p+1-N)] \qquad (12.43)$$

Similarly, the estimation of the cross-correlation vector between input and reference is:

$$R_{yx}(n) = \sum_{p=1}^{n} y(p)X(p) \qquad (12.44)$$

whenever a new data set $\{x(n+1), y(n+1)\}$ becomes available the new coefficient vector $A(n+1)$ can be recursively computed from $A(n)$. The definition equations (12.43) and (12.44) lead to the recursions:

$$R_N(n+1) = R_N(n) + X(n+1)X^t(n+1)$$
$$R_{yn}(n+1) = R_{yn}(n) + X(n+1)y(n+1) \qquad (12.45)$$

Now:

$$R_N(n+1)A(n+1) = R_{yn}(n+1) = R_{yn}(n) + X(n+1)y(n+1)$$

and:

$$R_N(n+1)A(n+1) = R_N(n)A(n) + X(n+1)y(n+1)$$

and finally:

$$R_N(n+1)A(n+1) = [R_N(n+1) - X(n+1)X^t(n+1)]A(n) + X(n+1)y(n+1)$$

Hence the recursion:

$$A(n+1) = A(n) + R_N^{-1}(n+1)X(n+1)[y(n+1) - A^t(n)X(n+1)] \qquad (12.46)$$

It is interesting to observe that the term:

$$e(n+1) = y(n+1) - A^t(n)X(n+1) \qquad (12.47)$$

is the error at the output of the system, computed at time $(n + 1)$ with the coefficients $A(n)$ available at time n. It is called the *a priori* error.

The algorithms in which the coefficients are computed each time according to equation (12.46), are the least squares algorithms.

Simplified but very useful algorithms are obtained when the matrix $R_N^{-1}(n)$ is replaced by the diagonal matrix δI_N, where δ is a scalar called the adaptation step size. Accordingly the coefficients are updated by:

$$A(n + 1) = A(n) + \delta X(n + 1)e(n + 1) \tag{12.48}$$

The algorithm is called the gradient algorithm, because the vector $-X(n + 1)e(n + 1)$ is actually the gradient of the function $\frac{1}{2}e^2(n + 1)$, the instantaneous value of the quadratic error [4]. Thus, the change in the coefficient values is carried out in the direction of the error gradient, but with opposite sign, to move towards the minimum. The procedure is similar to steepest descent in optimization.

In a stationary environment the coefficient vector converges, in the mean, to the theoretical solution. Substituting equation (12.47) into equation (12.48) yields:

$$A(n + 1) = [I_N - \delta X(n + 1)X^t(n + 1)]A(n) + \delta X(n + 1)y(n + 1) \tag{12.49}$$

Now, defining

$$R_N = E[X(n)X^t(n)]; \quad R_{yx} = E[y(n)X(n)] \tag{12.50}$$

The optimum set of coefficients is reached when n goes to infinity:

$$E[A(\infty)] = A_{opt} = R_N^{-1}R_{yn} \tag{12.51}$$

The matrix R_N is the $N \times N$ signal autocorrelation matrix and R_{yn} is the vector of the N first elements of the inter-correlation between input and reference. Finally, the gradient algorithm converges in the mean to the optimal solution A_{opt}, hence also the denomination of stochastic gradient which is used sometimes. The minimization criterion is the least mean squares (LMS) criterion.

The circuit for the resulting adaptive filter is shown in Figure 12.4. The variations of the coefficients are calculated by multiplying for each value of the difference $e(n)$ and summing.

The choice of the value δ in equation (12.48) leads to a compromise between the adaptation speed and the value of the residual error when the adaptation is obtained. These two properties will be studied later, but it is first necessary to study the convergence conditions.

12.6 CONVERGENCE CONDITIONS

A convergence condition can be derived by considering the *a posteriori* error defined by:

$$\varepsilon(n + 1) = y(n + 1) - A^t(n + 1)X(n + 1)$$

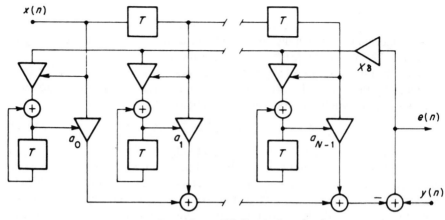

Fig. 12.4   Adaptive FIR filter in direct form

The adaptive filter works properly if the adaptation is efficient, namely if the *a posteriori* error is, in the mean, smaller than the *a priori* one:

$$E[|\varepsilon(n+1)|] < E[|e(n+1)|] \qquad (12.52)$$

Substituting the coefficient updating equation (12.48) into the definition of $\varepsilon(n+1)$ yields:

$$\varepsilon(n+1) = [1 - \delta X^t(n+1)X(n+1)]e(n+1) \qquad (12.53)$$

The two errors are in fact proportional. Since:

$$X^t(n+1)X(n+1) = \sum_{i=0}^{N-1} x^2(n+1-i)$$

the following convergence condition is obtained, assuming independence between the error and input signals:

$$0 < \delta < \frac{2}{N\sigma_x^2} \qquad (12.54)$$

where σ_x^2 is the input signal power.

In practice, it might be insufficient to impose condition (12.52) in the mean, and a peak factor for $X^t(n+1)X(n+1)$ might have to be introduced. Moreover, the above independence assumption is not strictly verified. Therefore it is wise to take a margin of a few units for the upper bound of δ in applications. In the case of Gaussian signals it can be shown that convergence is guaranteed as soon as the adaptation step is smaller than

$$\frac{1}{3}\frac{2}{N\sigma_x^2}$$

After convergence, the optimum coefficient values are given by equation (12.51).

The minimum value $E_{\min}$ of the squared error corresponding to the set of optimum values of the coefficients can also be expressed as a function of the signals $y(n)$, $x(n)$ and their intercorrelation.

For the set of coefficients A_{opt} the mean square error $E_{\min}$ is

$$E_{\min} = E[e^2(n)]$$

with:

$$e(n) = y(n) - A_{\text{opt}}^t X(n) = y(n) - X^t(n)A_{\text{opt}}$$

However, one has

$$e^2(n) = y^2(n) - 2y(n)X^t(n)A_{\text{opt}} + A_{\text{opt}}X(n)X^t(n)A_{\text{opt}}$$

Thus, allowing for equation (12.51) after transposition,

$$E_{\min} = E[y^2(n)] - A_{\text{opt}}^t R_N A_{\text{opt}} \tag{12.55}$$

or, again:

$$E_{\min} = E[y^2(n)] - A_{\text{opt}}^t R_{yx} \tag{12.56}$$

Alternatively, in another form:

$$E_{\min} = E[y^2(n)] - R_{yx}^t R_N^{-1} R_{yx} \tag{12.57}$$

In this expression the order N of the filter appears as the dimension of the square autocorrelation matrix of the signal $x(n)$ and of the intercorrelation column vector between $x(n)$ and the reference $y(n)$.

The stability being ensured, it is interesting to evaluate the adaptation speed and to determine the time constant of the adaptive filter [5].

12.7 TIME CONSTANT

This is the parameter which defines the adaptation speed of the filter. Let us first consider a filter with a single coefficient $a_1(n)$. In these conditions equation (12.49) can be written:

$$a_1(n+1) = [1 - \delta x^2(x+1)]a_1(n) + \delta x(x+1)y(n+1)$$

The coefficient of this filter has the mean value $b = 1 - \delta\sigma_x^2$. For a global assessment of the characteristics that filter can be viewed as a first-order IIR section with a fixed coefficient equal to b. Then for small δ, expression (6.5) yields the time constant:

$$\tau \approx \frac{1}{\delta\sigma_x^2}$$

The result can be generalized to a filter with N coefficients, under certain conditions.

For a given value $A(n)$ of the set of coefficients with index n, it is possible by using equations of the previous section to show the divergence from the optimum to be:

$$E(n) = E_{\min} + (A_{opt} - A(n))^t R_N (A_{opt} - A(n)) \qquad (12.58)$$

Assume $[\alpha(n)]$ is the column vector:

$$[\alpha(n)] = M(A_{opt} - A(n)) \qquad (12.59)$$

where M is the square matrix of order N, which appears in the decomposition of R_N given by equation (12.28).

Equation (12.58) can then be written in a concise form:

$$E(n) - E_{\min} = [\alpha(n)]^t \operatorname{diag}(\lambda_i)[\alpha(n)]$$

or, by multiplying out,

$$E(n) - E_{\min} = \sum_{i=0}^{N-1} \lambda_i \alpha_i^2(n) \qquad (12.60)$$

In its adaptation phase an adaptive filter changes from an initial value for the coefficients which can be zero, for example, to a value close to the optimum.

By assuming the coefficients for values with index $n \leq 0$ to be zero, and by assuming that the vector R_{yx} is constant, using equation (12.51) we derive:

$$M R_{yx} = M E[y(n)X(n)] = \operatorname{diag}(\lambda_i) M A_{opt}$$

or,

$$\operatorname{diag}(\lambda_i)[\alpha(0)] = M R_{yx} \qquad (12.61)$$

Then, allowing for equation (12.49) which gives the change in the coefficients at time index $n\,(n > 0)$, we can write:

$$M E[A(n + 1)] = M E[A(n)] - \delta \cdot \operatorname{diag}(\lambda_i) M E[A(n)] + \delta \operatorname{diag}(\lambda_i) M A_{opt}$$

and, using the definition of the vector $[\alpha(n)]$:

$$E[\alpha(n + 1)] = (I - \delta \operatorname{diag}(\lambda_i)) E[\alpha(n)]$$

which results in:

$$E[\alpha(n)] = [I - \delta \operatorname{diag}(\lambda_i)]^n [\alpha(0)] \qquad (12.62)$$

In fact the elements of the vector $[\alpha(0)]$ are determined by equation (12.61) with the hypotheses retained. Under these conditions the quadratic error $E(n)$ at time n is obtained by substituting into equation (12.60) the values computed that way for the elements of $[\alpha(n)]$. Thus:

$$E(n) = E_{\min} + \sum_{i=0}^{N-1} \lambda_i \alpha_i^2(0)(1 - \delta \cdot \lambda_i)^{2n}$$

For small values of δ, the mean square error can be approximated by a weighted summation of exponentials, with time constants τ_i, such that

$$\tau_i \simeq \frac{1}{2\delta\lambda_i}; \quad 0 \leqslant i \leqslant N-1$$

To find the time constant τ_e of the set $|e(n)|$, $E(A(n)) - E_{min}$ must be approximated by an exponential. Then,

$$\sum_{i=0}^{N-1} \lambda_i \alpha_i^2(0)\left(1 - \frac{2n}{\tau_e}\right) \simeq \sum_{i=0}^{N-1} \lambda_i \alpha_i^2(0)(1 - 2n\delta\lambda_i)$$

or

$$\tau_e \simeq \frac{\sum_{i=0}^{N-1} \lambda_i \alpha_i^2(0)}{\sum_{i=0}^{N-1} \lambda_i \alpha_i^2(0)\lambda_i} \tag{12.63}$$

By making the following approximation, which is valid for values of λ_i which are not too dispersed:

$$\frac{\sum_{i=0}^{N-1} \lambda_i \alpha_i^2(0)\lambda_i}{\sum_{i=0}^{N-1} \lambda_i \alpha_i^2(0)} \simeq \frac{1}{N} \sum_{i=0}^{N-1} \lambda_i$$

and, allowing for equation (12.26), one finally obtains:

$$\tau_e \simeq \frac{N}{\delta\sum_{i=0}^{N-1} \lambda_i} = \frac{1}{\delta\sigma_x^2} \tag{12.64}$$

This equation thus gives an estimate of the time constant of the adaptive filter which is inversely proportional to the adaptation step in the coefficients. It is interesting to note that if the power of the input signal σ_x^2 is fixed or estimated continuously, with an adaptation step as follows:

$$\delta = \Delta/\sigma_x^2$$

one obtains:

$$\tau_e \simeq \frac{1}{\Delta} \tag{12.65}$$

The convergence condition (12.54) provides an upper bound for the adaptation step and if δ exceeds that bound the output error will grow. A geometric illustration, based on the error surface representation, assumed symmetric, shows that the fastest adaptation is obtained with half that bound, $\delta = 1/N\sigma_x^2$. That result can be proven analytically using the approach described in the next section. In these conditions the time constant satisfies the following inequality:

$$\tau_e \geqslant N \tag{12.66}$$

It must be pointed out that, according to equation (12.62), the convergence time depends on the smallest of the eigenvalues λ_i. Thus, overall, the adaptive filter works all the better when the ratio of the extreme eigenvalues is close to unity, that is when the input signal $x(n)$ is close to a white noise.

In order to complete the examination of this adaptive filter the residual error after adaptation has still to be evaluated [6].

12.8 RESIDUAL ERROR

After a transition phase corresponding to the convergence, the coefficients of the adaptive filter continually vary about their optimal value, because the adaptation step δ remains constant. This, incidentally, is the permanent adaptation condition of the system. As a result, the residual error E_R, defined as the limit of the expectation of the quadratic error $E(n)$ as n goes to infinity, is greater than the minimum value E_{min}.

The residual error E_R is evaluated by considering the vector $[\alpha(n)]$ defined by equation (12.59) and its evolution described by the following recursion:

$$[\alpha(n + 1)] = [\alpha(n)] - \delta M X(n + 1)e(n + 1)$$

In order to estimate the variances of the elements of the vector $[\alpha(n)]$ it is convenient to consider the matrix $[\alpha(n)][\alpha(n)]^t$, whose main diagonal is made of those elements squared:

$$[\alpha(n + 1)][\alpha(n + 1)]^t = [\alpha(n)][\alpha(n)]^t - 2\delta M X(n + 1)e(n + 1)[\alpha(n)]^t$$
$$+ \delta^2 e^2(n + 1)M X(n + 1)X^t(n + 1)M^t \qquad (12.67)$$

In terms of $[\alpha(n)]$, the error is expressed by:

$$e(n + 1) = y(n + 1) - A_{opt}^t X(n + 1) + X^t(n + 1)M^t[\alpha(n)]$$

The evolution of the adaptive system is governed by the two above equations. In order to obtain useful results it is necessary to make several simplifying assumptions [6, 7].

The following variables are assumed to be independent:

(1) The output error when the coefficients take on their optimal values;
(2) The data vector: $X(n + 1)$;
(3) The coefficient deviation: $A(n) - A_{opt}$.

The consequence is the following equality:

$$E[[y(n + 1) - A_{opt}^t X(n + 1)]X^t(n + 1)M^t[\alpha(n)]] = 0 \qquad (12.68)$$

Taking the expectation of both sides of equation (12.67) yields

$$E\{[\alpha(n + 1)][\alpha(n + 1)]^t\} = [I_N - 2\delta \operatorname{diag}(\lambda_i)]E\{[\alpha(n)][\alpha(n)]^t\}$$
$$+ \delta^2 E(n) \operatorname{diag}(\lambda_i) \qquad (12.69)$$

After the transient phase, when n goes to infinity:

$$E\{[\alpha(\infty)][\alpha(\infty)]^t\} = \frac{\delta}{2}E(\infty)I_N \qquad (12.70)$$

Because of the definition (12.59) of $[\alpha(n)]$ we have also:

$$E\{[A_{\text{opt}} - A(\infty)][A_{\text{opt}} - A(\infty)]^t\} = \frac{\delta}{2}E(\infty)I_N \tag{12.71}$$

Therefore, after convergence the coefficient deviations are independent and they have all the same variance.

For variable coefficients equation (12.60) takes the form

$$E(n) = E_{\min} + \sum_{i=0}^{N-1'} \lambda_i E[\alpha_i^2(n)]$$

Thus, the residual error $E_R = E(\infty)$ is given by:

$$E_R = \frac{E_{\min}}{1 - (\delta/2)N\sigma_x^2} \tag{12.72}$$

It is worth pointing out that the above equation leads to the stability condition (12.54) derived by a different approach.

In practice, due to the margin generally taken on the step size δ, the following approximation holds:

$$E_R \simeq E_{\min}\left(1 + \frac{\delta}{2}N\sigma_x^2\right) \tag{12.73}$$

In terms of the time constant, with equation (12.64), we have:

$$E_R \simeq E_{\min}\left[1 + \frac{NT}{2\tau_e}\right] \tag{12.74}$$

The relation between the time constant and the residual divergence is thus clearly seen. T is the sampling period, taken here as being equal to unity.

The increase in residual error due to the step size δ can be viewed as a gradient noise.

The operation of an adaptive filter can be illustrated by the second-order predictor defined by the following equations:

$$e(n+1) = x(n+1) - a_1(n)x(n) - a_2(n)x(n-1)$$
$$a_1(n+1) = a_1(n) + \delta x(n)e(n+1)$$
$$a_2(n+1) = a_2(n) + \delta x(n-1)e(n+1)$$

Taking $x(n) = \sin n\pi/4$ as a signal, with zero values for the coefficients at the origin, the evolution of the error and that of the coefficients is shown in Figure 12.5. The optimum coefficient values correspond to an FIR filter having a zero on the unit circle at the frequency $f_s/8$, to cancel the input signal. The case of linear prediction is treated in more detail in Section 12.13.

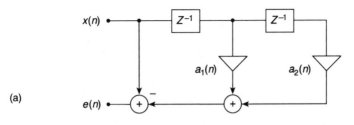

(a)

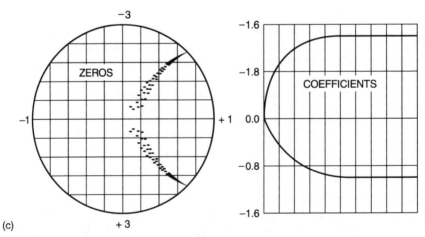

(b)

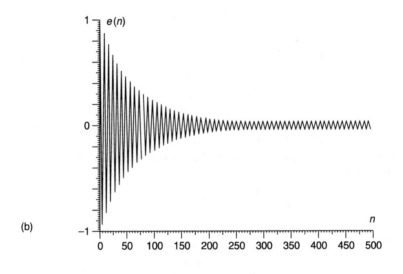

(c)

Fig. 12.5   Second-order prediction filter: (a) arrangement of the second-order predictor, (b) output error, (c) development of the coefficients

12.9 PARAMETERS OF COMPLEXITY

The complexity parameters of adaptive filters are the same as those for filters with fixed coefficients, and the most important are the multiplication rate, the number of bits of the coefficients and the internal memories. The limitations on the number of bits of the coefficients and internal data contribute to the increase in the residual error E_{RT}. The specifications are generally given in terms of a minimum gain of the system. That is, the ratio of the power of the reference signal σ_y^2 to the total residual error E_{RT} should exceed a specified value G^2:

$$\frac{\sigma_y^2}{E_{RT}} \geq G^2 \qquad (12.75)$$

The time constant τ_e, if it is imposed, should be compatible with the minimum gain of the system so that the system can be realized using the gradient technique.

The values of the parameters G and τ_e of the adaptive filter permit the calculation of the number of bits of the coefficients and internal data for each structure; the order of the filter is chosen to be sufficiently large for the minimum divergence E_{min} to be acceptably small and for equation (12.75) to be satisfied.

In the case of the FIR filter realized as a direct structure as in Figure 12.4, rounding is generally carried out on the output of the multipliers and the scheme becomes that of Figure 12.6.

The noise generated by rounding the internal data with the quantization step q_2 corresponds to the addition of the power $N(q_2^2/12)$ to the minimal error power E_{min}.

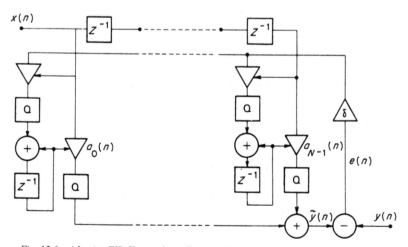

Fig. 12.6   Adaptive FIR filter with coefficient and internal data wordlength limitations

If the coefficients are quantized with the step q_1, the round-off errors make a vector which has to be incorporated into the coefficient evolution equation. Assuming that round-off errors are independent of other signals, the additional term $(q_1^2/12)I_N$ has to be introduced in equations (12.69) and equation (12.70) becomes:

$$E\{[\alpha(\infty)][\alpha(\infty)]^t\} = \frac{\delta}{2}E(\infty)I_N + \frac{1}{2\delta}\frac{q_1^2}{12}\text{diag}\left(\frac{1}{\lambda_i}\right) \qquad (12.76)$$

Finally, the total residual error E_{RT} is given by:

$$E_{RT} = \frac{1}{1-(\delta/2)N\sigma_x^2}\left[E_{min} + N\frac{q_2^2}{12} + \frac{N}{2\delta}\frac{q_1^2}{12}\right] \qquad (12.77)$$

which, for small step size δ, can be approximated by:

$$E_{RT} \simeq E_{min}\left(1 + \frac{\delta\cdot N\sigma_x^2}{2}\right) + \frac{N}{2\delta}\frac{q_1^2}{12} + N\frac{q_2^2}{12} \qquad (12.78)$$

The relative values of the four terms in this expression can be chosen to suit each application. One common option is to consider E_{min} as the most important term and to assume that the additional residual error caused by the adaptation step δ is equal to the noise introduced by the internal rounding, that is,

$$\tfrac{1}{2}E_{min}\frac{\delta\cdot N\sigma_x^2}{2} = \frac{N}{2\delta}\frac{q_1^2}{12} = N\frac{q_2^2}{12} \qquad (12.79)$$

If b_c is the number of bits of the coefficients and a_{max} is the amplitude of the largest coefficient, then

$$q_1 = a_{max}2^{1-b_c}$$

Under these conditions,

$$2^{2b_c} = \frac{2}{3}\frac{a_{max}^2}{\delta^2 E_{max}\sigma_x^2}$$

With the assumption that E_{min} is the dominant term in equation (12.78), that is,

$$G^2 E_{min} \simeq \sigma_y^2$$

and by introducing the time constant, one has approximately

$$b_c \simeq \log_2(\tau_e) + \log_2(G) + \log_2\left(a_{max}\frac{\sigma_x}{\sigma_y}\right) \qquad (12.80)$$

The term $[a_{max}(\sigma_x/\sigma_y)]$ depends on the system gain, the signals and the filter order. It can be bounded, assuming the powers of the sequences $y(n)$ and $\tilde{y}(n)$ are similar:

$$\sigma_y^2 \approx \sigma_{\tilde{y}}^2 = A^t R_N A$$

Thus:

$$\lambda_{min} \sum_{i=0}^{N} a_i^2 < \sigma_y^2 < \lambda_{max} \sum_{i=0}^{N-1} a_i^2$$

and:

$$\left[\frac{\sigma_x^2}{N\lambda_{max}}\right]^{1/2} < \frac{\sigma_x a_{max}}{\sigma_y} < \left[\frac{\sigma_x^2}{\lambda_{min}}\right]^{1/2}$$

In fact, the upper bound above is often impractical, and chosen estimations have to be worked out in applications. However, if the input signal is close to a white noise, clearly the term $(\sigma_x a_{max}/\sigma_y)$ is close to unity; with some signals (for example, those with narrow spectrum) it can take on large values.

Equation (12.80) shows that the number of bits of the coefficients is directly related to the time constant and to the gain of the system.

In a similar way equation (12.79) permits the number of bits of the internal data b_i to be determined by setting:

$$q_2 = \max\{|x(n)|, |y(n)|\}2^{1-b_i}$$

With the assumption that $\sigma_x^2 \geqslant \sigma_y^2$, which corresponds particularly to the cases of linear prediction and echo cancelling, and taking the value 4 as the peak factor of the signal $x(n)$ as in the Gaussian case, this becomes:

$$q_2 = 4\sigma_x 2^{1-b_i}$$

whence

$$2^{2b_i} = 2^4 \times \frac{4}{3}\frac{1}{E_{min}\cdot\delta}$$

By introducing the parameters for the system, we obtain:

$$b_i \simeq 2 + \log_2\left(\frac{\sigma_x}{\sigma_y}\right) + \log_2(G) + \tfrac{1}{2}\log_2(\tau_e) \tag{12.81}$$

Equations (12.80) and (12.81) can assist systems designers in selecting from the available options.

12.10 NORMALIZED ALGORITHMS AND SIGN ALGORITHMS

The time constant of an adaptive filter and its residual error are related to the power of the input signal $x(n)$. When this power can vary by large amounts, the adaptation can be modified as follows:

$$A(n+1) = A(n) + \frac{\delta}{X^t(n+1)X(n+1)}X(n+1)e(n+1) \tag{12.82a}$$

This is described as a normalized algorithm. It can be shown that it leads *a posteriori* to a zero error.

In practice, rather than calculate the scalar product $X^t(n+1)X(n+1)$, a recursive estimate of the power can be made and this leads to:

$$P_x(n+1) = (1-\varepsilon)P_x(n) + \varepsilon x^2(n+1)$$

$$A(n+1) = \frac{A(n) + \delta X(n+1)e(n+1)}{P_x(n+1)} \qquad (12.82b)$$

The parameter ε of the recursive estimate is chosen as a function of the variations of the signal power. It must be at least of order $1/N$, where N is the number of coefficients for the filter.

In certain applications it is important to minimize the operations, and simplified algorithms are then used in which the variations of the coefficients are a function of the sign of the terms $e(n)$ and $x(n)$, or even the products $e(n)x(n-i)$; these are the sign algorithms. The reduction in complexity thus obtained is paid for by a degradation in some aspects of system performance [8].

Consider the following coefficient adaptation algorithm:

$$a_i(n+i) = a_i(n) + \Delta.e(n+1)\text{sign}[x(n+1-i)] \qquad (12.83)$$

For non-zero x:

$$\text{sign}[x] = \frac{x}{|x|}$$

If the amplitude distribution of $x(n)$ is symmetrical, in a coarse approximation, $|x(n)|$ can be replaced by σ_x. Therefore, expression (12.83) is similar to equation (12.82) with:

$$\delta \approx \frac{\Delta}{\sigma_x}$$

It appears that it is more interesting, for complexity-reduction purposes, to take the sign of $x(n)$ rather than the sign of $e(n)$. Further in that direction, the coefficient variations can be limited to constant values:

$$a_i(n+1) = a_i(n) + \Delta\,\text{sign}\,[e(n+1)]\,\text{sign}\,[x(n+1-i)] \qquad (12.84)$$

Then, to a first approximation, the coefficient variations in equation (12.84) are comparable with those in equation (12.82) with:

$$\delta = \frac{\Delta}{\sigma_e\sigma_x} \qquad (12.85)$$

Starting from zero values for the coefficients, in the convergence phase of the filter, it can be taken that $\sigma_e \simeq \sigma_y$ and the time constant τ_s for the sign algorithm can be expressed by:

$$\tau_s \simeq \frac{1}{\Delta}\frac{\sigma_y}{\sigma_x} \qquad (12.86)$$

After convergence, one can take $\sigma_e^2 = E_{\min}$. If the variation step is sufficiently small, the residual error E_{RS} in the sign algorithm is

$$E_{RS} \simeq E_{\min}\left(1 + \frac{N\Delta}{2}\frac{\sigma_x}{\sqrt{E_{\min}}}\right) \qquad (12.87)$$

The residual error is thus found to be larger than with the gradient algorithm. This leads to low values for Δ. It should also be noted that, remembering equation (12.85), the stability condition (12.51) is represented by the inequality

$$\Delta \leqslant \frac{2}{N}\frac{\sqrt{E_{\min}}}{\sigma_x} \qquad (12.88)$$

which can be taken as a convergence condition and can result in very small values of Δ. In practice, equation (12.84) is usually modified to become

$$a_i(n+1) = (1-\varepsilon)a_i(n) + \Delta\,\text{sign}\,\{e(n+1)x(n+1-i)\} \qquad (12.89)$$

The constant ε, positive and small, introduces a leakage function, which is needed, for example, in transmission systems which must tolerate a certain error rate. Under these conditions the coefficients are bounded by:

$$|a_i(n)| \leqslant \frac{\Delta}{\varepsilon}; \quad 0 \leqslant i \leqslant N-1 \qquad (12.90)$$

This modification leads to an increase in the residual error. The coefficients are biased, and instead of (12.51) for small ε and Δ we can write:

$$E[A(\infty)] = \left[\frac{\varepsilon}{\delta}I + R_N\right]^{-1}\cdot E[y(n)X(n)] \qquad (12.91)$$

The corresponding increase in residual error can be calculated by an expression similar to equation (12.58).

The choice of the constants ε and Δ is made as a function of the performance to be achieved in each case.

The adaptive filters considered above are of the direct structure FIR type. It is a simple and robust approach and is commonly used. However, as with fixed coefficient filters, other structures can be employed.

12.11 ADAPTIVE FIR FILTERING IN CASCADE FORM

In certain modelling problems, particularly in automatic control, it is important to know the roots of the Z-transfer function of the adaptive filter [9]. It is then convenient to use a cascade arrangement of L second-order sections $H_i(Z)$, $1 \leqslant i \leqslant L$, such that:

$$H_i(Z) = 1 + a_1^i Z^{-1} + a_2^i Z^{-2}$$

From the results of Chapter 6, if the roots Z_1^i and Z_2^i are complex, then

$$Z_2^i = \overline{Z_1^i}; \quad a_2^i = |Z_1^i|^2; \quad a_1^i = -2\,\mathrm{Re}\,(Z_1^i)$$

Consider an adaptive filter whose transfer function $H(Z)$ is

$$H(Z) = \prod_{i=1}^{L} (1 + a_1^i Z^{-1} + a_2^i Z^{-2}) \tag{12.92}$$

Beginning with a given set of values of the coefficients, variations proportional to the gradient of the error function $E(A)$ must be applied so as to minimize the mean square error. Using the definition (12.2) of $E(A)$, this leads to

$$\frac{\partial E}{\partial a_k^i} = -\frac{2}{N_0} \sum_{n=0}^{N_0-1} [y(n) - \tilde{y}(n)] \frac{\partial \tilde{y}(n)}{\partial a_k^i}; \quad k = 1, 2; \quad 1 \leqslant i \leqslant L \tag{12.93}$$

In order to calculate the term g_k^i such that:

$$g_k^i(n) = \frac{\partial \tilde{y}(n)}{\partial a_k^i}$$

one can use the expression for $y(n)$ obtained by an inverse Z-transform on the transform $X(Z)$ of the set $x(n)$. Hence,

$$\tilde{y}(n) = \frac{1}{2\pi j} \int_{\Gamma} Z^{n-1} \prod_{i=1}^{L} (1 + a_1^i Z^{-1} + a_2^i Z^{-2}) X(Z)\,dZ$$

where Γ is a suitable integration contour. Hence,

$$\frac{\partial \tilde{y}(n)}{\partial a_k^i} = \frac{1}{2\pi j} \int_{\Gamma} Z^{n-1} Z^{-k} \prod_{\substack{l=1 \\ l \neq i}}^{L} (1 + a_1^l Z^1 + a_2^l Z^{-2}) X(Z)\,dZ$$

or

$$\frac{\partial \tilde{y}(n)}{\partial a_k^i} = \frac{1}{2\pi j} \int_{\Gamma} Z^{n-1} Z^{-k} \frac{H(Z)}{1 + a_1^i Z^{-1} + a_2^i Z^{-2}} X(Z)\,dZ \tag{12.94}$$

Thus, to form the term $g_k^i(n)$ it is sufficient to apply the set $\tilde{y}(n)$ to a recursive section whose transfer function is the inverse of that of the initial section of order i. This recursive section has the same coefficients, but with the opposite sign. The corresponding circuit is given in Figure 12.7. The variations in the coefficients are calculated by the expressions

$$da_k^i(n) = \delta g_k^i(n)[y(n) - \tilde{y}(n)]; \quad k = 1, 2; \quad 1 \leqslant i \leqslant L \tag{12.95}$$

The filter obtained in this way is more complicated than that in the previous section, but it offers a very simple method of finding the roots, which, due to the presence of a recursive part, should be inside the unit circle in the Z-plane to ensure the stability of the system. The techniques derived for FIR filters also apply to IIR filters.

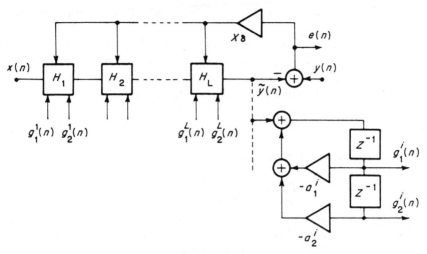

Fig. 12.7 Adaptive FIR filter in cascade form

12.12 ADAPTIVE IIR FILTERING

The coefficients of an IIR filter can be calculated for time specifications by using the least mean squares technique in an iterative procedure, as was done in Section 7.3. Algorithms for adapting the coefficients to the time-evolution of the system can also be deduced.

A linear system can be modelled by a purely recursive IIR filter with a Z-transfer function, $G(Z)$, such that:

$$G(Z) = a_0 \frac{1}{1 + \sum_{k=1}^{K} b_k Z^{-k}} \tag{12.96}$$

In this case the model is said to be autoregressive (AR) [10]. This important and convenient approach is also appropriate if the best representation of the system corresponds to a Z-transfer function, $H(Z)$, which is the quotient of two polynomials

$$H(Z) = \frac{N(Z)}{D(Z)}$$

where $N(Z)$ has all its roots inside the unit circle and thus has minimum phase. In this case, for a suitable integer M, we can write:

$$\frac{1}{N(Z)} \simeq 1 + \sum_{i=1}^{M} c_i Z^{-i}$$

It is then sufficient to let the degree K of the denominator of the function $G(Z)$ take a value sufficient for representing $H(Z)$. The presence of roots inside the unit circle of the system results in an increase in the number of poles of the model [1].

The general IIR filter corresponds to an autoregressive model with a moving average (ARMA). This is the most general approach for modelling a linear system. For the IIR filter whose coefficients must be calculated over a set of N_0 indices in order to approximate a set $y(n)$, the output is written

$$\tilde{y}(n) = \sum_{l=0'}^{L} a_l x(n-l) - \sum_{k=1}^{K} b_k \tilde{y}(n-k) \qquad (12.97)$$

The error function $E(A, B)$ is expressed by:

$$E(A, B) = \frac{1}{N_0} \sum_{n=0}^{N_0-1} [y(n) - \tilde{y}(n)]^2 \qquad (12.98)$$

Starting from a set of values for the coefficients, this function can be minimized using the gradient algorithm if the coefficients are given increments proportional to the gradient of $E(A, B)$ and of the opposite sign. The presence of a recursive part causes complications. Calculation of the gradient leads to the following expressions:

$$\frac{\partial E}{\partial a_l} = -\frac{2}{N_0} \sum_{n=0}^{N_0-1} [y(n) - \tilde{y}(n)] \frac{\partial \tilde{y}(n)}{\partial a_l}; \quad 0 \leqslant l \leqslant L$$

$$\frac{\partial E}{\partial b_k} = -\frac{2}{N_0} \sum_{n=0}^{N_0-1} [y(n) - \tilde{y}(n)] \frac{\partial \tilde{y}(n)}{\partial b_k}; \quad 1 \leqslant k \leqslant K$$

with

$$\frac{\partial \tilde{y}(n)}{\partial a_l} = x(n-l) - \sum_{k=1}^{K} b_k \frac{\partial \tilde{y}(n-k)}{\partial a_l} \qquad (12.99)$$

$$\frac{\partial \tilde{y}(n)}{\partial b_k} = -\tilde{y}(n-k) - \sum_{k=1}^{K} b_k \frac{\partial \tilde{y}(n-k)}{\partial b_k} \qquad (12.100)$$

To show the method of realizing equations (12.99) and (12.100), we can write

$$H(Z) = \frac{\sum_{l=0}^{L} a_l Z^{-l}}{1 + \sum_{k=1}^{K} b_k Z^{-k}} = \frac{N(Z)}{D(Z)}$$

Then

$$\tilde{y}(n) = \frac{1}{2\pi j} \int_{\Gamma} Z^{n-1} H(Z) X(Z) \, dZ$$

and consequently:

$$\frac{\partial \tilde{y}(n)}{\partial a_l} = \frac{1}{2\pi j} \int_{\Gamma} Z^{n-1} Z^{-l} \frac{1}{D(Z)} X(Z) \, dZ \qquad (12.101)$$

$$\frac{\partial \tilde{y}(n)}{\partial b_k} = \frac{1}{2\pi j} \int_{\Gamma} Z^{n-1} Z^{-k} \frac{(-1)}{D(Z)} H(Z) X(Z) \, dZ \qquad (12.102)$$

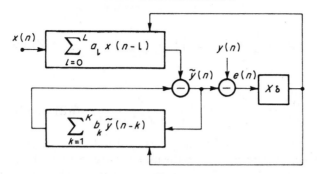

Fig. 12.8 Adaptive IIR filter

The gradient is thus calculated from the set obtained by applying $x(n)$ and $\tilde{y}(n)$ to the circuits corresponding to the transfer function $1/D(Z)$.

To simplify the implementation, the second terms in equations (12.99) and (12.100) can be ignored, which leads to the following increments:

$$da_l(n) = \delta \cdot [y(n) - \tilde{y}(n)]x(n - l); \qquad 0 \leqslant l \leqslant L \qquad (12.103)$$

$$db_k(n) = -\delta \cdot [y(n) - \tilde{y}(n)]\tilde{y}(n - k); \quad 1 \leqslant k \leqslant K \qquad (12.104)$$

The corresponding circuit is shown in Figure 12.8.

For each value of the index n, the coefficients a_l and b_k are incremented by a quantity which is proportional to the product of the error $e(n) = y(n) - \tilde{y}(n)$ with $x(n - l)$ and $\tilde{y}(n - k)$, respectively.

The stability and parameters of this type of filter can be studied as for FIR filters [11, 12]. However, it is not difficult to control the stability of an IIR filter when it is constructed as a cascade of second-order sections, which, as indicated in previous chapters, also offers other advantages.

Consider a filter constructed as a cascade of second-order elements and with a transfer function $G(Z)$. In the autoregressive case, we have

$$G(z) = a_0 \prod_{i=1}^{L} \frac{1}{1 + b_1^i Z^{-1} + b_2^i Z^{-2}} \qquad (12.105)$$

To control the stability of such a filter it is sufficient, following Section 6.7, to ensure that the following conditions are fulfilled:

$$|b_2^i| < 1; \quad |b_1^i| < 1 + b_2^i; \quad 1 \leqslant i \leqslant L \qquad (12.106)$$

As before, the calculation of the gradient of the error function requires knowledge of the term g_k^i, where

$$g_k^i(n) = \frac{\partial \tilde{y}(n)}{\partial b_k^i}$$

Since

$$\tilde{y}(n) = \frac{1}{2\pi j} \int_{\Gamma} Z^{n-1} a_0 \prod_{i=1}^{L} \frac{1}{1 + b_1^i Z^{-1} + b_2^i Z^{-2}} X(Z)\, dZ$$

we obtain:

$$\frac{\partial \tilde{y}(n)}{\partial b_k^i} = -\frac{1}{2\pi j} \int_{\Gamma} Z^{n-1} \frac{Z^{-k}}{1 + b_1^i Z^{-1} + b_2^i Z^{-2}} G(Z) X(Z)\, dZ$$

This expression indicates that the terms $g_k^i(n)$, with $k = 1, 2$ and $1 \leqslant i \leqslant L$, are obtained by applying the set $\tilde{y}(n)$ to the ith recursive section. The corresponding circuit is given in Figure 12.7, where the non-recursive elements of the second order are replaced by recursive ones. The stability of the system is tested by equation (12.106) for each value of the index n.

The method which has been discussed is also applicable to an ARMA model, but in this case the circuits are rather more complicated.

The techniques used in the previous sections involve overall minimization of the mean square error. It is interesting to give an example of sequential or stage-by-stage minimization, which is presented below in the context of linear prediction.

12.13 LINEAR PREDICTION

Linear prediction is an inverse modelling operation, which is an important field of application for adaptive filtering. For a given set $x(n)$, the power of the set $e(n)$ such that:

$$e(n) = x(n) - \tilde{x}(n) = x(n) - \sum_{i=1}^{N} a_i x(n - i) \tag{12.107}$$

is to be minimized.

In this expression, the term $\tilde{x}(n)$ represents a prediction of the input set $x(n)$, which is obtained by a linear operation on the terms for preceding values of the index. The set $e(n)$ is called the prediction error.

The circuit is shown in Figure 12.9. On a domain of N_0 indices, the coefficients a_i $(1 \leqslant i \leqslant N)$ of the prediction filter must be determined so as to minimize the mean square error $E(A)$ defined by:

$$E(A) = \frac{1}{N_0} \sum_{n=0}^{N_0 - 1} e^2(n)$$

Using the definition equation (12.107), the terms:

$$\frac{\partial E}{\partial a_k} = -\frac{2}{N_0} \sum_{n=0}^{N_0 - 1} x(n - k) e(n); \quad 1 \leqslant k \leqslant N$$

must be zeroed, and the coefficients must be chosen so that the following

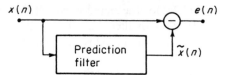

Fig. 12.9   Linear prediction system

equations are true:

$$\sum_{n=0}^{N_0-1} x(n-k)x(n) = \sum_{i=1}^{N} a_i \sum_{n=0}^{N_0-1} x(n-k)x(n-i); \quad 1 \leqslant k \leqslant N \quad (12.108)$$

Using the elements $r(p)$ of the autocorrelation function of the signal $x(n)$, we arrive at

$$r(p) = \sum_{i=1}^{N} a_i r(p-i), \quad 1 \leqslant p \leqslant N \quad (12.109)$$

The results of the previous sections can be applied directly, with $y(n) = x(n)$. Thus, using (12.56), the least mean square error E is expressed by:

$$E = r(0) - \sum_{i=1}^{N} a_i r(i) \quad (12.110)$$

The coefficients of the prediction filter can be determined using equation (12.109):

$$\begin{bmatrix} a_1 \\ a_2 \\ \vdots \\ a_N \end{bmatrix} = R_N^{-1} \begin{bmatrix} r(1) \\ r(2) \\ \vdots \\ r(N) \end{bmatrix} \quad (12.111)$$

The filter obtained that way minimizes the power of the prediction error signal $e(n)$. Its transfer function is:

$$H(Z) = 1 - \sum_{i=1}^{N} a_i Z^{-i}$$

and it is minimum phase; its zeros are inside or on the unit circle. The proof is as follows. Assume a zero z_0 is outside the unit circle, $|z_0| > 1$, then $H'(Z)$ such that:

$$H'(Z) = \frac{H(z)}{(1 - z_0 z^{-1})(1 - \bar{z}_0 z^{-1})} \times \left(1 - \frac{z^{-1}}{z_0}\right)\left(1 - \frac{z^{-1}}{\bar{z}_0}\right)$$

yields a smaller prediction error because:

$$|H'(e^{j\omega})| = \frac{1}{|Z_0|^2} |H(e^{j\omega})|$$

According to expression (4.24) for the signal power at the output of a filter, the power at the output of $H'(Z)$ is smaller than that at the output of $H(Z)$ when the same signal is fed to both filters. This contradicts the definition of the prediction filter.

As concerns the computation of the coefficients, equation (12.111) leads to $N^3/3$ multiplications and divisions. Instead, an iterative procedure can be used; it needs about N^2 operations and, moreover, at lends itself to an efficient implementation.

The Levinson-Durbin procedure [10] provides a solution for the system of equations (12.105) by recursion over N stages. It begins by taking the power of the error signal as

$$E_0 = r(0)$$

At the ith stage $(1 \leqslant i \leqslant N)$ the following calculations are performed:

$$k_i = \frac{1}{E_{i-1}} \left[r(i) - \sum_{j=1}^{i-1} a_j^{i-1} r(i-j) \right]; \quad 1 \leqslant i \leqslant N$$

$$a_i^i = k_i \qquad\qquad\qquad\qquad\qquad\qquad (12.112)$$

$$a_j^i = a_j^{i-1} - k_i a_{i-j}^{i-1}; \quad 1 \leqslant j \leqslant i-1$$

$$E_i = (1 - k_i^2)E_{i-1} \qquad\qquad\qquad\qquad (12.113)$$

At the Nth stage, the N coefficients a_i are obtained by:

$$a_i = a_i^N; \quad 1 \leqslant i \leqslant N$$

The term E_i corresponds to the power of the residual error with a predictor of order i. However, the values of the coefficients k_i obtained at the preceding stages are not changed in the ith stage. The procedure is sequential and the model is improved as the number of stages (and hence the number of coefficients) increases, because, from equation (12.113), the power of the error is reduced at each stage if $|k_i| < 1$.

The coefficients k_i completely define the filter and indicate the method of realization. The error signal at stage i is formed by the set $e_i(n)$, where

$$e_i(n) = x(n) - \sum_{j=1}^{i} a_j^i x(n-j)$$

The transfer function of the corresponding filter is expressed by $A_i(Z)$, where

$$A_i(Z) = 1 - \sum_{j=1}^{i} a_j^i Z^{-j}$$

Using equation (12.112) this becomes:

$$A_i(Z) = A_{i-1}(Z) - k_i Z^{-i} A_{i-1}(Z^{-1})$$

From the results of Section 8.5, by assuming

$$B_{i-1}(Z) = Z^{-(i-1)} A_{i-1}(Z^{-1})$$

we obtain:

$$A_i(Z) = A_{i-1}(Z) - k_i Z^{-1} B_{i-1}(Z) \tag{12.114}$$

The function $B_i(Z)$ has the corresponding set $b_i(n)$, with

$$b_i(n) = b_{i-1}(n-1) - k_i e_{i-1}(n) \tag{12.115}$$

Finally, the coefficients k_i result in a lattice structure as in Figure 12.10

The convergence of the procedure is ensured if

$$|k_i| < 1; \quad 1 \leqslant i \leqslant N \tag{12.116}$$

The filter can be made adaptive by dividing the set $x(n)$ into sets of N_0 terms which are used to calculate the coefficients. Such a method is applied, for example, to the modelling of a speech signal sampled at 8 kHz with $N_0 = 200$ and a filter of order $N = 10$. The filter coefficients are then updated 40 times per second.

The lattice filter in Figure 12.10 can also be adapted by a gradient algorithm for each value of the index. Indeed, using the equations

$$e_i(n) = e_{i-1}(n) - k_i b_{i-1}(n-1)$$
$$b_i(n) = b_{i-1}(n-1) - k_i e_{i-1}(n) \tag{12.117}$$

one can write the gradients as

$$\frac{\partial e_i^2(n)}{\partial k_i} = -2e_i(n) b_{i-1}(n-1)$$

$$\frac{\partial b_i^2(n)}{\partial k_i} = -2b_i(n) e_{i-1}(n) \tag{12.118}$$

and the following variations can be applied to the coefficients by assuming that the functions $E[e_i^2(n) + b_i^2(n)]$ for $1 \leqslant i \leqslant N$ are to be minimized:

$$k_i(n+1) = k_i(n) + \delta_i(e_i(n) b_{i-1}(n-1) + b_i(n) e_{i-1}(n)) \tag{12.119}$$

As the power of the signals $e_i(n)$ and $b_i(n)$ decreases with the index i the variation step δ_i must be related to this power in order to obtain a certain homogeneity

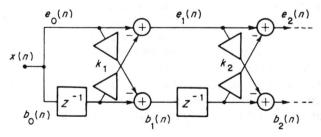

Fig. 12.10   Linear prediction in a lattice structure

of the time constants. This produces complications in realization. Other structures for adaptive filters, such as the FIR structure in Figure 12.4 or the simplified mixed FIR–IIR structure in Figure 12.8, produce simpler realizations.

The complexity parameters are determined from the results in Section 12.9. In this case, the gain of the system is the prediction gain G such that

$$G^2 = \frac{E[x^2(n)]}{E} = \frac{r(0)}{E}$$

that is,

$$G^2 = \frac{1}{1 - \sum_{i=1}^{N} a_i \dfrac{r(i)}{r(0)}} \tag{12.120}$$

By using $[r_{1x}]$ to denote the N-element column vector

$$[r_{1x}] = \begin{bmatrix} r(1)/r(0) \\ r(2)/r(0) \\ \vdots \\ r(N)/r(0) \end{bmatrix}$$

and by using equation (12.57), the prediction gain can be written in a different form:

$$G^2 = \frac{1}{1 - [r_{1x}]^t R_N^{-1} [r_{1x}]} \tag{12.121}$$

Thus, the prediction gain appears as a fundamental signal parameter.

12.14 CONCLUSIONS

Several techniques for designing and producing adaptive filters have been presented in this chapter. They are based on the gradient algorithm, which is the simplest and most robust approach for changing the coefficients. The direct FIR structure has been studied in detail by developing the parameters of adaptation (the time constant and the residual error) and the parameters of complexity (multiplication rate and the number of bits of the coefficients and internal data). This structure is the one most commonly used in practice. In some specific cases different structures, such as IIR, mixed FIR–IIR or lattice structures, can offer significant advantages. Analysis of the stability conditions in these structures and the study of the adaptation and complexity parameters can be made by a method similar to that given for the FIR structure.

The gradient algorithm results in a relatively slow change in the values of the coefficients of the filter, especially when a low residual error is required and when it is used in its most reduced form, the sign algorithm. In order to find the most rapid rate of adaptation all the coefficients can be recalculated periodically by using fast iterative procedures. The lattice structure is well suited to this approach and allows real time analysis or modelling of signals such as speech with circuits of moderate complexity.

The gradient algorithm can be improved; for example, by using different adaptation steps for the coefficients, which are obtained from statistical estimates of the signal characteristics.

It is possible to envisage criteria which, for certain applications, are more appropriate than the minimization of the mean square error and algorithms which are more efficient than the gradient algorithm can be elaborated [1]. However, putting these algorithms into operation is generally more complicated and problems of sensitivity to any imperfection in the realization can occur.

In conclusion, FIR and IIR structures operating according to the least mean squares error criterion and using the gradient algorithm in its simplest form, the sign algorithm, offer a simple and effective compromise for adaptive filtering applications.

REFERENCES

[1] M. BELLANGER, *Adaptive Filters and Signal Analysis*, Marcel Dekker, New York, 1987.

[2] S. HAYKIN, *Adaptive Filter Theory*, Prentice-Hall, Englewood Cliffs, NJ, 1986.

[3] W. A. GARDNER, Learning characteristics of stochastic gradient descent algorithms: a general study, analysis and critique. *Signal Processing*, No. 6, North-Holland, Amsterdam, 1984, pp. 113–33.

[5] B. WIDROW, J. McCOOL, M. G. LARIMORE and C. JOHNSON, Stationary and non-stationary learning characteristics of the LMS adaptive filter. *Proceedings of the IEEE*, **64**(8), 1976.

[6] J. E. MAZO, On the independance theory of equalizer convergence. *BSTJ*, **58**(5), 963–93, 1979.

[7] O. MACCHI, Le filtrage adaptatif en télécommunications. *Annales des télécomm.* **36**(11/12), 1981.

[8] T. A. C. M. CLAASEN and W. F. G. MECKLENBRAUKER, Comparison of the convergence of two algorithms for adaptive FIR digital Filters. *IEEE Transactions*, **ASSP29**(3), 1981.

[9] L. B. JACKSON and S. L. WOOD, Linear prediction in cascade form. *IEEE Transactions*, **ASSP26**(6), 1978.

[10] J. MAKHOUL, Linear prediction, a tutorial review. *Proceedings of the IEEE*, **63**(4), 1975.

[11] I. D. LANDAU, An extension of a stability theorem applicable to adaptive control. *IEEE Transactions*, **AC25**, Aug 1980.

[12] C. F. N. COWAN and P. M. GRANT, *Adaptive Filters*, Prentice Hall, Englewood Cliffs NJ, 1985.

EXERCISES

1 Calculate the autocorrelation function $r(p)$ of the signal $x(n)$ by applying the following equation:

$$r(p) = \frac{1}{N_0} \sum_{n=0}^{N_0-1} x(n)x(n-p)$$

for the function $x(n)$ such that:

$$x(n) = \sin(2\pi(n/N_0))$$

with $N_0 = 16$.

What are the characteristics of the autocorrelation function obtained?

Perform the same calculations for the function $x(n)$ such that:

$$x(n) = \sin\left(2\pi\frac{n}{N_0}\right) \quad \text{for } 0 \leqslant n \leqslant 15$$

$$x(n) = 0 \quad\quad\quad \text{for } n < 0 \text{ and } n \geqslant 16$$

Compare the results of the two calculations.

2 Calculate the eigenvalues of the autocorrelation matrices of order $N = 2$ and $N = 3$ for a signal $x(n)$ as a function of the elements $r(0)$, $r(1)$ and $r(2)$ of its autocorrelation function.

How do these eigenvalues change when $r(1)$ and $r(2)$ tend towards zero?

How are the eigenvalues modified when white noise of power σ^2 is added to the signal?

3 The signal $x(n) = m + e(n)$, where m is a constant and $e(n)$ is a white noise of power σ_e^2, is applied to a recursive estimator, as in Section 12.4 whose output is:

$$y(n) = (1 - b)y(n - 1) + bx(n)$$

assuming $x(n) = 0$ for $n < 0$, compute $y(n)$. If $b = 0.8$ how many samples are needed for $y(n)$ to approach m, in the mean, within 1%?

Compute the output mean square error, $E[[y(n) - m]^2]$, for $n > 0$. What is the limit when n goes to infinity? Study the evolution and the choice of coefficient b for the three cases: $\sigma_e \approx m$; $\sigma_e > m$; $\sigma_e < m$.

Compare the performance of the recursive estimator with that of the non-recursive estimator defined by:

$$y(n) = \frac{1}{n+1} \sum_{i=0}^{n} x(i)$$

4 A transmission channel has an impulse response composed of three terms:

$$h(0) = 1; \quad h(1) = 1.2; \quad h(2) = 0.5$$

Using a direct approach, calculate the coefficients of an equalizing FIR filter with five coefficients approximating the transfer function inverse to that of the transmission channel.

The following signal with period $N_0 = 8$ is applied to the transmission channel:

$$u(0) = 1$$

$$u(n) = 0; \quad 1 \leqslant n \leqslant 7$$

An iterative procedure using the gradient algorithm (12.46) is used to compute the five coefficients of an FIR equalizer, starting from zero values. What is the value of the adaptation step δ which yields the smallest time constant; compute the residual error. (Answer the same questions with sign algorithm.)

5 The impulse response of a given system takes the values

$$h(0) = 1; \quad h(1) = 1.3; \quad h(2) = 0.6; \quad h(3) = 0.2$$

Assume the system is to be modelled with an IIR filter having $N = 2$ coefficients. For a periodic signal and with period $N_0 = 8$ such that

$$x(0) = 1$$
$$x(n) = 0; \quad 1 \leqslant n \leqslant 7$$

calculate the values of the coefficients which minimize the mean square error, using the simplified iterative procedure given in Section 12.12. Study the convergence of the procedure and its stability.

6 A sinusoidal signal $x(n)$ is applied to a second-order FIR prediction filter:

$$x(n) = \sin 2\pi(n/8)$$

Calculate the coefficients a_1 and a_2 of this filter and locate the zeros in the Z-plane.

Starting from zero initial coefficients, examine the trajectory of the zeros of this filter for an adaptation step $\delta = 0.1$.

Give the new values of the prediction coefficients when discrete white noise of power σ^2 is added to the signal.

7 Use the procedure given in Section 12.13 to determine a lattice prediction filter with $N = 5$ coefficients. The first values of the autocorrelation function are given by:

$$r(n) = 1/(n + 1); \quad n \geqslant 0$$

Comment on the results.

Give the number of arithmetic operations needed if these values are assumed to have been obtained using 128 data values.

CHAPTER 13

CIRCUITS AND FACTORS OF COMPLEXITY

Digital signal processing equipment uses the techniques and circuits generally used in data processing machines. There are, however, some special features, of which the most notable is the large number of arithmetic computations required. It follows that the complexity of the equipment is very dependent on the number of arithmetic operations to be performed during the elementary time period. The various operations and manipulations are carried out on binary numbers which can be represented in various ways.

13.1 BINARY REPRESENTATIONS

Since the operations involved include addition and subtraction, a suitable binary representation should permit the inverse N' of any number N to be defined in such a way that their sum is zero. The simplest representation involves reserving one bit for the sign. This is the sign and absolute value representation. The arithmetic operations are carried out on absolute values, and the sign is treated separately. Distinguishing one bit in the number in this way introduces complications which are avoided in complementary representations.

Given a binary number N, the inverse number N' (with respect to addition) can be obtained by inverting all the bits in the number N and then adding one least significant bit. This ensures that the sum $N + N'$ is zero. If M is the total number of bits in N, the N' is given by:

$$N' = 2^M - N \tag{13.1}$$

This is called the 2s complement representation of N.

Another representation is obtained if the least significant bit is not added after inversion of the bits. This is called the 1s complement representation of N and is the number N'' given by:

$$N'' = 2^M - N - 1 \tag{13.2}$$

In both representations the $+$ sign is written as 0 and the $-$ one is written as 1. The sign is treated in all operations as being part of the number. For a negative number, the 2s complement representation is obtained by adding 1 to the 1s complement. Figure 13.1 illustrates these properties.

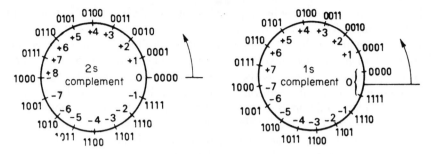

Fig. 13.1 1s and 2s complement representation

In both representations, the arithmetic operations are applied to all bits, including the sign bit, but the algorithms are different:

(1) In 1s complement arithmetic, for an addition, the second term is added to the first. For a subtraction, the 1s complement of the second term is added to the first. If a carry bit results, it is transferred to the last position.

Example

```
  0 0 1 0 1 1    (11)      1 1 0 1 0 0     (−11)
+ 0 0 1 1 0 1   +(13)    + 1 1 0 0 1 0    +(−13)
───────────             ─────────────
  0 1 1 0 0 0    (24)    1 1 0 0 1 1 0    −(  24)
                          └─────────→ 1


                            1 0 0 1 1 1
```

(2) In 2s complement arithmetic, for an addition, the second term is added to the first. For a subtraction, the 1s complement of the second term is added to the first, then a 1 is added to the last position. Carry bits are ignored.

Example

```
   0 0 1 0 1 1     (11)       0 0 1 0 1 1    (11)
 + 1 1 0 0 1 1    +(−13)     − 1 1 0 0 1 0   −(13)
 ───────────                 ───────────
   1 1 1 1 1 0     −2         1 1 1 1 0 1    −2
                                      + 1
                             ───────────
                             1 1 1 1 1 0
```

In order to compare these three binary representations—sign and absolute value, 2s complement and 1s complement—their respective advantages in arithmetic operations must be examined.

The sign and absolute value representation is of obvious importance for multiplication, as it is sufficient to multiply the absolute values in a product and to perform an 'exclusive-OR' logic operation on the signs. Nevertheless, the addition of such numbers involves a complicated addition–subtraction unit because the number which has the lower absolute value must be subtracted from the one with the larger absolute value when the signs are different.

2s complement addition is straightforward. Further, this representation has one notable property caused by the cyclic continuity shown in Figure 13.1 When several numbers are added to produce a final sum the partial results may exceed the maximum capacity in the representation without any effect on the final result, provided that it lies within the range of definition. For example, in the 4-bit representation of Figure 13.1 we can write: $7 + 4 - 5 = 6$. This property can allow a saving of several bits in the capacity of an accumulator. Multiplication introduces conditional operations in 2s complement arithmetic. One technique currently in use is to convert to the sign and absolute value representation to multiply the numbers, and then to return to the 2s complement form. Another approach is to use the Booth algorithm, which is a generalization of the subtraction rule for negative multipliers. In effect, when a set of P non-zero bits appears in the multiplier in positions $i, i + 1, \ldots, i + p - 1$. the corresponding multiplication can be written as

$$m(2^{i+P} - 2^i) = m2^i(2^P - 1) \tag{13.3}$$

where m represents the multiplicand. Thus, as multiplication is ultimately performed by a series of additions and shifts, Booth's algorithm corresponds to the following rule.

If, when scanning the multiplier from the least significant towards the more significant bits, one encounters:

(1) A 1–0 transition, an addition is made, and followed by a shift;
(2) A 0–1 transition, a subtraction is made and followed by a shift;
(3) No transition, only a shift is performed,

it can be seen that this algorithm involves a set of conditional operations.

The 1s complement representation has no particular advantage for either addition or multiplication, and so is the least-used representation. As there are often more additions to be performed overall than there are multiplications, the 2s complement representation is most frequently used.

The M bits of a number can be represented by convention as a whole number or as a fractional number of modulus less than 1. One can also operate on fractional numbers defined to a constant factor. These are called 'fixed-point' representations. A representation can also be used where the position of the decimal point moves and is constantly followed by the machine. This is the 'floating-point' representation. It permits the dynamic range of the numbers to be considerably increased and binary representation to be optimized at the price of some circuit complexity [1].

13.2 TYPES OF CIRCUIT

In order to illustrate the types of circuit required for the implementation of digital processing operations considerations will first be given to the general diagram of a canonic filter circuit given in Figure 13.2. The arithmetic operations are multiplication by the coefficients a_i and b_i and addition of the results. Each number $y(n)$ produced by the filter for the given configuration requires $(2N + 1)$ multiplications and $2N$ additions. If the rate at which the number $y(n)$ are obtained is denoted by f_s, the arithmetic unit should be capable of making M_R multiplications per second and A_R additions per second, where

$$M_R = (2N + 1)f_s$$
$$A_R = 2Nf_s$$

The general circuit diagram illustrates another basic function—the memory.

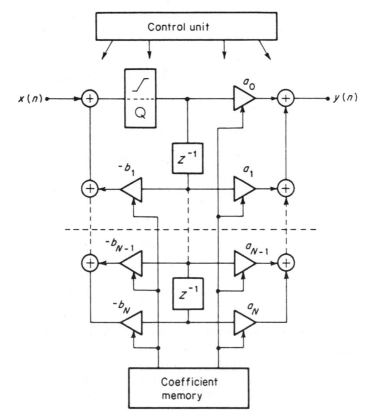

Fig. 13.2   Filter circuit diagram

Intermediate results, which in the case in the figure are the numbers

$$z(n) = x(n) - \sum_{i=1}^{N} b_i z(n - i)$$

are stored temporarily in memories called data memories, which are represented by Z^{-1} to reflect their role in the transfer function $H(Z)$. The coefficients a_i and b_i are provided by another memory which may have fixed contents in the case of fixed coefficient filters, and could be either mask-programmed or reprogrammable.

The data memories in realizable systems have a limited capacity, i.e. the intermediate results $z(n)$ have a limited number of bits. It is obvious that even if the input numbers $x(n)$ have a limited number of bits, the number required for $z(n)$, obtained by a weighted summation, will increase indefinitely. It is necessary, therefore, to introduce limitation, preferably by rounding, to reduce the power of the noise generated in this way.

On the other hand, to avoid overflow and to reduce the risk of auto-oscillations, limitation has to be introduced on the high-amplitude side using a logic saturation device. In Figure 13.2 these operations are placed after the input adder.

If the data memories have B_D bits per number and the coefficient memories B_C bits per coefficient, the total capacities MM_D and MM_C are

$$\begin{aligned} MM_D &= NB_0 & \text{bits} \\ MM_C &= (2N + 1)B_C & \text{bits} \end{aligned} \qquad (13.4)$$

One final element is necessary for the realization of a filter, the control unit which ensures the sequencing of the operations and controls the functioning of the other elements. This unit depends very much upon the architecture of the system.

To summarize, the following types of circuit are necessary:

(1) Arithmetic circuits for addition and multiplication
(2) Data memories;
(3) Coefficient memories;
(4) Auxiliary logic circuits such as for rounding and logic saturation;
(5) Circuits for the control unit.

These are the most commonly used functions in digital processing. It should be noted that division has not been mentioned as division by a factor other than two is rarely needed in equipment designed for signal processing. Division by powers of two is performed by shifting the numbers in the registers.

Each type of circuit will be discussed in turn, beginning with the arithmetic circuits [2, 3].

13.3 SERIAL MULTIPLIERS

Arithmetic circuits have to perform three operations—multiplication, addition and subtraction. These three operations use the same basic circuit, the adder.

The full adder is a circuit which, given a set of three binary digits, A, B and R_i, produces two binary digits S and R_{i+1} according to the table in Figure 13.3, which also shows the logic circuit for producing it using 'NAND' gates. The digits A and B represent the two numbers to be added, and R_i is the carry from the preceding stage. The digit S is the result of the addition and R_{i+1} is the carry to the next stage. As subtraction is the addition of a negative number this can also be performed by the circuit by inverting one input.

Multiplication is the most complicated arithmetic operation currently used in signal processing. This has been the subject of considerable attention and many varied techniques are available. It can be achieved as a set of additions. Multiplication of a given number D with B_D bits by a coefficient C with B_C bits provides a product P which has, at most, $B_P = B_D + B_C$ bits. This product P is obtained by making at most either $B_C - 1$ additions of the number D, suitably shifted, or $B_D - 1$ additions of the number C, suitably shifted.

Example

$$D = 13 = 1\ 1\ 0\ 1$$
$$C = 9 = 1\ 0\ 0\ 1$$

$$
\begin{array}{r}
1\ 1\ 0\ 1 \\
0\ 0\ 0\ 0 \\
0\ 0\ 0\ 0 \\
1\ 1\ 0\ 1 \\
\hline
\end{array}
$$

$$P = 117 = 1\ 1\ 1\ 0\ 1\ 0\ 1$$

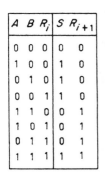

A	B	R_i	S	R_{i+1}
0	0	0	0	0
1	0	0	1	0
0	1	0	1	0
0	0	1	1	0
1	1	0	0	1
1	0	1	0	1
0	1	1	0	1
1	1	1	1	1

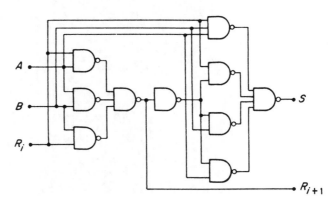

Fig. 13.3   Full adder circuit

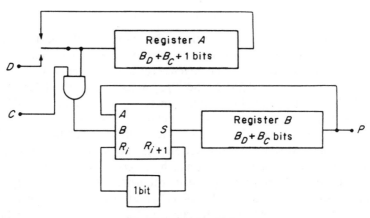

Fig. 13.4   Serial multiplier

The approach producing the simplest circuit uses only one adder and one accumulator. It is the serial multiplier represented in Figure 13.4.

The multiplicand D is introduced to the register A which has $B_D + B_C + 1$ bits, and which is looped to itself once the multiplicand is completely loaded. During this time the register B, with $B_D + B_C$ bits, is empty. At each rotation of the data in register A access to an adder is provided through a logic gate controlled by one bit of the coefficient C. The other input to the adder is the contents of accumulator register B, whose input is the output S of the adder.

At each set of $B_D + B_C + 1$ elementary time intervals the coefficient bit is changed and the contents of the register B (which has only $B_D + B_C$ bits) is shifted by one bit. The coefficient C is presented with the least significant bits

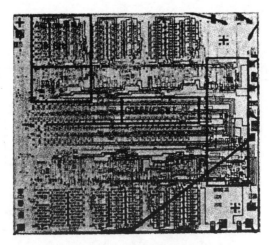

Fig. 13.5   Layout of serial multiplier in a filter

first, and in order to perform a multiplication a total of $B_C(B_D + B_C + 1)$ elementary time intervals are required. The carry R_{i+1} of the adder is connected to the input R_i via a one-bit memory. The unit formed by the complete adder and the one-bit memory for the carry is often called a 'serial adder'.

This form of multiplier is well suited for implementation using monolithic integrated circuits. Figure 13.5 shows an integrated PMOS circuit which includes six multipliers of this type, appearing in a repetitive structure and involving a serial adder and a 16-bit accumulator. The coefficients and the control logic can also be seen. This circuit can operate with a 2 MHz clock.

The serial multiplier is very simple in concept, but its computing power is limited, since, with clock frequency f_H, it can only perform M_R multiplications per second, where

$$M_R = \frac{f_H}{(B_D + B_C + 1)B_C} \qquad (13.5)$$

Its use is restricted to cases where low computing powers are required. Greater efficiency is achieved by means of the serial–parallel structure.

13.4 SERIAL–PARALLEL MULTIPLIERS

In this type of multiplier one of the factors (for instance, the data) is presented in series and the other in parallel, as in the circuit in Figure 13.6. The circuit contains B_C delay elements and B_C adders, with the latter providing the binary figures of the product, with the least significant bits first, following the multiplication rule. The product has $B_D + B_C$ bits. The data are presented with the least significant bits first, and to avoid overlaps B_C zeros must be inserted between them. The first adder in the chain can be used for a supplementary addition involving a number outside the product. The circuit thus involves B_C identical elements. $B_D + B_C$ elementary time intervals are required for each

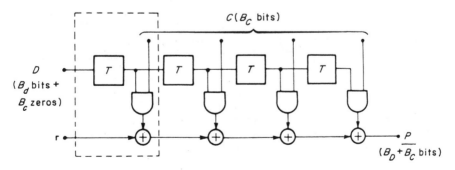

Fig. 13.6 Serial–parallel multiplier

multiplication and the computation rate with a clock frequency f_H is

$$M_R = \frac{f_H}{B_D + B_C} \tag{13.6}$$

This rate can be increased if limitation of the result to the B_D most significant bits is acceptable. The truncation is obtained by introducing an output gate for each adder and by controlling the propagation of the products with a suitable control signal. If B_C zeros are introduced between the data then the multiplication time is reduced to B_D clock periods.

An increase in the operating rate of the circuit can also be obtained by arranging not to have B_C adders in cascade, as this causes the propagation times to be added. This can be achieved by introducing a memory element between pairs of adders.

One common variation is to use the serial–parallel multiplier with a serial input, which allows serial presentation of the two factors. The fundamental element of the multiplier is constructed in such a way that the coefficient and the control signal for the truncation of the product can be applied in series. The corresponding circuit is shown in Figure 13.7.

The factors D and C are applied simultaneously (least significant bit first), and the rounding control signal, S, is an impulse of width equal to one elementary

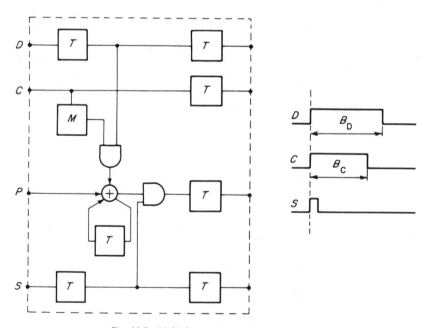

Fig. 13.7 Multiplier section with serial accesses

time unit applied simultaneously with the least significant bit of the factors. The product P has B_D bits and is presented with a delay of $2B_C$ bits.

The operation of the complete circuit, formed from a cascade of elements as shown in Figure 13.7, can be seen easily by writing the detailed timing diagram.

By way of example, a multiplier using this structure has been realized as an integrated circuit with 16 sections and offering two configurations, with 8- or 16-bit coefficients.

The number of bits B_D of the data and of the product is fixed by a control signal and should be greater than 16 or 8, depending on the configuration. With a clock frequency of $f_H = 2560\,\text{kHz}$, the computation rate is $128\,\text{kmult/s}$ (thousands of multiplications per second) for 20×16 bits, or $256\,\text{kmult/s}$ for 20×8 bits. The possibility is available of adding or subtracting an external number R to the product. The integrated circuit in NMOS technology is shown in Figure 13.8, where one can see the two sets of 8 elementary serial multiplier units, the adder–subtractor for the external number, and the control logic.

The computation rates offered by this circuit are very suitable for applications in telecommunications. A different structure has to be employed for higher rates.

13.5 PARALLEL MULTIPLIERS

The structure permits the highest calculation rate. In this circuit, the numbers D and C are applied in parallel, with the product P also being produced in parallel form. Such a multiplier involves $B_D \times B_C$ basic elements, each of which is formed from an adder and a switching circuit as shown in Figure 13.9.

The arrangement of such elements to form a multiplier is shown in Figure 13.10. The time τ_M required to perform a multiplication is approximately

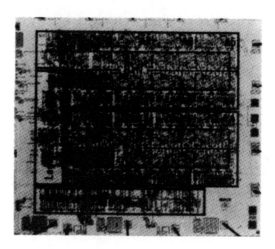

Fig. 13.8 Integrated circuit multiplier with serial access in NMOS technology

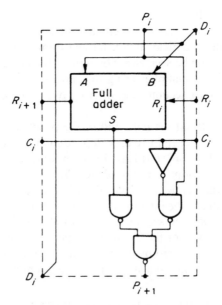

Fig. 13.9   *Parallel multiplier section*

equal to the product of the propagation delay τ_e of a basic element and the number of elements $(B_D + B_C)$. Thus:

$$\tau_M = (B_D + B_C)\tau_e$$

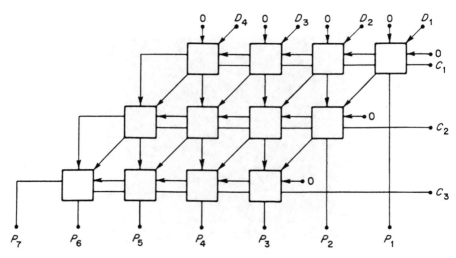

Fig. 13.10 4×3 *parallel multiplier*

The computation rate is then given by:

$$M_R = \frac{1}{B_D + B_C} \frac{1}{\tau_e} \qquad (13.7)$$

This is inversely proportional to the propagation delay of the elementary unit, which is itself a function of the technology. With bipolar technology, for example, a calculation rate of $M_R = 10^7$ mult/s can be achieved with $B_D = B_C = 16$ bits.

The clock frequency f_H of a system using this type of multiplier should be such that:

$$f_H < \frac{1}{\tau_M} \qquad (13.8)$$

The multiplication is then performed within an elementary time interval.

13.6 COMPARISON OF DISTRIBUTED AND RESIDUE ARITHMETIC

The characteristics of the types of multiplier which have been discussed above are summarized in Figure 13.1, which shows the number of adders that each involves, the number of elementary time periods required for a multiplication and the computation rate in multiplications per second. It is assumed that the data and the coefficients have B bits and the products $2B$ bits. The relative advantages of each type can be seen, but even so, it is not easy to compare the computation rates of the serial–parallel and parallel types. To do this, it is necessary to compare the propagation time of the circuit shown in Figure 13.7, a function of the logic elements involved and the propagation time τ_e of the elementary unit of the parallel multiplier. The advantage of the parallel structure, with comparable technology, lies in the fact that the propagation times are accumulated without being quantized as in the serial structure.

Type	Serial	Serial parallel	Parallel
Adders	1	B	B^2
Elementary propagation times	$B(2B+1)$	$2B$	1
Calculation rate	$\dfrac{f_H}{B(2B+1)}$	$\dfrac{f_H}{2B}$	$\dfrac{1}{2B}\dfrac{1}{\tau_e}$

Fig. 13.11 Characteristics of types of multiplier

Flexibility in use is an important factor in practice, and the serial structure offers the greatest flexibility as the numbers of bits of the two factors are controllable from outside the multiplier. In the serial–parallel type this only applies to one factor. The parallel type involves fixed structures, which can result in non-optimal use of the circuits.

In every case the multiplier is a complicated unit. To simplify it, one could consider changing the binary representation of the coefficients and using, for example, the signed digit representation, in which the digits are $+1, 0$, or -1. This allows sets of consecutive 1s to be suppressed so that the number of zeros in the coefficients is greater than half the total number of bits (up to about two-thirds). This leads to a reduction in the number of adders, but at the cost of additional logic circuits.

An unusual and interesting approach to avoid the use of multipliers, and well suited to a recursive second-order section, corresponds to a distributed arithmetic. This is in contrast to an arithmetic where the operations are localized, such as the multiplier.

In the general second-order filter section the output set $y(n)$ is related to the input set $x(n)$ by the equation:

$$y(n) = a_0 x(n) + a_1 x(n-1) + a_2 x(n-2) - b_1 y(n-1) - b_2 y(n-2)$$

By assuming that the data contains B bits and that, for example, $x(n)$ is written as

$$x(n) = \sum_{i=0}^{B-1} 2^{-i} x_i(n) \quad \text{with} \quad x_i(n) = 0 \text{ or } 1$$

we have

$$y(n) = \sum_{i=0}^{B-1} 2^{-i} [a_0 x_i(n) + a_1 x_i(n-1) + a_2 x_i(n-2) - b_1 y_i(n-1) - b_2 y_i(n-2)]$$

$$(13.9)$$

The expression in the brackets above can only have $2^5 = 32$ possible values:

$$0, a_0, a_1, \ldots, (a_0 + a_1 + a_2 - b_1 - b_2)$$

As the coefficients are fixed, these 32 numbers can be stored in a 32-word memory. By presenting the data in series (least significant bit first), and by using them to address the memory there appears at the output a sequence of numbers that only have to be properly shifted and accumulated in order to obtain the required number $y(n)$. The design is shown in Figure 13.12. The output of the memory is connected to an accumulator which is followed by a series–parallel converter. The structure obtained is not canonic in the data memory, and the number of bits for the numbers stored in the ROM memory is a function of the number of bits of the coefficients.

The method can be generalized for any weighted summation [4]. However, the size of the memory limits its use to applications with a small number of terms. In other cases the function has to be decomposed.

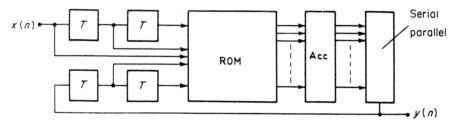

Fig. 13.12   Second-order section in distributed arithmetic

This technique reduces, in fact, to replacing complicated multiplier circuits with other similarly complicated memory circuits and address decoders. It derives benefit from the extensive work that has been expended on memory circuits and can lead to effective implementations.

Without eliminating multiplications completely, it is possible to perform an equivalent operation with several circuits operating in parallel; this reduces delay due to carry propagation. With residue arithmetic, a number X is represented by the set of remainders r_i obtained when it is divided by a set of numbers m_i called moduli. For example, with $N = 3$ and moduli 5, 7 and 8, the number 13 gives the following remainders:

$$r_1 = 13 \bmod 5 = 3$$
$$r_2 = 13 \bmod 7 = 6$$
$$r_3 = 13 \bmod 8 = 5$$

Hence the number 13 is represented by the set of remainders $(3, 6, 5)$.

Inversely, the initial number is calculated from the remainders by performing a weighted sum of these remainders, modulo the product of the moduli. The weighting coefficients C_i are obtained as follows:

$$C_i = (M/m_i) \times [(M/m_i)^{-1} \bmod m_i]$$

with:

$$M = \prod_{i=1}^{N} m_i$$

and:

$$X = \left(\sum_{i=1}^{N} C_i r_i \right) \bmod M$$

This is the Chinese remainder theorem. Taking the previous example:

$$13 = (56 \times 3 + 120 \times 6 + 105 \times 5) \bmod (5 \times 7 \times 8)$$

The value of this representation lies in the fact that the operations of addition and multiplication are performed on each remainder independently of the others.

Example

	Residue arithmetic		
Decimal	mod 5	mod 7	mod 8
13	3	6	5
$\times 9$	$\times 4$	$\times 2$	$\times 1$
117	2	5	5

The multiplication is thus distributed among independent parallel operations on the remainders. It can produce an acceleration which may be considerable for high precision. Hence in 32-bit arithmetic a residue multiplier can be an order of magnitude faster than a binary multiplier with an equivalent number of logic gates. In evaluating the overall complexity, however, it is necessary to take account of the conversion between binary and residue arithmetic [5].

13.7 MEMORIES AND AUXILIARY CIRCUITS

The memories perform a basic function, either in storing the coefficients or in temporarily saving the data and the partial results.

The coefficients, when they are fixed, are stored in read-only (ROM) memories, or, more conveniently, in programmable (PROM) or reprogrammable (EPROM) memories. The data change with each processing cycle and are stored in random access (RAM) memories or in shift registers. The RAM memory comprises two principal subsets, the address decoding circuits and the memory circuits. Shift registers are memory locations connected in series in which data are propagated at the frequency of an externally applied clock. They are particularly well suited for serial processing. The amount of material necessary for one bit of memory largely depends on the technology. In general, however, it is much less than is needed for a full adder.

The auxiliary circuits are principally to limit the number of bits of the numbers to be stored in the memory. Rounding is performed by an adder as a complementary function. In effect, it is performed by simple truncation after adding a 'one' at the level of the first truncated bit. For logic saturation different circuits must be used. The most complicated case is with serial logic, where the numbers are presented in series with the least significant bit first. Saturation information appears with the bit following the sign bit. Under these conditions a complete number must be stored in the memory, as shown in Figure 13.13. The output from the circuit is either the delayed input number or the number corresponding to saturation taking the sign into account. Its control is provided by an overflow detection circuit.

The control unit does not generally contain special digital processing circuits. It is formed of various logic circuits selected from the manufacturers' catalogues:

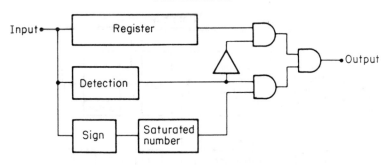

Fig. 13.13   Saturation circuit for serial logic

gates, counters, flip–flops, registers, etc. It size and its effect on cost are closely related to the architecture of the system.

13.8 PROCESSORS

The digital processing operations can also be performed using general-purpose programmable machines—digital computers, minicomputers, microcomputers and microprocessors. The set of operations to be performed has to be described in the instruction program, and nothing distinguishes digital signal processing from any other application. This approach is important in systems where there is already a computer available for other purposes, or in producing a digital filter, for example, with a reduced development budget and on a short time scale. The use of microprocessors for example is common [6].

However, in general, the calculating capacity of microprocessors remains weak compared with that of specialized circuits and this represents a considerable obstacle to their use in signal processing. Furthermore, their very general and flexible structure does not correspond exactly to requirements. In fact, digital processing uses a reduced number of simple and repetitive operations as in the typical following sequence:

> read a pair of numbers into memory, multiply these numbers, add the result to a previous result available in a working register, put the result in memory

Hence the data and instruction memories can be separated and the control unit can be completely isolated from the arithmetic unit, which is not needed for address calculation.

These considerations lead to the design of processors for signal processing which have a high calculating power and a reduced and well-adapted instruction set. An example of the architecture is provided in Figure 13.14.

The circuit contains a data bus and an instruction bus. The data and coefficient memory can be partly RAM and partly ROM. A multiplier is associated with an accumulator which is directly accessible to the data memory. The program

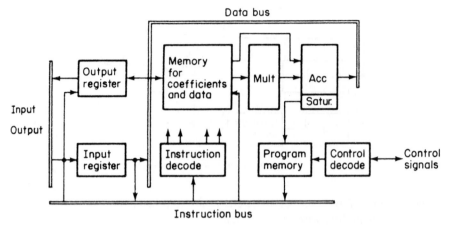

Fig. 13.14   Example of the architecture of an integrated signal processor

memory contains the series of instructions to be executed and is externally controllable. Such a circuit can be input/output compatible with a general-purpose microprocessor, in which case it is effectively a peripheral adapted for signal processing.

Linking of operations can be facilitated by an architecture with more than two buses. Hence Ref. [7] describes a circuit with four buses, one instruction bus and three data buses, for which the operations corresponding to one second-order filter section require eight instructions. The structure of this type of processor represents a compromise between computing power and flexibility of use.

13.9 CHOOSING THE COMPLEXITY PARAMETERS

The circuits used in signal processing have been described in the above sections, and show many variations, for example, in the multipliers. We can now list and compare the elements which contribute to the complexity of the system.

Overall, the complexity of a signal-processing machine can be defined as the quantity of material necessary for its production subject to a given set of constraints (related to the effort necessary for its design and operation). It can vary considerably, depending on the theoretical approach taken, the circuits used and the architecture selected. It is not possible to consider all the theoretical possibilities for a given application. For example, for a filtering function there are approaches based on FIR, multirate or IIR elements realized in direct form, cascade or lattice structures, with various types of circuits and architectures. The designer of a system is forced to make *a priori* choices on the basis of simple parameters which allow a quick but reliable evaluation of the different possible solutions. Based on the descriptions of the circuits made in the above sections, the following principal parameters for complexity can be listed:

(1) The number of multiplications to be made per second;
(2) The number of additions per second;
(3) The size of the memory;
(4) The number of circuits needed for the control unit.

It is immediately obvious that the number of additions to be made per second has a much smaller effect on the complexity than the number of multiplications. To illustrate this, it is sufficient to note that if the same number of additions and multiplications are to be performed per second, then approximately B and B^2 adders will be required, respectively, for numbers with B bits.

Comparison between memory size, parameters and number of multiplications per second, is less easy to establish. Reference must be made to circuits operating under the actual conditions of a well-designed system, i.e. to circuits which are compatible and homogeneous. The element delays, memory access times and multiplier propagation times must be of the same order in order to fully exploit the capabilities of the circuits.

As previously indicated, programmable logic devices generally have a low computing power with regard to their memory capacity, and the number of multiplications performed per second is often the deciding parameter for the type of machine to be used.

Only the data memory has been considered above; the coefficient memory can be considered either as an addition to the data memory or as making a small contribution if its contents are mask programmed (ROM).

The control unit, in a wired logic machine, can represent 10–30% of the circuits, according to the system architecture and processing algorithm. With programmable logic, the instruction memory can require a larger capacity than the data and coefficient memory. The control unit can be made more complicated to allow better use of the capabilities of the arithmetic unit, as in the case of multiplexing; its complexity is therefore related to the number of multiplications to be performed per second.

Comparing the contributions to the complexity of a system made by the arithmetic unit, the memory and the control unit, it appears that the predominant parameter is the number of multiplications performed per second. Analysis can be performed for types of linear processing such as filters and spectral analysis by calculating the discrete Fourier transform, or for non-linear processes such as frequency detection and adaptive filters; the conclusions are generally the same.

Finally it seems that the number of multiplications to be performed per second is generally the parameter most representative of the complexity of a processing system. It is easily accessible and will remain as the principal parameter. As a secondary parameter, the data memory capacity should be considered and sometimes the control unit. This justifies the interest in techniques for reducing calculation time, particularly multirate filtering.

When making these evaluations the term 'operation' often designates combined multiplication and accumulation and it is measured in million operations per second (MOPS); similarly, million instructions per second (MIPS) is a common unit in data processing.

To illustrate this discussion, it may help to consider an expression for complexity evaluation produced by a working group of the International Telecommunications Union (ITU) to compare implementations of low-speed speech coders at the start of the 1990s. The complexity parameter C is given by the expression:

$$C = \text{MOPS} + 0.2\text{RAM} + 0.05\text{ROM}$$

where RAM = size of data memory in kilobytes
     ROM = size of program memory in kilobytes

For a mobile radio speech encoder with a rate around 4.6 kbit/s, the distribution is as follows:

Number of operations = 9.624
RAM = 8.34 kilobytes
ROM = 11.07 kilobytes
Complexity C = 11.846.

Note that this evaluation is intended primarily for programmable device realization. It illustrates well the relative importance of the principal processing parameters at a given stage of the technology.

To complete the definition of a system, the architecture remains to be determined.

13.10 SYSTEM ARCHITECTURE

The most important element in defining a system is the arithmetic unit, or, more accurately, the multiplier, since the multiplication rate is a crucial complexity parameter. Indeed, with any given technology the computing capacity of serial, serial–parallel and parallel multipliers is imposed, and should correspond to the requirements of each individual application. Nevertheless, there still remains a wide range of possibilities. For example, one type of multiplier or another can be chosen, depending on whether the calculations are made by a single unit or are shared between several units. In fact, when choosing the architecture of a system there are two main choices as the numbers inside the system can be presented either in series or in parallel.

Parallel presentation has the advantage of simplifying the design and implementation of the equipment because a large amount of information is accessible simultaneously, and the number of events in a basic operating period of the system (for example, a signal sampling period), is reduced in comparison with the series presentation. It lends itself to bus organization, which is the structure of general-purpose computers. Another important point in practice is that of the evolution of the system. Parallel presentation offers the widest options for the evolution of the equipment and for modification without excessive redesign effort. This is because of the small number of sequential operations and the ease of access to the information. Equipment for testing the system and displaying internal signals which is suitably adapted to this structure is available on the market. In contrast, the parallel structure complicates the connections between the logic circuits and the wiring between circuit boards because the numbers are transferred from one circuit to another by a number of wires at least equal to the number of bits, while with a series presentation one wire is sufficient in principle. As a result, there is in general an advantage in the series presentation on grounds of economy, reliability in operation and power consumption (as bus structures consume more energy).

The basic system design options are summarized in Table 13.1. In systems which are to be optimized, as is generally the case with telecommunications equipment, it is desirable to select arithmetic circuits whose computation rate is suited to the natural flow of data in the system. In particular, it is desirable to avoid having to multiplex strings of data so that the multipliers are used to their full capacity. The memories should be chosen to be compatible in their organization and access time with the structure of the multiplier, whether they are RAM or shift registers. With regard to the control unit, while it would have to provide fewer signals with the parallel presentation it can be realized without excessive complication with a series presentation and judicious sequencing of the operations.

In conclusion, the architecture of the system depends essentially on the application involved and on the constraints imposed during the design of the equipment. The parallel type of architecture is a good approach when flexibility is required and when the system must be developed rapidly. In contrast, the series-type design is probably preferable if a degree of optimization is required in addition to the normal requirements of meeting the basic needs of the particular problem. Both approaches can also be combined. In fact research on architecture is very active, stimulated by the progress of technology. A major area is that of systolic machines. The structure consists of basic elements which

Table 13.1. BASIC OPTIONS IN SYSTEM DESIGN

Arithmetic complexity	Circuit architecture	Number representation
Rate of operation (MOPs)	serial ⟶	serial
Range of function	systolic ⟶	parallel
	parallel ⟶	parallel
	bus	

are locally interconnected and arranged in regular patterns. Data flows propagate in the structures like waves [8].

REFERENCES

[1] J. ALLEN, Computer architectures for signal processing. *Proc. IEEE*, **63**(4), 1975.
[2] S. L. FREENY, Special purpose hardware for digital filtering. *Proceedings of the IEEE*, **63**(4), 1975.
[3] M. DAVIO, J. P. DESCHAMPS and A. THAYSE, *Digital systems with algorithm implementation*, John Wiley, London, 1983.
[4] A. PELED and B. LIU, A new hardware realization of digital filters. *IEEE Transactions*, **ASSP22**, Dec 1974.
[5] F. J. TAYLOR, Residue arithmetic: a tutorial with examples. *Computer*, **17**(5), 50–62, 1984.
[6] KEN MARRIN, DSP: A technology in search of applications. *Computer Design*, 15 November 1986, 59–77.
[7] A. ALIPHAS and J. A. FELDMAN, The versatility of digital signal processing chips. *IEEE Spectrum*, June 1987, 40–45.
[8] S. Y. KUNG, On supercomputing with systolic wavefront array processors. *Proceedings of the IEEE*, **72**(7), 867–84, 1984.

EXERCISES

1 Give the 7-bit, sign and absolute value representation of the following numbers, then give their 2s complement and their 1s complement representations:

$$a = 9; \quad b = -7; \quad c = -5; \quad d = 4$$

In these three representations perform the operations:

$$a + b; \quad b + c; \quad a + c - d; \quad a \cdot d; \quad a \cdot b; \quad b \cdot c; \quad b \cdot d$$

What particular property is presented by the operation:

$$(a \cdot b + c) + a?$$

Explain why it is beneficial to use the 2s complement representation in this case.

2 Show that the serial multiplier can also be produced with a register A as in Figure 13.4, whose capacity is equal to $(B_D + B_C - 1)$ bits. Give the timing diagram and compare it with the diagram for the circuit in Figure 13.4. What is the easiest form to use?

3 Give the timing diagram of the serial–parallel multiplier in Figure 13.6. The circuit is modified by transferring the delay elements between the adders. The bits of the factor D are applied simultaneously to all gates controlled by the bits of the factor C. This arrangement has the advantage of no longer involving adders connected directly in cascade. Show that the circuit thus obtained fulfils the function of the serial–parallel multiplier. How should the bits of factor C be presented? Can a rounded product be easily obtained?

4 Give detailed logic circuits to achieve the functions required for logic saturation and rounding operations for numbers presented in series and in parallel. Compare the circuit volumes.

5 Compare the computing powers of the multiplier circuits available from the various semiconductor manufacturers and establish a classification as a function of the ratio of computing power to power consumption.

6 Assume that a low-pass filter is to be devised with a sampling frequency of 128 kHz. The pass band extends from 0 to 3400 Hz, the tolerable ripple is 0.1 dB. The stop band begins at 4600 Hz and the attenuation should be greater than 40 dB. An elliptic IIR filter of order 6 is selected. How many calculations and memories are required if the coefficients have 8 bits and the data 16 bits? Which circuits can be used for this application? How many circuits of the type given in Figure 12.8 are needed? Are microprocessors available which can perform these operations? Suggest a design for producing this filter with a parallel architecture and a bus structure.

7 Calculate the number of coefficients of a FIR filter which satisfies the characteristic of the filter in Exercise 6. Compare the volumes of the calculations and the memories required.

The sampling frequency at the output of the filter can be reduced to 8 kHz. Suggest an approach using multirate filters and give the volumes of calculations and memories required.

8 Determine the complexity of a fast Fourier transform computer of order 64, produced following the diagram in Figure 3.3. The operating frequency of the system is 1 kHz. Give the number of calculations to be performed per second. How much memory is required for data, coefficients and products with a length of 16 bits? How many multipliers of the type shown in Figure 12.8 are needed?

Give an architecture in which the numbers are presented in series to perform these calculations with the smallest number of multipliers of this type.

Answer the same questions for a parallel multiplier capable of performing a 16×16 bit multiplication in 200 ns.

CHAPTER 14

APPLICATIONS IN TELECOMMUNICATIONS

Digital signal processing can be applied to a very wide range of fields, from process control in manufacturing to the analysis of biological signals (such as electrocardiograms and electroencephalograms) or analytic reconstruction (tomography) in medicine, and to geophysics, radar and sonar systems. However, telecommunications is a very important field of application, and this is the area which has provided the major stimulus for research and development and the advancement of the technology.

The world telecommunications network is in the process of being digitized, with the aim of improving the overall quality of connections and introducing new services while providing substantial financial savings. As signals are carried in this network in digital form it is quite natural to consider their processing in digital form at the terminal devices [1, 2].

Before presenting some typical cases, the advantages and limitations of digital techniques will be summarized.

14.1 ADVANTAGES AND LIMITATIONS OF DIGITAL TECHNIQUES

The most important practical advantages of processing a signal in digital form are as follows:

(1) *Absence of drift in the characteristics*. The processing characteristics are fixed rigorously, for example, by the binary coefficients stored in the memories. They are thus independent of the external environment and of parameters such as temperature. Further, ageing has no effect.

(2) *Reproducibility*. The tolerances on the values of the components have no effect on the performance of the system, at least as long as correct operation is maintained. It follows that no adjustments need be made during fabrication, and no realignment is needed over the lifetime of the equipment. This represents a considerable advantage over analogue techniques and, moreover, allows safety margins to be reduced in the design of the equipment, which can lead to important simplifications.

(3) *Improved quality level*. The quality of the processing is limited only by economic constraints. Indeed, no matter how low the specified level of

394

degradations, it can be achieved by increasing the number of bits in the representation of data and coefficients. Further, an increase of 1 bit in the representation results in a 6 dB improvement in the signal to round-off noise ratio. Additionally, some characteristics which cannot be achieved rigorously in analogue equipment can be realized digitally. One example is phase linearity.

(4) *Possibilities of developing new functions*, such as adaptive filters, programmable filters and complementary filters. This illustrates the flexibility of digital techniques. The coefficients which define the processing are easily modified if they are held in an active memory (RAM). Adaptation of the filter response to match, for example, the evolutions of a transmission channel (that is, realization of an adaptive filter or equalizer) simply requires the development of a suitable algorithm. The configuration of the filter can also be changed by an instruction external to the system, in which case it is a programmable filter. Another case is the analysis and reconstruction of signals, where there is a need for two filter functions to be absolutely identical and to remain so, or to be complementary. Digital techniques are well suited to these applications.

(5) *Possibilities of multiplexing.* The same equipment can be shared between several signals, with obvious financial advantages for each function.

In view of these advantages and new opportunities the limitations must also be stated. First, digital systems, and in particular, filters, are active devices which, although performing the same functions as passive circuits, use more energy and are less reliable. On this latter point some compensation is obtained from the facility for automatic supervision and monitoring of digital systems.

On the other hand, the frequency range of digital filters is limited by the technology to values that correspond to the maximum calculation capacities which can be realized and exploited. For simple filtering operations, it is difficult to exceed a sampling frequency of several hundred megahertz at the present state of the technology. In fact a sampling frequency of several tens of megahertz appears to be a reasonable limit to the use of digital filters.

In applications where the signals are not already in digital form, analogue–digital and digital–analogue converters must be introduced, and this adds to the complexity of the system.

Thus, the preferred application area for digital techniques is with frequencies less than 10 MHz, with complex functions to be produced, where high quality is required and where a digital environment can be created. These conditions correspond to those at the terminals of a telecommunications network.

A common function with a number of applications is the detection of a single frequency in a spectrum, as occurs in spectral analysis, in telephone multiplexing, in monitoring systems and in the transmission of signalling signals.

14.2 FREQUENCY DETECTION

Assume that the amplitude of a signal component with frequency f_0 is to be determined when the signal is sampled at frequency $f_s > 2f_0$. The set of operations to be performed corresponds to the scheme in Figure 14.1.

The signal is applied to a narrow band-pass filter centred on the frequency f_0. Rectification is then performed by taking the absolute value of the numbers obtained. This set of absolute values is applied to a low-pass filter which provides the desired value of the amplitude. If the frequency component f_0 which is to be detected is present, threshold logic provides the logic information.

This process can be analysed as follows. Assume $s_0(t)$ is the signal to be detected, with

$$s_0(t) = A \sin(\omega_0 t)$$

Taking the absolute value of the numbers which represent samplings of this signal is equivalent to multiplying by a square wave $i_p(t)$ in phase with s_0, and of unit amplitude. Using equation (1.6) we can write:

$$i_p(t) = 2 \sum_{n=0}^{\infty} h_{2n+1} \sin\left[(2n+1)\omega_0 t\right] \tag{14.1}$$

with:

$$h_{2n+1} = (-1)^n \frac{\sin\left[\pi(2n+1)/2\right]}{\pi(2n+1)/2} = \frac{1}{\pi(2n+1)/2}$$

The signal $s_0^*(t)$ obtained after rectification is

$$s_0^*(t) = 2A \sum_{n=0}^{\infty} h_{2n+1} \sin\left[(2n+1)\omega_0 t\right] \sin(\omega_0 t)$$

or:

$$s_0^*(t) = A h_1 + A \sum_{n=1}^{\infty} (h_{2n+1} - h_{2n-1}) \cos(2n\omega_0 t) \tag{14.2}$$

To obtain the amplitude A, terms of the infinite sum have to be eliminated. Above a certain order the parasitic products have frequencies greater than half the sampling frequency $f_s/2$ and are folded in the useful band. The specifications of the low-pass filter, and, in particular, the stop band edge, have to be chosen so as to eliminate the largest parasites. Those occurring in the pass band result in fluctuations in the measurement of A.

In this approach it is advantageous to use an IIR band pass filter and an FIR low-pass filter, because the amplitude measurement can be made at a frequency less than f_s. There is another method available, which involves only

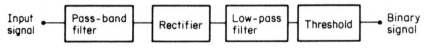

Fig. 14.1   Frequency detection by band-pass filtering

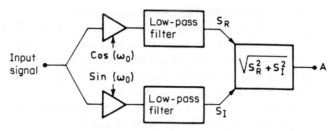

Fig. 14.2   Frequency detection by complex filtering

multirate filters. This is based on modulation by two carriers in quadrature at the frequency f_0 and is shown in Figure 14.2.

The component to be detected is

$$s(t) = A \sin (\omega_0 t + \phi)$$

where ϕ represents the phase of the component relative to the carrier. After low-pass filtering in the two branches to eliminate the unwanted modulation products the following signals are obtained:

$$S_R = \frac{A}{2} \sin \phi; \quad S_1 = \frac{A}{2} \cos \phi \qquad (14.3)$$

The required amplitude is

$$A = 2 \sqrt{(S_R^2 + S_1^2)}$$

Accurate evaluation of $X = \sqrt{(S_R^2 + S_1^2)}$ is difficult and one is generally satisfied with an approximation X', which depends on the phase ϕ.

Table 14.1 gives various approximations and the corresponding relative errors. These errors can be reduced by multiplication by a scaling factor C, i.e. by calculating the value x'_C:

$$X'_C = C \sqrt{(S_R^2 + S_1^2)}$$

Detection of a modulated frequency generally requires less calculation than the method which uses a band-pass filter, but it does require the availability of suitable carrier signals. The operation of detecting a frequency is used in signal

Table 14.1 APPROXIMATION OF $x = \sqrt{(S_R^2 + S_1^2)}$

X'	$\max\left(\dfrac{X' - X}{X}\right)$	C	$\max\left(\dfrac{X'_C - X}{X}\right)$
$|S_R| + |S_1|$	0.41421	0.20711	0.17157
$\max(|S_R|, |S_1|)$	0.29289	0.85355	0.17157
$\max(|S_R|, |S_1|) + \frac{1}{2} \min(|S_R|, |S_1|)$	0.11803	1.05803	0.05573

transmission systems and forms the basis of receivers of multifrequency codes.

Multifrequency code is used to transmit maintenance and tariff information over the connections between telephone exchanges and subscribers with push-button telephones. The subscriber's keypad uses two dialling tones; one provides a group of four high frequencies, thus forming a set of 16 possible codes. The values of these frequencies are given in Table 14.2.

The multifrequency code receiver detects the presence of these two tones and indicates the represented code. These functions can be realized using digital circuits.

14.3 PHASE-LOCKED LOOP

Phase-locked loops are used for clock recovery in terminals and receivers [3]. The principle is illustrated in Figure 14.3. When the loop is in equilibrium, the frequency produced by the voltage-controlled oscillator is equal to the frequency of the input signal and the phase detector produces a signal whose continuous component is extracted by the narrowband low-pass filter. The phase detector can be a modulator which forms the product of the oscillator output and the signal input. If the nominal frequency of the oscillator is equal to the input frequency, the signals are in quadrature and the continuous component at the output of the modulator is zero. If not then the phase difference with respect to the quadrature signal produces a continuous component which shifts the oscillator frequency by the required amount for the frequencies to become equal. The bandwidth of the loop filter determines the capture range, the response time and the residual noise level.

This operation can be reproduced entirely digitally. However, there is additional flexibility in respect of where the phase calculation is performed. The

Table 14.2 MULTIFREQUENCY CODE

Code	Frequencies (Hz)			
LF	697	770	852	941
HF	1209	1336	1477	1633

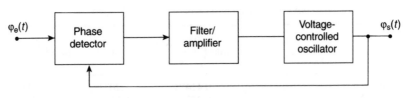

Fig. 14.3   The principle of a phase-locked loop

digital oscillator can be realized by means of a phase accumulator connected to a memory which provides samples of the sinusoid. The input phase values can thus be directly processed at the input and output of the loop, and the phase difference can be obtained by simple subtraction. A model corresponding to a second-order loop is given in Figure 14.4. It is a control loop with two coefficients K_1 and K_2 corresponding to the proportional and integral control terms respectively.

The voltage-controlled oscillator is represented by the integrator which provides the output phase $\varphi_s(n)$. The transfer function between the output and the input can be written:

$$H(Z) = \frac{\Phi_s}{\Phi_e} = \frac{K_1 Z^{-1} + (K_2 - K_1)Z^{-2}}{1 - (2 - K_1)Z^{-1} + (1 - K_1 + K_2)Z^{-2}} \qquad (14.4a)$$

This is the transfer function of a low-pass filter with a value of 1 at zero frequency. The characteristics of the filter are determined by the two parameters K_1 and K_2. The region of stability is examined by using the results of Section 6.7 and putting $b_1 = K_1 - 2$ and $b_2 = 1 - K_1 + K_2$. This gives $1 - K_1 + K_2 < 1$ and $|K_1 - 2| < 2 - K_1 + K_2$. In the plane of the coefficients K_1 and K_2, the stability domain is a triangle.

The transfer function between the phase shift and the input can be written:

$$\frac{\Phi_e(Z) - \Phi_s(Z)}{\Phi_e(Z)} = \frac{(1 - Z^{-1})^2}{1 - (2 - K_1)Z^{-1} + (1 - K_1 + K_2)Z^{-2}} \qquad (14.4b)$$

This shows that such a loop is capable of following a phase variation.

14.4 PCM TIME-DIVISION MULTIPLEXING

Pulse code modulation (PCM) is used for the time-division multiplexing which permits the transmission of several telephone communications using two conductors for each transmission direction. The International Telegraph and Telephone Consultative Committee (CCITT) has recommended a multiplexing hierarchy and defined the characteristics of the equipment to be used [4].

The first level of multiplexing has 30 channels and the digital signals are transmitted at a rate of 2048 kbit/s. At the transmitter the telephone signal is

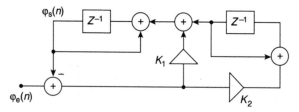

Fig. 14.4 Model of a second-order loop

sampled at 8 kHz and coded into 8 bits following the segmented A-law described in Section 1.11. This is therefore the basic coding of the telephone network and it must be included in terminals connected to this network. After transmission the signal is converted from digital to analogue form at the receiver.

To avoid aliasing when the signal is sampled components with frequencies greater than 4600 Hz have to be eliminated, and attenuation should be greater than 30 dB. The pass band of the corresponding low-pass filter should extend to 3400 Hz, with a ripple of less than 0.25 dB to 3000 Hz. A filter of the same type should be introduced at the receiver in order to attenuate the image bands of the signal around multiples of 8 kHz.

The transmitter and receiver filters of the PCM terminals can be produced in digital form according to the scheme in Figure 14.5. The telephone signal is coded at 32 kHz and 12 bits. A digital filter eliminates components higher than 4000 Hz and allows the sampling frequency to be reduced to 8 kHz.

Consequently, it is advantageous to use an FIR-type filter or a cascade of half-band filters. After filtering, a logic compressor circuit converts the linear coding law to the 8-bit segmented A-law, and supplies the PCM signal in the normalized form. The receiver extends the 8-bit code to a 12-bit linear code followed by filtering with an increase of the sampling frequency to 32 kHz and digital–analogue conversion to restore the telephone signal.

The number of internal memory bits should be calculated so that the round-off noise, when added to the quantization distortions produced by the coding, the rounding operation in the compression unit and the rounding operation before decoding, remains within the limits imposed by the recommendations concerning the quality of the PCM channels. The filtering conditions are satisfied with a linear phase FIR filter with 32 coefficients. The internal data have 16 bits and the coefficients can be limited to 8 bits.

Under these conditions the number of multiplications M_R to be performed per second for both transmission and reception and for a total of 30 channels is

$$M_R = 8000 \left(\frac{N}{2} \right) \times 30 \times 2 = 7.68 \times 10^6 \, \text{mult/s}$$

The amount of memory needed for the data and coefficients MM_D and MM_C are

$$MM_D = N \times 16 \times 30 \times 2 = 30\,720 \, \text{bits}$$

$$MM_C = \frac{N}{2} \times 8 \times 2 = 256 \, \text{bits}$$

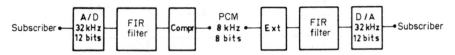

Fig. 14.5   Digital filters in PCM channel

As well as the digital filtering, it should be noted that an additional stage of analogue filtering is required before conversion from analogue to digital in the transmitter and after conversion from digital to analogue in the receiver, to eliminate components at frequencies in the region of about 32 kHz. This can be achieved simply using a passive network of resistances and capacitances.

In this application the use of digital filters involves complication of the A/D and D/A converters which should operate at a rate greater than 8 kHz. This point should be allowed for when making overall cost estimates. An alternative approach in this case is to use switched capacitor devices, as discussed in Section 8.3, because it is then sufficient to convert into digital form the amplitudes of the signal samples from the filter. That is, the rate of A/D and D/A conversion is 8 kHz with filtering based on these devices. Digital processing can also be applied to reducing the bit rate necessary for transmission of a telephone channel.

14.5 ADAPTIVE DIFFERENTIAL PCM (ADPCM)

A telephone channel coded into 8 bits as described in the previous section, with a sampling frequency of 8 kHz, requires a bit rate of 64 kbit/s. With this type of coding the signal can have a flat spectrum or be a sine wave with the maximum permissible amplitude at any frequency in the channel band. The spectral distribution of the quantization distortion can be regarded as uniform. However, the most common signal in telecommunications, the speech signal, has a spectral density which decreases rapidly with frequency from a value of less than 1 kHz. Under these conditions coding techniques other than PCM can be used to obtain a reduction in the bit rate.

Differential PCM coding is based on a linear prediction of the signal to be encoded, following the technique presented in Chapter 10. It is the difference $e(n)$ between the signal $x(n)$ and the prediction $\tilde{x}(n)$ which is coded and transmitted at each sampling period. As the error signal $e(n)$ has a spectral distribution which is almost uniform it makes much better use of the potential of the coder than the initial speech signal.

Figure 14.6 shows the principle of differential PCM coding. The set $e'(n)$ is the result of adding the quantization error to $e(n)$ and $x'(n)$ is the set issued by the decoder. The signal $e(n)$ is expressed by:

$$e(n) = x(n) - \tilde{x}(n) = x(n) - \sum_{i=1}^{N} a_i x(n - i)$$

The prediction filter P has a transfer function $P(z)$ such that:

$$P(Z) = \sum_{i=1}^{N} a_i Z^{-i} \tag{14.5}$$

The order N of the filter and the coefficients a_i $(1 \leqslant i \leqslant N)$ should be chosen to minimize the power of the signal $e(n)$. Under these conditions, for a given value of N, the coefficients are calculated as indicated in Section 12.12, from the elements $R(k)$ $(0 \leqslant k \leqslant N)$ of the autocorrelation function of $x(n)$. The following normalized values have been suggested for speech signals:

$$R(0) = 1; \quad R(1) = 0.8644; \quad R(2) = 0.5570; \quad R(3) = 0.2274$$

They show a strong correlation between neighbouring samples. The corresponding coefficients have values of:

$$a_1 = 1.936; \quad a_2 = -1.553; \quad a_3 = 0.4972$$

The values $N = 3$ or $N = 4$ are generally chosen for the order of the prediction filter [5].

In the actual implementation of differential PCM systems for telecommunications several improvements are made to the basic principle in Figure 14.6, in order to reach a high level of performance with low-complexity equipment [5, 7]:

(1) The prediction is carried out using the sequence $e_q(n)$ transmitted after quantization, which brings a reduction in quantization distortion power. Moreover, the transmitter and receiver then operate on the same source of information and nothing else has to be transmitted if some adaptive procedures are introduced.
(2) The quantizer is made adaptive by relating the quantization step to an evaluation of the signal power. That is, to benefit from the fact that telephone signals have either a fixed or a slowly varying power. Speech, for example, is a non-stationary signal, which, however, can be considered as almost stationary over short periods of time (of the order of 10 ms).
(3) The prediction is made adaptive to cope with the spectra of the various telephone signals and follow the short-term variations of the speech spectrum.

These are called Adaptive Differential PCM (ADPCM) systems.

The block diagram of the linear prediction filter in the transmitter and the reconstruction filter in the decoder is given in Figure 14.7. The Z-transfer function of the transmitter $H(Z)$ is expressed by:

$$H(Z) = 1 - \sum_{i=1}^{N} a_i Z^{-i} = \frac{1}{1 + (\sum_{i=1}^{N} a_i Z^{-i}/1 - \sum_{i=1}^{N} a_i Z^{-i})} \qquad (14.6)$$

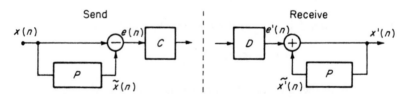

Fig. 14.6   Principle of differential PCM coding

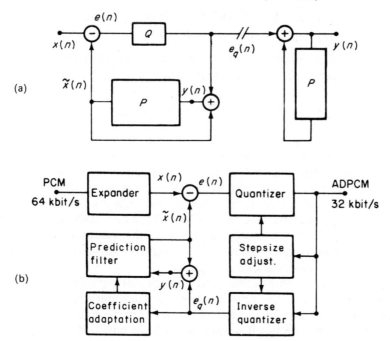

Fig. 14.7 (a) Filters in a differential PCM; (b) block diagram of a PCM-ADPCM converter

The prediction $\tilde{x}(n)$ is computed using the reconstructed sequence $y(n)$. The quantizer Q introduces a quantization distortion which is simply added to the input signal and can be assumed to have a flat spectrum. By doing so, the quantization distortion is reduced by a factor approaching the prediction gain G, with respect to the scheme in Figure 14.6.

The signal-to-noise ratio, SNR, of the whole system can be expressed by:

$$SNR = \frac{\sum_n x^2(n)}{\sum_n [y(n) - x(n)]^2} = \frac{\sum_n x^2(n)}{\sum_n e^2(n)} \frac{\sum_n e^2(n)}{\sum_n [y(n) - x(n)]^2}$$

For stationary signals we can write:

$$SNR = \frac{E[x^2(n)]}{E[e^2(n)]} \frac{E[e^2(n)]}{E[(y(n) - x(n))^2]} \tag{14.7}$$

The first term is the prediction gain. If the error sequence $e(n)$ is assumed to have a flat spectrum, and if the quantizing distortion $q(n) = e(n) - e_q(n)$ is sufficiently small, then the prediction gain G is expressed by:

$$G^2 = \frac{1}{2\pi} \int_{-\pi}^{\pi} \frac{d\omega}{|1 - \sum_{i=1}^{N=1} a_i e^{-ji\omega}|^2} \tag{14.8}$$

The following set of equalities holds:

$$q(n) = e(n) - e_q(n) = x(n) - \tilde{x}(n) - e_q(n) = x(n) - y(n)$$

Assuming a transmission without errors between the send and receive sides, the signal-to-noise ratio of the system is expressed in decibels by:

$$SNR = 20 \log G + 10 \log \frac{E[e^2(n)]}{E[q^2(n)]} \tag{14.9}$$

In order to maximize the second term in the above expression optimum quantizers can be used, as in Section 1.12.

The Z-transfer function $H(Z)$ of the transmitter can be of the FIR, IIR or mixed FIR–IIR types. Using the techniques described in Chapter 12, the prediction filter and the receiver filter can be made adaptive.

The block diagram of a 64-kbit/s PCM to 32-kbit/s ADPCM converter is shown in Figure 14.7. The quantization step of the quantizer is adjusted so as to remain proportional to the root mean square of the signal $x(n)$, which is estimated from the quantizer output sequence.

The transmission quality of this system is such that negligible degradation is introduced for most of the signals carried by telephone channels. With an adaptive prediction filter of adequate order (for example, tenth-order FIR) the prediction gain for speech can range from 6 dB for unvoiced signals to 16 dB for voiced ones, with an overall subjective value of about 13 dB; for high-speed data modems (4800 bit/s) it is about 4 dB. The signal-to-noise ratio of the optimum 4-bit quantizer is near 20 dB. With these figures the signal-to-noise ratio of the ADPCM system is sufficient to transmit speech and data.

An algorithm for PCM-ADPCM conversion has been standardized by the CCITT. It is described in Rec. G721 [4].

14.6 FREQUENCY-DIVISION MULTIPLEXING— TRANSMULTIPLEXER

Frequency-division multiplexing (FDM) of telephone channels consists of distributing the spectra of the channels over a frequency band at a spacing of 4 kHz. For example, 12 multiplexed channels occupy the band 60–108 kHz. As with time multiplexing, the International Telegraph and Telephone Consultative Committee recommends a hierarchy for frequency-division multiplexing.

In order to achieve this by analogue techniques, modulation is combined with filtering to minimize the cost of the filters. Quality constraints are very severe for these systems, particularly those relating to the cross-talk between the channels, and the filters have very high performance specifications.

Digital processing is well suited to this type of application, particularly when interconnection has to be made between a time-division multiplex and a frequency-division one because the signals are already available in digital form at one side. This device is called a transmultiplexer.

The achievement of frequency multiplexing by digital techniques is dependent upon their reproducibility and stability. In essence, both multiplexing and demultiplexing of the telephone channels can be obtained by applying a bank of filters to sampled signals. By way of illustration, we will consider 60 telephone PCM channels to be multiplexed over the frequency band 8–248 kHz.

Each of the signals of a telephone channel is available in the form of real numbers which represent samples taken at a rate of 8 kHz. The corresponding spectrum is given in Figure 14.8. This is a real signal spectrum which is periodic in frequency. A displacement of the whole of this spectrum by 4 kHz corresponds in fact to a reversal in each 4 kHz band and can be obtained very easily by simple inversion of the sign of every other sample, as this operation is actually a modulation of a 4 kHz frequency carrier. Thus, in each 4 kHz band a suitable signal spectrum is available for the multiplexing.

The bank of filters used to multiplex 60 signals comprises 64 real filters, because it is necessary to provide a transition band for the filtering after digital–analogue conversion, and a frequency translation of the multiplexed signal will take place. Further, this number allows the use of FFT algorithms to calculate the DFT which the bank of filters contains. The sampling frequency of the real signal after multiplexing is

$$f_s = 8 \times 64 = 512 \, \text{kHz}$$

The filters in the bank are based on a low-pass filter whose pass band is limited to 1700 Hz, as shown in Figure 14.8.

The subassemblies are determined using the technique described in Section 10.8, allowing for the fact that the translations of the basic filter must be odd multiples of $f_s/2N$, where $N = 128$ is both the total number of filters and the number of branches of the polyphase network. Thus, allowing for the centring of the filters, the Z-transfer function of index m is obtained from that of the basic filter element by changing the variable Z to:

$$Z \exp\left(j2\pi \frac{2m+1}{2N} \right)$$

thus:

$$B_m(Z) = Z^{-n} \exp\left(-j2\pi \frac{2m+1}{2N} n \right) H_n(Z^N) \tag{14.10}$$

In the matrix factorization which describes the transfer function of the filter bank, the DFT which occurs under these conditions is the odd DFT described

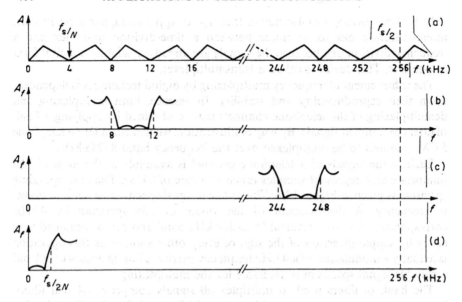

Fig. 14.8   *Filter bank for frequency multiplexing 60 PCM channels. (a) PCM signal spectrum: (b) and (c) filter frequency responses: (d) basic filter response*

in Section 3.3.1. In fact, it is advantageous to shift the sampling of the basic filter in order to use the doubly odd DFT described in Section 3.3.2. The functions $H_n(Z^N)$ which form the polyphase network are written as in Section 11.8, in the form:

$$H_n(Z^N) = \frac{N_n(Z^N)}{D(Z^N)}$$

They all have the same denominator. By using the calculation techniques given for the FIR and IIR filters, and in particular by using the technique described in Section 7.3.2, it is possible to calculate the basic filter in such a way as to reveal a symmetry in the numerator of its Z-transfer function. This symmetry gives the functions $N_n(Z^N)$ and $N_{N-1-n}(Z^N)$ the same coefficients, except in reverse order. Under these conditions, if the signals to be multiplexed are real it is possible to halve the number of calculations to be performed in this part of the system by using suitable structures [8]. The block diagram of this system is shown in Figure 14.9. The 60-channel PCM signals sampled at 8 kHz are applied to the bank of filters which provides the frequency multiplexed signal in the 8–248 kHz band, sampled at 512 kHz.

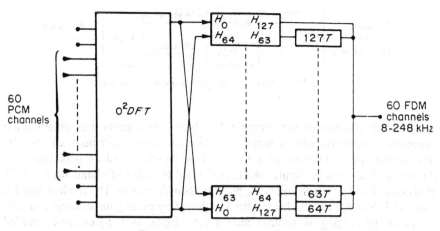

Fig. 14.9   Filter bank block diagram

To meet the quality requirements of FDM systems, and particularly the noise power in a channel, evaluation of the round-off noise in the DFT and the polyphase network results in a 20-bit representation of the numbers in the internal memories. The constraints on cross-talk between channels produce a basic filter of order 6, with coefficients represented by 16 bits. Evaluation of the degradation introduced by limiting the number of bits of the coefficients in the FFT computer results in a value equal to at least 14 bits.

14.7 CODING OF SOUND

Filter banks are the basis of digital sound compression, since they allow the characteristics of the ear to be exploited, particularly the effect of masking. The algorithm represented in Figure 14.10, and standardized as ITU-T/G722, permits transmission of sound on the telephone channel at 64 kbit/s. The audio signal has a bandwidth extending to 7 kHz, it is sampled at 16 kHz and coded in 14 bits. A bank of two QMF filters enables two sub-bands to be obtained sampled at 8 kHz which are then coded in ADPCM at rates of 48 and 16 kbit/s for the low and high bands respectively. One multiplexing operation, with possible insertion of data, provides a transmission rate of 64 kbit/s.

Compression of high-quality sound for digital broadcasting or recording is defined by the ISO/CEI 1 1172-2 standard [9]. It is based on a bank of 32 filters

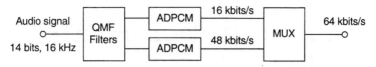

Fig. 14.10   Coding of an audio signal in two sub-bands

with 512 coefficients of the pseudo-QMF type. The signals thus obtained are quantized separately, with a number of bits for each sub-band such that the quantization noise remains at a level below the threshold of masking. This threshold, illustrated in Figure 14.11, is defined for each sub-band from a DFT analysis at 1024 points by applying psycho-acoustic results. The method enables a rate of 128 kbit/s to be achieved for a high-quality monophonic sound channel.

Speech processing is another area where digital techniques have enabled considerable progress to be made in compression, recognition and synthesis [10]. Synthesis is considered briefly in the next section.

14.8 SPEECH SYNTHESIZER

A speech synthesizer produces a signal from binary numbers obtained by analysis and stored in a memory. It is generally used in man–machine communication.

When the speech signal to be synthesized has a long duration (for example, of several minutes) it is preferably in order to reduce the size of the memory holding the information to use coding techniques which lead to low bit rates. From this point of view, linear prediction coding is important because of the ease with which it can be realized and because of the standard of quality that it permits. This approach is based on a greatly simplified representation of the

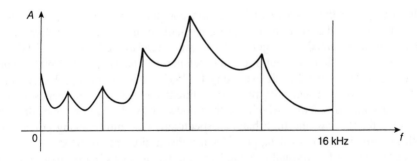

Fig. 14.11   Example of a masking curve for a sound signal

mechanism of speech production which consists of applying periodic vibrations or a noise source to a vocal tract modelled by a variable filter, depending upon whether the sounds to be produced are voiced or unvoiced [10]. The variable filter can be a purely recursive filter which has a transfer function $H(Z)$:

$$H(Z) = G \frac{1}{1 - \sum_{i=1}^{N} a_i Z^{-i}} \qquad (14.11)$$

The scheme of the synthesizer is given in Figure 14.12.

The filter can be implemented using a structure deduced by analogy with sound pipes, and which is the lattice structure of Section 8.5. An example of such a circuit is given by Figure 8.21. The calculations of the N partial correlation coefficients k_i involved is presented in Section 12.13.

The coefficients k_i $(i = 1, N)$ are stored in the memory with the smallest wordlength compatible with the level of quality required for the synthesized speech. This allows the quality level of a speech signal to be obtained. This memory also contains the gain G of the filter and the information for the type of source to be applied. The noise source is produced using a pseudo-random number generator, the principle of which was introduced in Section 1.8. The periodic vibrations are impulses and the speech signal is obtained after digital–analogue conversion.

14.9 MODEMS

Data transmission over telephone lines at rates as high as 4.8 or 9.6 kbit/s requires correction of the line distortions. This is achieved by digital filters with variable coefficients, which are called adaptive equalizers, and operate following, for example, the principle presented in Section 12.9.

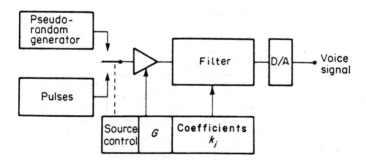

Fig. 14.12   Linear prediction speech synthesizer

The modulation techniques used for synchronous data transmission over voice-grade telephone channels have been standardized by the International Telegraph and Telephone Consultative Committee [11]. They include phase modulation, phase and amplitude modulation and quadrature amplitude modulation. The transmitted signal $s(t)$ generally has a symmetrical spectrum around the frequency f_m, which is equal to the carrier frequency for double-sideband systems. If the baud rate, or symbol rate, is $f_b = 1/T_b$, the line signal $s(t)$ is expressed by:

$$s(t) = \sum_{i=-\infty}^{\infty} h(t - iT_b)[x_i \cos(2\pi f_m t) + y_i \sin(2\pi f_m t)] \qquad (14.12)$$

where $h(t)$ is the impulse response of the spectral shaping filter and the set (x_i, y_i) is related to the input binary sequence d_i by standardized coding rules. The set of all possible values of the pair (x_i, y_i) determines the signal constellation of the system and is usually given a two-dimensional representation; for each x_i and y_i the set of possible values is very restricted.

In a 4.8 kbit/s data modem using 8-phase modulation the parameter values are:

$$f_b = 1600\,\text{Hz}; \quad f_m = 1800\,\text{Hz}; \quad \text{values of } x_i \text{ and } y_i = 0, \pm 1, \pm \sqrt{2}$$

and for 9.6 kbit/s with phase and amplitude modulation:

$$f_b = 2400\,\text{Hz}; \quad f_m = 1700\,\text{Hz}; \quad \text{values of } x_i \text{ and } y_i = 0, \pm 1, \pm 3, \pm 5$$

The principle of a receiver for data transmission at 9.6 kbit/s is shown in Figure 14.13. The line signal is digitized by an analogue–digital converter, with a sampling frequency of 9.6 kHz. After a frequency shift by the carrier frequency f_m the complex signal obtained is applied to a low-pass filter and to a complex equalizer which corrects the phase and amplitude distortions of the telephone channel. Then the original binary data signal at 9.6 kbit/s is regained by means of decision and decoding circuitry, which also provides the information for the filter coefficient adjustment. This principle lends itself to various realizations, in particular the order of the operations can be modified.

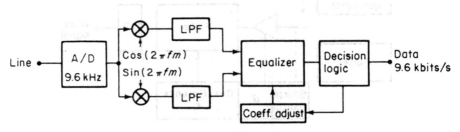

Fig. 14.13   *Data modem receiver*

Note that besides equalization and filtering, which form a dominant part of the receiver, many other functions can be realized within the wide range of available modems [12].

14.10 ECHO CANCELLATION

Another application of adaptive filtering, echo cancellation aims to reduce the annoyance caused by a range of imperfections that create echoes in the telephone network [13]. Firstly, there are echoes arising from acoustic coupling between the microphone and loudspeaker in a telephone set; an adaptive echo canceller can be used in this case to provide a hands-free set. Secondly, electrical echoes are produced on the transmission lines in the form of reflected signals due to impedance mismatching and imperfections in the hybrid transformers which perform two-wire to four-wire conversions. In the case of speech, these signals are reflected back to the subscriber who is speaking and they become annoying as the distance between the subscribers increases; this is the case with long-distance transmissions, particularly by satellite. These echoes must also be eliminated for high-speed data transmission in simultaneous bidirectional mode and for two-wire connection of subscribers to the integrated services digital network (ISDN).

The insertion of an echo canceller is shown in Figure 14.14 for the case where the transmission channel uses digital PCM. Transmission over the network is supported differently for the two directions, so transmission on the subscriber line is in bidirectional mode on a single two-wire connection; the signal received from the network, digital in Figure 14.14, is applied to a hybrid transformer after conversion to analogue form. If matching is not perfect then part of the signal,

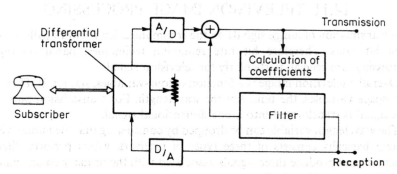

Fig. 14.14 PCM channel with echo cancellation

the echo, is introduced into the transmission chain. This parasitic signal can be cancelled by the addition of a signal provided by a digital filter of the FIR type operating on the received signal. This is an example of modelling. The coefficients are calculated from the transmitted signal following the principle explained in Section 12.5.

The number of coefficients N is determined by the characteristics of the transmission circuits, the length of the connection and the maximum delay encountered. ITU-T Recommendation G.168 specifies a maximum delay of 64 ms, which corresponds to $N = 256$ coefficients for a sampling frequency of 8 kHz. It also specifies an echo attenuation of greater than 30 dB over an input level range of 30 dB and a convergence time of less than 1 s.

Because the input signal is speech, it is necessary to use a standardized algorithm as indicated in Section 12.10. The data items are linearly coded 12-bit numbers following the A-law of Section 1.11. Specification of the adaptation parameters leads to coefficient values which have of the order of 16 bits.

Detection of double speech and non-linear processing are also important in this telephone echo canceller, for overall system quality and for subscriber perception. Double speech is produced when the speech of the local subscriber is superimposed on the echo in the reference signal. To cancel the echo, this disturbing signal entails a change in the coefficients which must be stopped as rapidly as possible. Adaptation is therefore stopped as soon as detection has occurred. Detection can be simply by comparing the levels of the three signals: input, reference and output error. Level estimation can be made recursively from the amplitudes. Non-linear processing consists of cancelling the output error signal when its level falls below a predetermined threshold.

For data transmission in simultaneous bidirectional mode on two-wire lines, the echo cancellers have different complexity characteristics. Overall there are fewer bits for internal data and more bits for the coefficients.

14.11 TELEVISION IMAGE PROCESSING

The transmission of image signals on the communication network calls for very high bit rates. Therefore bit rate reduction techniques, based on digital processing, are crucial, particularly for television signals.

Overall a television image is a function of four-variables, $s(x, y, t, \lambda)$. There are two space variables, the time and the wavelength. For transmission purposes, this signal is transformed into a one-dimensional signal.

The wavelength variable can be dropped by considering that the human visual system basically consists of three types of receivers, which perform filtering functions and produce three signals associated with the primary colors, namely Red, Green and Blue (R, G, B).

The television scanning process converts these three-dimensional signals into a one-dimensional signal. The images are scanned 25 times per second, with

625 lines per image. In fact, the odd- and even-numbered lines form two consecutive frames which are multiplexed in time. Hence the recurrence of 50 interleaved frames per second.

For transmission, the primary components R, G, B are replaced by linear combinations called luminance Y and colour differences or chrominance U and V.

$$Y = 0.30R + 0.59G + 0.11B$$
$$U = R - Y = 0.70R - 0.59G - 0.11B$$
$$V = B - Y = -0.30R - 0.59G + 0.89B$$

Digitization is performed with a frequency of 13.5 MHz for the luminance and 6.75 MHz for the chrominance signals. Since the analogue-to-digital conversion is of 8 bits, the corresponding data rate rises to 216 Mbit/s. This format corresponds to CCIR Recommendation 601 of the ITU and is described as Type 422. It leads to images presented in the form of tables of 8-bit numbers containing 720 points per line and 576 useful lines, in the case of 625-line scanning. One image therefore corresponds to 414 720 bytes for the luminance and 207 360 bytes for each chrominance component.

Bit rate reduction techniques rely on the fact that good modelling is provided by the output of a first-order IIR filter to which white Gaussian noise is applied [14]. The corresponding two-dimensional autocorrelation function can be written:

$$V(x, y) = r_0 e^{-(\alpha x + \beta y)}$$

where α and β are positive constants. For the associated spectrum this gives:

$$S(\omega_1, \omega_2) = r_0 \left(\frac{4\alpha\beta}{(\alpha^2 + \omega_1^2)(\beta^2 + \omega_2^2)} \right) \qquad (14.13)$$

The greatest compression in the representation of a signal is obtained with a transformation based on the eigenvectors of the autocorrelation matrix. In the case of first-order signals, this transformation is well approximated by the discrete cosine or sine transformation, presented in Sections 3.3.3 and 3.3.4. In the image compression standards it is the DCT applied to blocks of 8×8 elementary image points, or pixels, which has been retained. The standards formulated for videophones, image storage and digital television make use of the following three techniques [15]:

(1) Estimation of movement in order to be able to minimize the difference between the current image and the preceding one;
(2) The discrete cosine transform to minimize spatial redundancy;
(3) Variable length statistical coding (VLC).

The general arrangement of an image encoder is given in Figure 14.15. The

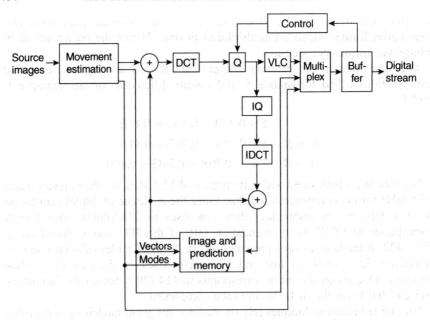

Fig. 14.15   General arrangement of a moving-image encoder

digitizer Q operates above thresholds which can be set by a control device, permitting a constant bit rate to be obtained with the help of a buffer memory. For commercial quality television, the bit rate can be reduced to around 4 Mbit/s, a compression factor of the order of 50 [16].

Digital filters are used for interpolation and subsampling operations during changes of image format or movement estimation. These are separable filters. Digital compression of multimedia signals, speech, image and sound permits considerable reduction of the bit rates required for broadcasting programmes, and associated with the techniques of digital transmission, it offers the possibility of high spectral efficiency by transmitting several programmes in the channels used for a single analogue programme. This results in an economic advantage which can be large, as in satellite television broadcasting.

Techniques with high spectral efficiency make intensive use of digital processing and they make the best use of the characteristics of the channels. In this way, multicarrier techniques can lead to capacities of several bits/s per hertz on channels which are of limited quality or susceptible to interference.

14.12 MULTICARRIER TRANSMISSION—OFDM

The objective of multicarrier transmission techniques is to approach the theoretical capacity of a channel, on the one hand by limiting the effect of

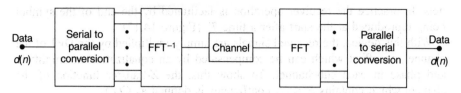

Fig. 14.16   The principle of OFDM transmission

distortion and on the other by adjusting the data rate in accordance with the spectral density of the noise. In fact, by dividing a given channel into tens, hundreds or thousands of subchannels, the effect of distortion on each subchannel is made negligible and the data rate it is able to support can be allocated to each subchannel. One simple and effective approach to implementing this procedure involves the fast Fourier transform; known as OFDM (orthogonal frequency division multiplexing), its principles are illustrated in Figure 14.16. The flow of data to be transmitted is converted into N elementary flows, each N times smaller, which are applied to the input of an inverse DFT computer. In accordance with the definition of the inverse DFT given in Chapter 2, this operation corresponds to a modulation of N carriers by the elementary flows, at frequency multiples of f_s/N, and the addition of the set of signals modulated in this way. The rate of OFDM symbols is thus f_s/N. On reception, after passing through the channel, a direct DFT performs the range of demodulations and reconstructs the original data, which merely requires to be serialized to recover the initial flow.

This simple principle is a direct illustration of the definition of the DFT and its inverse. However, to operate correctly, it requires several precautions and adaptations.

Referring back to Section 2.4, in particular Figure 2.8, notice that orthogonality of the signals is valid only for frequencies which are at the centre of the interval of length $f_s/2N$ allocated to each subchannel; and notice how the subchannels have an area of overlap and that the amplitude of overlap reduces with increasing frequency difference. On the edges of the transmission channel, the frequency responses of the subchannels are not symmetric and this can lead to interference. It is therefore necessary to avoid using extreme subchannels and to provide a margin of at least a few subchannels on each side of the chosen frequency band.

In the time domain, a practical transmission channel has an impulse response of duration τ. To avoid superposition of two consecutive OFDM symbols on reception, the symbols must be separated by a sufficient time; that means it is necessary to introduce a guard time $T_g > \tau$. During this guard time it is necessary to prolong the OFDM symbol, to introduce the circular convolution mentioned in Section 2.1 and hence to avoid interference between the subchannels. In practice the receiver operation is facilitated by the end of the symbol being reproduced at the start after a time T_g (Figure 14.17).

With this device, the received signals are simply multiplied by the DFT of the channel, an effect which can be compensated by an equalization in amplitude and phase in each subchannel. To show this, the Z-transfer function of the channel which contains $P \leq N_g$ coefficients is defined as $C(Z)$:

$$C(Z) = \sum_{p=0}^{P} C_p Z^{-p} \tag{14.14}$$

If $x(n)$ is the transmitted signal, the received signal $y(n)$ can be written:

$$y(n) = \sum_{p=0}^{P} C_p x(n - p)$$

As $x(n)$ is expressed in terms of the data d_k by

$$x(n) = \sum_{k=0}^{N-1} d_k e^{j(2\pi/N)kn} \tag{14.15}$$

The double summation is obtained for $y(n)$:

$$y(n) = \sum_{p=0}^{P} C_p \sum_{k=0}^{N-1} d_k e^{j(2\pi/N)k(n-p)} \tag{14.16}$$

Putting:

$$H_k = \sum_{p=0}^{P} C_p e^{-j(2\pi/N)kp}$$

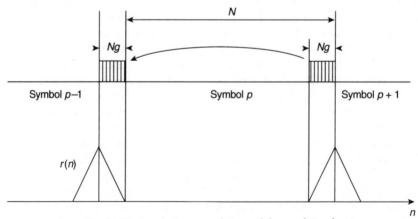

Fig. 14.17 Introducing a guard time and the correlation function

finally gives

$$y(n) = \sum_{k=0}^{N-1} (d_k H_k) e^{j(2\pi/N)kn} \qquad (14.17)$$

The circular convolution property of the DFT is again found and the receiver provides the transmitted data multiplied by the channel spectrum H_k.

The redundancy of the transmitted signals can be exploited by the receiver for synchronization. In fact, by calculating the following correlation function:

$$r(n) = \sum_{i=n-N_g+1}^{n} y(i)y^*(i-N) \qquad (14.18)$$

peaks appear as shown in Figure 14.17 which characterize the start of each symbol and allow the temporal analysis window of the receiver to be adjusted and a contribution made to synchronization of the clocks.

Synchronization in time and frequency is a difficult problem in systems with a large number of carriers, and special reference symbols are introduced or some subchannels are reserved for fixed signals called pilots.

Figure 14.18 shows the block diagram of a digital television receiver for terrestrial broadcasts [17]. The analogue interfaces carry the signal in the band 0.76–8.37 MHz and the analogue-to-digital conversion is performed at $f_s = 18.28$ MHz. Then a real-to-complex conversion using a quadrature filter is performed and the sampling frequency is reduced to 9.14 MHz. Before calculation of the FFT at $N = 8192$ points, a complex multiplier performs frequency adjustment on the spectrum of the signal. Temporal synchronization controls the positioning of the window of the FFT. The transmitted signal

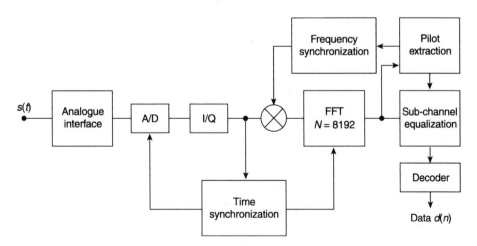

Fig. 14.18 Receiver for terrestrial digital television

contains 6817 active carriers, and 177 of them are dedicated to pilot signals which permit exact synchronization of the receiver, an estimate of the frequency response of the channel for equalization and a measure of the distortion in each subchannel. The guard time can reach 20% of the symbol duration. This system must permit transmission rates up to 32 Mbit/s in a channel with 8 MHz spacing, or 4 bits/s/Hz.

Another example of OFDM is the ADSL (asymmetric digital subscriber line) system; this permits transmission to the subscriber using a pair of conductors at rates which can reach 6 Mbit/s above the telephone signal. The multicarrier signal occupies a band of 1 MHz with 256 carriers.

The advantages of the OFDM technique lie in its low sensitivity to channel distortion, a relative immunity to impulse noise by averaging effects of the DFT, the possibility of avoiding interference with a narrow spectrum, and adjustment of the data rate in accordance with the spectral density of the noise.

In contrast, OFDM requires special synchronization devices and it is sensitive to non-linearities. In particular, as the signal transmitted in the channel is the sum of a large number of random signals of the same distribution, there is a Gaussian amplitude distribution and a crest factor of 12 dB, which is a disadvantage in comparison with modulation techniques having a constant envelope, as used in radiocommunication for example.

The efficiency of OFDM is reduced by the guard intervals in time and frequency, the pilot signals and the reference symbols. These losses can be avoided at the cost of additional processing by adopting the transmultiplexer principle of Section 14.6, but this time using almost perfect banks of decomposition and reconstruction filters which limit overlapping of the subchannels to the immediate vicinity. With adaptive equalizers in each subchannel, the synchronization constraints are also reduced and this makes it possible to approach the theoretical limits of transmission capacity.

REFERENCES

[1] S. L. FREENY, J. F. KAISER and H. D. McDONALD, In *Applications of digital signal processing*, A. Oppenheim (ed.), Prentice Hall, Englewood Cliffs NJ, 1978, Ch. I.
[2] J. DUNN, Signal Processing: technology and prospects. *Electrical Communication*, **59**, No. 3, 252–9, 1985.
[3] F. M. GARDNER, *Phaselock techniques*, John Wiley, New York, 1979.
[4] CCITT, Réseaux Numériques: systèmes de transmission et équipements de multiplexage, Vol. III.3, Geneva, 1985.
[5] J. D. GIBSON, Adaptive prediction in speech differential encoding systems. *Proc. IEEE*, **68**, April 1980.
[6] D. J. GOODMAN and R. W. WILKINSON, A robust adaptive quantizer. *IEEE Transactions*, **COM23**(11), 1975.
[7] J. M. RAULIN, G. BONNEROT, J. L. JEANDOT and R. LACROIX. A 60-channel PCM-ADPCM Converter. *IEEE Trans.*, **COM-30**(4), 1982.

[8] G. Bonnerot, M. Coudreuse and M. Bellanger, Digital processing techniques in the 60 channel transmultiplexer. *IEEE Transactions*, **COM26**(5), 1978.

[9] ISO-CEI 11172, Technologies de l'information: codage des images animées et du son, Geneva, 1994.

[10] L. R. Rabiner and R. W. Schafer, *Digital Processing of Speech Signals*, Ed. A. Oppenheim, Prentice-Hall, Englewood Cliffs, NJ, 1978.

[11] CCITT, Data communication over the telephone network. *Red Book*, Vol. VIII-1. Geneva, 1985.

[12] P. J. Van Gerwen, N. Verhoeckx, H. Van Essen and F. Snijders, Microprocessor implementation of high speed data modems. *IEEE Trans.*, **COM 25**(2), February 1977.

[13] M. Erdreich, Le phénomène de l'écho dans les réseaux de télécommunications. *Commutation et Transmission*, **3**(4), 1981.

[14] F. Kretz, Codage MIC-différentiel à prédiction adaptative en télévision. *Annales des Télécom*, **37**, 1982.

[15] A. N. Netravali and B. Haskell, *Digital pictures: representation and compression*, Plenum, New York, 1988.

[16] RecUIT-T H262/ISO-CEI 13818-2, Information technology: generic coding of moving pictures and associated audio information—video, Geneva, 1995.

[17] T. de Couasnon, R. Monnier and J. B. Rault, OFDM for Digital TV Broadcasting. *Signal Processing*, **39**, 1–32, 1994.

EXERCISES: SOLUTIONS AND HINTS

CHAPTER 1

1.1 $I_L(t) = \dfrac{1}{2} + \dfrac{2}{\pi} \displaystyle\sum_{p=1}^{4} \dfrac{(-1)^{p+1}}{2p-1} \cos 2\pi(2p-1)\dfrac{t}{T}$.

1.2 $s(nT) = \sin(n\pi + \varphi) = (-1)^n \sin\varphi$.

     The possibility of reconstruction depends on φ. $\left(\varphi = \dfrac{\varphi}{2} \text{ yes; } \varphi = 0 \text{ no} \right)$.

1.3 $H(f_s/2) = \dfrac{2\sqrt{2}}{\pi}$ $(0,92\,\mathrm{dB})$.

1.4 $f_2 < f_s < 2f_1$.

1.5 $s(nT) = s_r(nT) + js_i(nT) = e^{j(\pi/2)n} \dfrac{\sin\dfrac{3\pi}{8}n}{\sin\left(\dfrac{\pi}{8}n\right)}$.

1.6   Maximum value of $s(n) = 8$;
$$s(n) = 0 \quad \text{for} \quad \varphi_k = -2\pi\dfrac{k}{8}n + k\pi$$

1.7 $f_s = 2\,\mathrm{MHz}; \quad \Delta f = 1\,\mathrm{kHz}$.

1.8 $p(1) = \dfrac{1}{\pi}\dfrac{1}{\sqrt{A^2 - s^2}};$ $r(\tau) = 2(f_2 - f_1)\dfrac{\sin\pi(f_2 - f_1)\tau}{\pi(f_2 - f_1)\tau}\cos\pi(f_2 + f_1)\tau$.

1.9   Periodic part; Fourier coefficient: $C_n = \dfrac{p\sin\dfrac{\pi}{2}n}{\pi n}$.

     Non periodic part; spectrum: $S_2(f) = p(1-p)T\dfrac{1 - \cos\pi fT}{\pi^2 f^2 T^2}$.

1.10 Signal to noise ratio in the band: $300\text{–}500\,\mathrm{Hz} = 75\,\mathrm{dB}$ ($f_s = 16\,\mathrm{kHz}$; gain $3\,\mathrm{dB}$).

1.11 Quantizing distortion: line at $\tfrac{3}{8}f_s$; power: 0.0195^2.

1.12 If the characteristic is centred: $a_1 = 0$ for $0 \leqslant |\alpha| \leqslant \frac{1}{2}$; $a_1 = \dfrac{4q}{\pi}\sqrt{1 - \dfrac{1}{4\alpha^2}}$ for $0 \leqslant |\alpha| \leqslant 1$.

Centring at $\dfrac{q}{2}$: $a_1 = \dfrac{2q}{\pi}$ for $0 \leqslant |\alpha| \leqslant 1$.

1.13 Without clipping (peak factor): 10 bits; with clipping at $1\% = 9$ bits.

1.14 Linear coding $(S/N)_{max} = 50\,dB$; with non linear coding it varies from 35 to 38 dB when the signal varies from $-36\,dB$ to $0\,dB$

1.15 Optimal values: $x_0 = 0$; $x_1 = 0.9816$: $y_1 = 0.4528$; $y_2 = 1.510$.

CHAPTER 2

2.1 The DFT of the second set is related to the DFT of the first by $X'(k) = e^{-jk(\pi/4)}X(k)$

2.2 Real multiplications: $M_R = 28$: real additions: $A_R = 84$.

2.3 The small differences come from aliasing and decrease when N increases.

2.4 Number of complex multiplications: 160, 96, 72. Additions: 384.

2.5 Maximum noise-power at any output: 28. $q^2/12$. With quantization at 8 bits of the coefficients: $|\varepsilon(i, k)| \leqslant 0.003$.

2.6 Recursion: $X_0 = x(N - 1)$; $X_m = x(N - 1 - m) + WX_{m-1}$ for

$$1 \leqslant m \leqslant N - 1.$$

$N - 1$ complex multiplications are needed.

2.7 Total roundoff noise: $N\dfrac{q^2}{12} + Nq^2$; signal-to-noise ratio degradation: $\Delta SB = 11.5\,dB$

(input noise $q^2/12$).

2.8 Recording: 20 000 samples; memory 160 $k$bits; cycle time per multiplication: 1 μs.

2.9 Cosine, Hamming and Blackman windows attenuate the secondary lobes but do not permit detection of weak components.

2.10 In figure 2.13 the multipliers are used for 50% of the time. Full efficiency is achieved through doubling the input memory with alternate reading.

2.11 Memories: 120, 30 and 6 numbers in the 3 stages respectively. Three complex multipliers are needed in 2 stages. Their efficiency can become 100% with intermediate memories.

CHAPTER 3

3.1 It is sufficient to verify that the products $I_3 \times A$ and $A \times I_3$ are different.

3.2 It is sufficient to verify the relations 3.3, 3.4, 3.5, 3.6.

3.3 Number of real multiplications in radices 2, 4 and 8: 384, 284 and 246.

3.4 The order 12 DFT requires 20 complex multiplications.

3.5 Use relations (3.18) and (3.21) to get the 2 factorisations.

3.6 The complex DFT of order 8 leads to 24 real multiplications. The odd transform leads to 26.

3.7 With that approach the operations in Δ_{12} vanish, which reduces the number of complex multiplications to 16.

3.8 Matrices of the transformation:

$$T = \begin{bmatrix} 1 & 1 & 1 & 1 \\ 1 & 4 & 16 & 13 \\ 1 & 16 & 1 & 16 \\ 1 & 13 & 16 & 4 \end{bmatrix}; \quad T^{-1} = 13 \begin{bmatrix} 1 & 1 & 1 & 1 \\ 1 & 13 & 16 & 4 \\ 1 & 16 & 1 & 16 \\ 1 & 4 & 16 & 13 \end{bmatrix}$$

CHAPTER 4

4.1 Response to the sequence a^n: $y(n) = a^{n-3} \dfrac{1 - a^{9-n}}{1 - a}$ for $5 \leqslant n \leqslant 8$.

4.2 Take the derivative, perform series expansion and an integration

$$\log(z - a) = -\sum_{n=1}^{\infty} \frac{1}{na^n} Z^n; \quad \frac{Z^{-1}}{(1 - aZ^{-1})(1 - bZ^{-1})} = \frac{1}{a - b} \sum_{n=0}^{\infty} \left(\frac{1}{b^n} - \frac{1}{a^n} \right) Z^{-n}.$$

with $|a| < 1; \quad |b| < 1$

4.3 $H(Z) = \dfrac{1}{(1 - re^{j\theta}Z^{-1})(1 - re^{-j\theta}Z^{-1})}$.

4.4 Output power: 21; $H(\omega) = 4.41 - 1.536 \cos \omega + 0.46 \cos 2\omega$.

4.5 System response:

$$y(n) = \frac{e^{jn\omega}}{1 - e^{-j\omega} + 0.8e^{-2j\omega}} + r^{n-1} \left[(a - 0.8b) \frac{r \sin(n+1)\theta}{\sin \theta} - 0.8a \frac{\sin(n\theta)}{\sin \theta} \right]$$

CHAPTER 5

5.1 The response is zero for $f = 0.288$; 0.347; 0.408; 0.469 maximal ripple: 0.08.

   Zeros of $H(Z)$: 0.606; 1.651; $0.4292 \pm j0.464$; $1.073 \pm j1.161$.

5.2 Coefficients: -0.012; 0; 0.042; 0; -0.093; 0; 0.314; 0.5. Zeros of $H(Z)$: 0.4816; 2.076; $0.3764 \pm j0.368$; $1.3583 \pm j1.328$; maximal ripple: 0.03.

5.3 $\delta = 0.017$, which is less than the above values.

5.4 With the sampling frequency $f_s/2$ at the output the numbers of memories and multiplications are divided by 2, through interleaving (see §10.5).

5.5 In the complex plane $H(Z)$ is rotated by π and $\pm \pi/2$, which yields a high-pass filter and a low-pass filter.

5.6 Coefficients: $N = 27$; computation accuracy: $b_c = 12$ bits; $b_i = 20$ bits.

5.7 Coefficients error function: -0.0065; 0.0034; -0.0015; -0.0019

$$e(0.1925) = e^{-j2\pi 8 f_0}(-0.0028)$$

$$b_c \simeq 1 + \tfrac{1}{2} \log_2 \left(\frac{f_e}{2\Delta f} \right) + \log_2 \left(\frac{1}{\min\{\delta_1, \delta_2\}} \right).$$

CHAPTER 6

6.1 Follow the derivation in section 6.1. The difference between filter delay and group delay illustrates the non linearity of the phase response.

6.2 Unit step response: $y(n) = \frac{1}{1.8}(1 - (-0.8)^{n+1}) + (-0.8)^{n+1} y(-1)$.

6.3 Poles: $P = 0.78 \pm j0.438$. The zeros do not add multiplications to the circuit.

6.4 $H_m = 85$; $\cos \omega_0 = 0.808$; $\|H\|_2 = 8.53$.
With zeros at $3f_s/8$: $\|H\|_2 = 25.8$.
'Limit-cycle frequency close to $f_s/10$; amplitude $\simeq 42q$. Large amplitude oscillations possible because (6.56) is not verified.

6.5 $b_c \geqslant 13$ bits; infinite attenuation point displacement: $df_i \leqslant 2.210^{-5} f_e$.

6.6 $H(Z) = \dfrac{0.796 - 1.42Z^{-1} + Z^{-2}}{1 - 1.42Z^{-1} + 0.796Z^{-2}}$; $\tau_g(\omega)$ calculated by (6.45).

Realization possible with 3 multiplications.

CHAPTER 7

7.1 First order section: $|H(\omega)|^2 = \dfrac{1.49 + 1.4 \cos \omega}{1.81 - 1.8 \cos \omega}$

$$\varphi(\omega) = \text{Arctg} \cdot \frac{1.6 \sin \omega}{1.63 - 0.2 \cos \omega}; \quad \tau(\omega) = \frac{1.6(0.37 \cos \omega - 0.2)}{(0.37 \cos \omega - 0.2)^2 + 2.66 \sin^2 \omega}$$

7.2 Characteristic frequencies: 0.162; 0.231; 0.538; 0.736.

7.3 Transfer function: $H(Z) = 0.094 \dfrac{(1 + Z^{-1})^4}{(1 + 0.039Z^{-2})(1 + 0.447Z^{-2})}$.

7.4 Transformation: $Z^{-1} = -\dfrac{Z^{-1} - \alpha}{1 - \alpha Z^{-1}}$; $\alpha = \dfrac{\sin \pi(f_1 - f'_1)}{\sin \pi(f_1 + f'_1)}$

with $f_1 = 0.1725$; $f'_1 = 0.1$; $\alpha = 0.3$.

7.5 Scale factors: $a_0^0 = 2^{-6}$; $a_0^1 = 2^{+2}$; $a_0^2 = 2$.
For $H(j) = 1$, we get: $a_0 = 0.515$ and $a_0^3 = 4.12$; accuracy: $b_i = 16$ bits.

7.6 Coefficient wordlength: $b_c \simeq 12$ bits. The optimum is obtained through systematic search about rounding. The critical pole $0.9235 \pm j0.189$ does not permit reducing b_c to 11 bits.

7.7 The filter in paragraph 7.2.3 can have limit cycles of amplitude less than $3q$ and frequency near $f_s/5$. The same applies to the filter in Figure 7.21.

7.8   The IIR filter requires 7 multiplications and 4 memories while the FIR counterpart requires 8 multiplications and 16 memories.

7.9   Transfer function:

$$H(Z) = 0.0625 \frac{1 - 1.165Z^{-1} + Z^{-2}}{1 - 1.404Z^{-1} + 0.84Z^{-2}} \frac{1 - 0.198Z^{-1} + Z^{-2}}{1 - 1.238Z^{-1} + 0.455Z^{-2}}$$

$$f_1 = 4832\,\text{Hz}; \quad f_2 = 7495\,\text{Hz}; \quad \Delta f_1 = 4\,\text{Hz}; \quad \Delta f_2 = -3\,\text{Hz}.$$

Scale factors: $a_0^0 = 2^{-3}$; $a_0^1 = 2^{-1}$; $a_0^2 = 1$.

7.10 Theoretic order: $N = 5.19$; for $N = 6$, δ_1 becomes very small. Coefficient wordlength: $b_c \simeq 11$ bits. Difference between input and internal data: 7 bits.

CHAPTER 8

8.1

$$S = \frac{1}{z+2}\begin{bmatrix} z & 2 \\ 2 & z \end{bmatrix}; \quad t = \begin{bmatrix} 1 - z/2 & z/2 \\ -z/2 & 1 + z/2 \end{bmatrix}$$

For LC circuits take $z = Lp + \dfrac{1}{Lp}$ or $z = \dfrac{LC}{Lp + \dfrac{1}{Cp}}$

8.2   The diagram is that in fig. 8.3 with $N = 6$ and $Y_6 = 0$. For

$$f_s = 40\,kHz: a_1 = a_4 = 0.205; \quad a_2 = a_3 = 0.085;$$

The coefficients are multiplied by 4 for $f_s = 10\,kHz$.

8.3   The circuit in fig. 8.2 can be used. The products sL_2/sL_3 and sL_2/sL_1 are implemented through direct connections between the input of the central path and the two adders. The coefficient values are: $a_1 = 5.510^{-4}$; $a_2 = 2.610^{-4}$; $a_3 = 2.410^{-4}$; $a_4 = 4.610^{-4}$; $L_2/L_1 = 0.097$; $L_2/L_3 + L_2/L_1 = 0.32$.

8.4   Coefficients:

$$\alpha_1 = 0.4425; \quad \alpha_3 = 0.1856; \quad \alpha_5 = 0.1793; \quad \alpha_7 = 0.7359$$
$$\beta_2 = 0.2255; \quad \beta_4 = 0.1781; \quad \beta_6 = 0.1944; \quad \alpha_7' = 0.7169$$

With $b_c = 5$ bits the wave filter has less ripple.

8.5   Lattice filter zeros: 0.6605; $0.6647 \pm j0.5020$; after rounding the k_i to 5 bits: 0.6661; $0.6377 \pm j0.5002$.

CHAPTER 9

9.1 $X(f) = \dfrac{1 - a\cos 2\pi f}{1 + a^2 - 2a\cos 2\pi f} + j\dfrac{-a\sin 2\pi f}{1 + a^2 - 2a\cos 2\pi f}.$

9.2   Calculate $X_1(\omega)$ through Hilbert transform; we can write:

$$X_R(\omega) = \frac{1}{2}\left[\frac{1}{1-p} + \frac{1}{1-\bar{p}}\right].$$

9.3 The nonzero terms in the sets $x_R(n)$ and $x_I(n)$ are interleaved. The operation performed is analytic filtering: $y(n) = \frac{1}{2}e^{-jn\frac{\pi}{5}}$.

9.4 Coefficient wordlength: $b_c \simeq 2 + \frac{1}{2}\log_2\left(\frac{f_e}{2\Delta f}\right) + \log_2\left(\frac{1}{\delta}\right)$.

9.5 Phase shifter order: $N \simeq \log\left(\frac{\pi}{\varepsilon}\right)\log\left(\frac{f_e f_e}{f_1 f_2}\right)$; coefficients: $b_c \approx \log_2\left(\frac{\pi}{\varepsilon}\right)$

$+ \log_2\left(\frac{f_e}{f_1}\right) + \log_2\left(\frac{f_e}{f_2}\right)$. For the example in paragraph 9.4: $N = 4.97$; $b_c \simeq 14$ bits.

9.6 Transfer functions:

$$H_m(Z) = 1 - 2Z^{-1} + 2Z^{-2} - Z^{-3} + 0.25Z^{-4}$$
$$H_L(Z) = 0.5 - 1.5Z^{-1} + 2.25Z^{-2} - 1.5Z^{-1} + 0.5Z^{-4}$$
$$H_M(Z) = 0.25 - Z^{-1} + 2Z^{-2} - 2Z^{-3} + Z^{-4}.$$

CHAPTER 10

10.1 Coefficient wordlengths: $b_c = 1, 2, 5, 6, 9, 10, 10, 11, 14$. For the half-band filter: $b_c \approx 2 + \log_2(1/\delta_m - \delta_0)$.

10.2 Filters in the cascade of 3 filters: $\Delta f = 0.4$ with $M = 2$; $\Delta f = 0.15$ with $M = 3$; $\Delta f = 0.025$ with $M = 8$. Roundoff noise at the output of a half-band filter: $2M\frac{q^2}{12}$.

After 3 filters: $P_N = 20\frac{q^2}{12}$.

10.3 The function can be carried out with a half-band filter ($M = 3$) and a low-pass filter with 54 coefficients, hence the computation rate of 264 kmult/s. A direct realization with 100 coefficients leads to 400 kmult/s.

10.4 The odd DFT corresponds to a frequency shift of: $f_s/2N$.

10.5 Polyphase network functions:

$$D(Z)(1 - 0.1354Z^{-1} + 0.069Z^{-2})(1 + 0.98Z^{-1} + 0.51Z^{-2})$$
$$N_1(Z) = 1 + 7.806Z^{-1} + 9.718Z^{-2} + 3.773Z^{-3} + 0.1883Z^{-4}$$
$$N_2(Z) = 3.713(1 + 2.908Z^{-1} + 2.035Z^{-2} + 0.317Z^{-3}).$$

The diagram is that of fig. 10.11.b; multiplication rate: $8f_s$.

CHAPTER 12

12.1 The first function $r(n)$ is periodic while the second $r'(n)$ is not. The ratio $\left|\frac{r'(n)}{r(n)}\right|$ varies roughly as $\left(1 - \frac{n}{16}\right)$.

12.2 The eigenvalues for $N = 3$ are the solutions of the equation:

$$(r_0 - \lambda)^3 - (r_0 - \lambda)(r_2^2 + 2r_1^2) + 2r_1^2 r_2 = 0$$

For a sinusoid: $r_0 = \frac{1}{2}$; $r_1 = \frac{1}{2}\cos\omega$; $r_2 = \frac{1}{2}\cos 2\omega$ then:

$$\lambda_1 = 0; \quad \lambda_2 = \frac{1}{2}(1 - \cos 2\omega); \quad \lambda_3 = \frac{1}{2}(2 + \cos 2\omega); \quad \text{with noise: } \lambda_i + \sigma^2.$$

12.3 Time constant $\tau = 1/\delta$. For $y(n)$ to approach m within 1% in average, 23 samples are needed. After the transition phase the quadratic residual error is given by (12.39). Recursive and non recursive estimators are equivalent for $n \simeq 2/\delta$.

12.4 Through polynomial division one gets:

$$H_5(Z) = 1 - 1.2Z^{-1} + 0.94Z^{-2} - 0.528Z^{-3} + 0.1636Z^{-4} + 0.068Z^{-5}.$$

The equalizer acts like a predictor and its coefficients are given by (12.111). Since $\sigma_x^2 = 0.3375$, $\delta_m \simeq 0.5$.

12.5 The prediction filter exhibits infinite attenuation at frequency $f_s/8$; Hence: $a_1 = \sqrt{2}$; $a_2 = -1$.
With the noise σ^2, one gets:

$$a_1 = \sqrt{2}\,\frac{1 + 2\sigma^2}{1 + 8\sigma^2 + 8\sigma^4} \simeq \sqrt{2}(1 - 6\sigma^2); \quad a_2 = -\frac{1}{1 + 8\sigma^2 + 8\sigma^4} \approx -(1 - 8\sigma^2)$$

CHAPTER 13

13.1 The operation $(ab + c) + a$ can be carried out with 7 bits in twos complement.

13.2 With a $(B_D + B_C - 1)$ bits shift register the operations are realized in reversed order and the coefficients are presented with most significant bit first.

13.3 With the delays between the address, the coefficient bits have to be reversed. Rounding can be performed through gates on the coefficient bits.

13.4 Less logic circuits are needed for the parallel type.

13.5 A factor of merit can be the power consumption to multiplication rate ratio (W/mult/s): in the order of $10^{-6} - 10^{-7}$.

13.6 Multiplication rate: 128.10^4 mult/s.

13.7 Direct filter order: 213. In multirate implementation 4 half-band filters ($M = 1, 2, 3, 8$) are needed, which leads to 244 kmult/s and 28 data memories.

13.8 Multiplication rate: 384 kmult/s. Data memory 8192 bits. A bus architecture as in fig. 13.14 is well suited.

INDEX

427